CLIMATE, CHANGE AND RISK

What are the prospects for changes in climatic hazards?

What impacts will result from present and future climatic hazards?

What methodologies are appropriate for understanding climatic risks as they change in the future?

The focus of the 'climate change debate' has begun to shift from 'what if?' to 'what should we do?' The scale of losses from weather-related disasters in the past decade has been unprecedented. The European windstorms of 1987 and 1990, Hurricane Andrew in 1992, the US Midwest flood in 1993 and heavy flooding in China in 1996 have caused considerable uncertainty about the behaviour of extreme weather events. Their potential impact on the insurance industry is even less well understood.

Climate, Change and Risk presents an overview of climatic hazards and climate change, focusing on societal responses, insurance and methodologies for analysis. Drawing on primary research from leading researchers world-wide, this volume explores the potential sensitivity to changes in weather hazards that might be expected with climate change.

Present hazard vulnerability and risk are often poorly understood and changes in disaster management are crucial. *Climate, Change and Risk* stresses that disaster managers must adapt their policies to new climatic conditions and continue monitoring trends to detect significant shifts in risk and respond appropriately.

This volume advances significantly our understanding of the rapidly emerging knowledge on extreme events related to climate variability and change. It should be read by climate specialists, natural hazard experts and climate policy-makers in all parts of the world.

Thomas E. Downing is the Programme Leader for Climate Impacts and Responses at the Environmental Change Unit, University of Oxford. **Alexander A. Olsthoorn** and **Richard S.J. Tol** are both researchers at the Institute for Environmental Studies, Vrije Universiteit Amsterdam.

CLIMATE, CHANGE AND RISK

Edited by Thomas E. Downing,
Alexander A. Olsthoorn and
Richard S.J. Tol

Routledge
Taylor & Francis Group

LONDON AND NEW YORK

First published 1999
by Routledge

2 Park Square, Milton Park, Abingdon, Oxon OX14 4RN
711 Third Avenue, New York, NY 10017, USA

*Routledge is an imprint of the Taylor & Francis Group,
an informa business*

First issued in paperback 2016

Transferred to Digital Printing 2005

Typeset in Galliard by
J&L Composition Ltd, Filey, North Yorkshire

British Library Cataloguing in Publication Data
A catalogue record for this book is available from the British Library

Library of Congress Cataloguing in Publication Data
Climate, change and risk/edited by Thomas E. Downing, Alexander A.
Olsthoorn & Richard S.J. Tol.
p.   cm.
Includes bibliographical references and index.
ISBN 0–415–17031–1 (cloth)
1. Climatic changes. 2. Climatic changes—Social aspects. 3. Natural
disasters. I. Downing, Thomas E., 1953–   II. Olsthoorn,
Alexander A., 1942– . III. Tol, Richard S.J., 1969– .
QC981.8.C5C5114 1999
363.34'92–dc21   97–51582
CIP AC

ISBN 978-0-415-17031-4 (hbk)
ISBN 978-1-138-99141-5 (pbk)

CONTENTS

v

CONTENTS

FIGURES

TABLES

CONTRIBUTORS

Robert J. Allan is Senior Research Scientist at CSIRO Division of Atmospheric Research, Aspendale, Victoria 3195, Australia

A. Paul Brignall
124 Naunton Crescent, Cheltenham GL53 7BE, UK

Cornelis Dorland is Researcher at the Institute for Environmental Studies, Vrije Universiteit, De Boelelaan 1115, 1081 HV Amsterdam, The Netherlands

Hadi Dowlatabadi is Director of the Center for Integrated Study of the Human Dimensions of Global Change, Department of Engineering and Public Policy, Carnegie Mellon University, Pittsburgh, PA 15213, USA

Thomas E. Downing is Senior Research fellow and Programme Leader at the Environmental Change Unit, 1a Mansfield Rd, Oxford OX1 3TB, UK

David Favis-Mortlock is Research fellow at the Environmental Change Unit, 5 South Parks Road, Oxford OX1 3UB, UK

Megan J. Gawith is Programme Officer, UK, Climate Impacts Programme, at the Environmental Change Unit, 1a Mansfield Rd, Oxford OX1 3TB, UK

John Handmer is Professor at the Flood Hazard Research Centre, Middlesex University, Queensway, Enfield EN3 4SF, UK

Paula A. Harrison is Research Scientist at the Environmental Change Unit, 1a Mansfield Rd, Oxford OX1 3TB, UK

Kevin J. Hennessy is Research Scientist at the CSIRO Division of Atmospheric Research, Aspendale, Victoria 3195, Australia

Matthijs Hisschemöller is Senior Researcher at the Institute for Environmental Studies, Vrije Universiteit, De Boelelaan 1115, 1081 HV Amsterdam, The Netherlands

Theodore S. Karacostas is Professor at the Department of Meteorology and Climatology, Aristotelian University of Thessaloniki, Thessaloniki 54006, Greece

Andreas Langen is a History and Geography student at Westfälische Wilhelms-Universität Munster, Schoßplaatz 2, D-48149 Münster, Germany

Frank P.M. Leek is Researcher at the Institute for Environmental Studies, Vrije Universiteit, De Boelelaan 1115, 1081 HV Amsterdam, The Netherlands

Kathy L. McInnes is Senior Research Scientist at CSIRO Division of Atmospheric Research, Aspendale, Victoria 3195, Australia

Heather McMaster is Researcher at the Climatic Impacts Centre, Macquarie University, North Ryde, NSW 2109, Australia

W. John Maunder is a consultant with W.J. Maunder Consultants Ltd., Tauranga, New Zealand

Alexander A. Olsthoorn is Senior Researcher at the Institute for Environmental Studies, Vrije Universiteit, De Boelelaan 1115, 1081 HV Amsterdam, The Netherlands

John L. Orr is Data Manager, UK Climate Impacts Programme, at the Environmental Change Unit, 1a Mansfield Rd, Oxford OX1 3TB, UK

Jean P. Palutikof is Reader at the Climatic Research Unit, University of East Anglia, Norwich, NR4 7TJ, UK

Edmund Penning-Rowsell is Professor and Director of the Flood Hazard Research Centre, Middlesex University, Queensway, Enfield EN3 4SF, UK

A. Barrie Pittock is Leader of the Climate Impact Group, CSIRO Division of Atmospheric Research, Aspendale, Victoria 3195, Australia

Roger S. Pulwarty is Research Scientist at the Cooperative Institute for Research in Environmental Sciences/Climate Diagnostics Center, University of Colorado, Boulder, CO 80309–0449, USA

Ramasamy Suppiah is Senior Research Scientist at the CSIRO Division of Atmospheric Research, Aspendale, Victoria 3195, Australia

Ros Taplin is Visiting fellow at the Climatic Impacts Centre, Macquarie University, North Ryde, NSW 2109, Australia

Sue Tapsell is Researcher at the Flood Hazard Research Centre, Middlesex University, Queensway, Enfield EN3 4SF, UK

Richard S.J. Tol is Researcher at the Institute for Environmental Studies, Vrije Universiteit, De Boelelaan 1115, 1081 HV Amsterdam, The Netherlands

Pier Vellinga is Professor and Director of the Institute for Environmental Studies, Vrije Universiteit, De Boelelaan 1115, 1081 HV Amsterdam, The Netherlands

Kevin J. Walsh is Senior Research Scientist at CSIRO Division of Atmospheric Research, Aspendale, Victoria 3195, Australia

J. Jason West is PhD student, Department of Civil and Environmental Engineering and Department of Engineering and Public Policy, Carnegie Mellon University, Pittsburgh, PA 15213, USA

Peter H. Whetton is Principal Research Scientist at CSIRO Division of Atmospheric Research, Aspendale, Victoria 3195, Australia

FOREWORD

Human society has always suffered from, and in some cases adapted to, extremes and severe weather events. As populations have grown, the supporting infrastructure has in some ways become more resilient, and in others more susceptible to the hazards of extreme events. In general, the trends have shown a decline in number of deaths with improved warning and preparedness systems, but an increase in economic losses. Indeed, on a global basis, economic losses from natural disasters have risen by about US$1 billion per year in the 1990s, in constant dollars. It should be noted, however, that there is some suggestion, in the incomplete records, of higher losses in the 1940s and 1950s than in the 1960s. Data compiled for the 1994 World Conference on Natural Disaster Reduction (a part of the International Decade for Natural Disaster Reduction) indicate that most of these losses are from climate-related disasters (storms, floods, droughts, forest fires, etc.) rather than from geological causes (earthquakes, volcanoes, tsunamis) and that losses from climatic disasters from the 1960s to 1992 were rising some three times as rapidly as those from geological disasters.

It is obvious that increased population, especially in vulnerable coastal regions and along rivers, and increased economic growth are major contributors to the rising toll of disaster losses. The urbanisation of parts of watersheds and the loss of woody vegetation cover in both urban and rural areas in most countries have contributed to increased floods and droughts. Yet it is difficult to explain the fifty-fold increase in economic losses on these grounds alone.

Some insurance companies, climate scientists and natural hazard experts have become increasingly convinced that a changing frequency of extreme climatic events is under way, especially over the past thirty years. Some global climate model (GCM) outputs, forced with increasing greenhouse gases and aerosols, suggest increasing rain intensities but longer dry spells between heavy rains, and more severe extra-tropical winter storms in the northern hemisphere. The IPCC concluded that there is no consistent global evidence of changes in extreme weather events in the generally poor records available. However, it goes on to cite a number of regional changes observed over the past century which are consistent with the GCM projections.

Climate extremes are the most obvious and important manifestations of anthropogenic climate change and of natural climate variability. Trends in such extremes have enormous implications in billions of dollars worth of buildings and drainage facilities. Increased frequency of heavy rains and flood events would require changes in flood-plain zoning and control structures, and affect soil erosion, sensitive ecosystems and insurance rates. More severe droughts can cause economic hardships, much human suffering and devastating effects on natural ecosystems.

This volume advances significantly our understanding of the rapidly emerging knowledge on extreme events as related to climate variability and change. It should be read by climate specialists, natural hazards experts and climate policy-makers in all parts of the world.

James P. Bruce

Co-chair, Intergovernmental Panel on Climate Change, Working Group III

PREFACE

The intent of this volume is to explore potential sensitivity to changes in weather hazards that might be expected with climate change. The emphasis is on identifying critical issues and data requirements, exploring different methodologies, and synthesising results across sectors and studies. The focus is insurance, especially the commercial insurance sector. Insurance is widely recognised as an effective response to risk, but there is considerable debate regarding its role in responding to climate change. The target audience for this volume is researchers and practitioners concerned with climate change adaptation, weather hazards and disaster management.

The chapters on Europe are the culmination of five years of research on climate change and extreme events. It began with a study by the Institute for Environmental Studies (IVM) at the Vrije Universiteit, Amsterdam, for the Netherlands National Research Programme on Global Air Pollution and Climate Change, with assistance from the Environmental Change Unit (ECU). This was followed by a European Commission (EC) study co-ordinated by the IVM, with contributions from the ECU, the Aristotelian University of Thessaloniki (AUT), the Flood Hazard Research Centre (FHRC) at Middlesex University, and the Climatic Research Unit (CRU) at the University of East Anglia (no. EV5V CT94 0391). An extended network of researchers, representatives from insurance companies and policy-makers participated in these two projects. Several workshops reviewed emerging research.

The EC project resulted in a final report prepared by the three editors of this book. The present volume includes revised and extended papers from that final report, three new chapters on Australia and the USA, and a fresh introduction.

The research also benefited from related work on European climate change and agriculture (CLAIRE, EV5V-CT92–0294), the economics of climate change (several projects for the EC, International Energy Agency Greenhouse Gas R&D, Nuclear Electric, and others). As we write, a new round of EC-funded research is about to get under way, focusing on societal responses to changing drought and flood risk in Spain, the Netherlands and the UK.

In Australia, research at CSIRO and the Climatic Impacts Centre at Macquarie University has largely been funded via the Australian Federal Department of the

xix

Environment, Sport and Territories as part of its core greenhouse research programme and specific projects concerned with the impacts of climate change. General support has also come from CSIRO and Macquarie University, and specific projects have been funded by the governments of Victoria, New South Wales, Queensland, Western Australia and the Northern Territory, the Asian Development Bank, the United Nations Environment Programme, and a number of private companies and associations.

In the USA, the National Science Foundation funded several centres for research on global change to further stimulate collaborative and integrated work. Colleagues at the Center for Integrated Study of the Human Dimensions of Global Change at Carnegie Mellon University and the Cooperative Institute for Research in Environmental Sciences at the University of Colorado contributed to this volume.

Against this background, the editors would like to thank all the people who have contributed funding, insight, data, analyses and personal support – the essential commodities of successful research. They are too numerous to mention everyone, but a few deserve particular credit: Andrew Sors, Angela Liberatore, Roberto Fantechi, and Denis Peter in the European Commission, the Directors of the IVM and ECU for encouragement, and Graham Francis and Kate Lonsdale for technical editing. And we thank the many authors for their insightful contributions.

A summary of the findings, updates when available, and additional information can be found on web sites maintained by the contributors. For example, look for links in web sites at the:

- Center for Integrated Study of the Human Dimensions of Global Change, Department of Engineering and Public Policy, Carnegie Mellon University: //hdgc.epp.cmu.edu
- Cooperative Institute for Research in Environmental Sciences, University of Colorado: http://cires.colorado.edu
- Climatic Impacts Centre, Macquarie University: //cic.mq.edu.au
- Climatic Research Unit, University of East Anglia: http://www.cru.uea.ac.uk
- Division of Atmospheric Research, CSIRO: //www.dar.csiro.au
- Environmental Change Unit, University of Oxford: http://www.ecu.ox.ac.uk
- Flood Hazard Research Centre, Middlesex University: //www.mdx.ac.uk/www/gem/index.htm
- Institute for Environmental Studies, Vrije Universiteit: http://www.vu.nl/ivm

T.E. Downing, A.A. Olsthoorn and R.S.J. Tol

ABBREVIATIONS

ACCP	Atlantic Climate Change Program
AET	actual evapotranspiration
AGCM	Atmosphere Only General Circulation Model
AIR	Applied Insurance Research
AOSIS	Alliance of Small Island States
APC countries	African, Pacific and Caribbean countries
AWC	available water-holding capacity
AWD	agricultural water deficit
BGS	British Geological Survey
BMRC	Bureau of Meteorology Research Centre
CCIRG	Climate Change Impacts Review Group
CEGCM	CSIRO ENSO General Circulation Model
COT	Crisis Research Team
CPB	Central Planning Bureau (Netherlands Bureau for Economic Policy Analysis)
CRED	Centre for Research on the Epidemiology of Disease
CSIRO	Commonwealth Science and Industrial Research Organisation
CVS	Centre for Insurance Statistics
CZMA	Coastal Zone Management Act
DoE	Department of the Environment (UK)
DRL	Dutch Red Cross
ECU	European Currency Unit
ENSO	El Niño–Southern Oscillation
EPIC	Erosion–Productivity Impact Calculator
FC	field capacity
FCCC	Framework Convention on Climate Change
FEMA	Federal Emergency Management Agency
G & E	Gooi en Eemlander
GA	General Accident
GAO	Government Accounting Office (USA)
GCM	General Circulation Model (or Global Climate Model)
GDP	Gross Domestic Product

GFDL	Geophysical Fluid Dynamics Laboratory model
GHASP	Geo-Hazard Susceptibility Package
GHG	greenhouse gas
GIS	Geographic Information System
GMT	global mean temperature
IDNDR	International Decade for Natural Disaster Reduction
IGBP	International Geosphere Biosphere Programme
IPCC	Intergovernmental Panel on Climate Change
IS92a	IPCC reference scenario a of greenhouse gas emissions, compiled in 1992
KNMI	Royal Netherlands Meteorological Office
MAFF	Ministry for Agriculture, Fisheries and Food (UK)
MAGICC	Model for the Assessment of Greenhouse Gas-Induced Climate Change
MAXCSWD	maximum annual potential cumulative soil water deficit
MPI	maximum potential intensity
MSLP	mean sea-level pressure
NCAR	National Center for Atmospheric Research (USA)
NDP	net domestic product
NEN	Netherlands Standardisation Institute
NFIA	National Flood Insurance Act
NFIP	National Flood Insurance Programme
NGRS	National Greenhouse Response Strategy
NOAA	National Oceanic and Atmospheric Administration
NRC	National Research Council
NS	Netherlands Railways
NSW	New South Wales
OECD	Organisation for Economic Co-operation and Development
P	precipitation
PET	potential evapotranspiration
PMF	probable maximum flood
PML	probable maximum loss
ppmv	parts per million by volume
QBO	quasi-biennial oscillation
RWS	Rijkswaterstaat (Dutch National Water Authority)
SGP	seasonal genesis parameter
SLPA	sea-level pressure anomalies
SLR	sea-level rise
SPARTECA	South Pacific Regional Trade and Economic Co-operation Agreement
SPCZ	South Pacific Convergence Zone
SST	sea surface temperature
SWB	soil water balance
SWD	soil water deficit

THI	temperature–humidity index
UKHI	UK Meteorological Office High Resolution Model
UKMO	United Kingdom Meteorological Office
UKTR	UK Meteorological Office transient simulation
UN	United Nations
UNDHA	United Nations Department of Humanitarian Affairs (formerly the UN Disaster Relief Organization (UNDRO))
UNDRO	United Nations Disaster Relief Organization
USAID	United States Agency for International Development
USDA/ARS	United States Department of Agriculture/Agricultural Research Service
USGS	United States Geological Survey
WISM	Weather Insurance Simulation Model
WMO	World Meteorological Organization
WTO	World Trade Organization; World Tourism Organization
ZWA	zonal wind anomalies

1

INTRODUCTION

*Thomas E. Downing, Megan J. Gawith, Alexander A. Olsthoorn,
Richard S.J. Tol and Pier Vellinga*

1.1 Climate and insurance

The scale of losses from weather disasters in the past decade or so has been
decidedly noticeable, if not unprecedented. Major losses include:

- European windstorms of 1987, which caused losses of US$3.7 million
  (Simons, 1992), and 1990, with losses of US$15 billion (Dorland *et al.*,
  Chapter 10, this volume).
- Hurricane Hugo, which set a new record for insured losses in 1989, $3
  billion, surpassed in 1992 by Hurricane Andrew, with insured losses in the
  USA of $17 billion (of total losses estimated at $30 billion) (Wilson, 1994;
  Pulwarty, Chapter 7, this volume).
- Heavy flooding in China in 1996 that affected over 250 million people and
  caused US$27 billion in economic losses (3.2 per cent of GDP) (Tong *et
  al.*, 1997), and $12 billion in the US floods in 1993 in the Midwest (M.F.
  Myers, personal communication).
- Subsidence in the UK due to the prolonged drought, which cost an estimated
  £100 million to UK insurers in 1975 and 1976 (Doornkamp, 1993) and over
  £2 billion from 1989 to 1996 (ABI, personal communication).
- Hailstorms in Australia in 1990 and 1996, with insured losses of A$384
  and 150 million respectively.

Since 1960, losses from windstorms alone have averaged between US$2 and
US$20 billion per year (see Table 1.1).[1] The population affected by disasters is
also rising, although the death toll has declined (Figure 1.1). The cluster of losses
in the 1990s prompted insurance and reinsurance companies, climatologists and
policy-makers (especially in vulnerable locales) to raise the question of whether the
effects of climate change were now being realised (see, for example, Dlugolecki *et
al.*, 1995; Doherty, 1997; Leggett, 1993; Greenpeace, 1994; Munich Re, 1993;
Tucker, 1997). Perhaps for the first time, private, commercial stakeholders
became concerned about potential climate change impacts and began addressing

1

Table 1.1 Major windstorms world-wide, 1960–92

	1960s	1970s	1980s	1990s
Average number per year	0.8	1.4	2.9	5.0
Average annual damage, US$ billion (1990)	2.3	3.4	3.8	20.2
Average annual insured losses, US$ billion (1990)	0.5	0.8	1.9	11.3

Notes: A major windstorm is defined as one costing more than US$500 million in total damages. The 1990s include only 1990, 1991 and 1992 (15 storms)

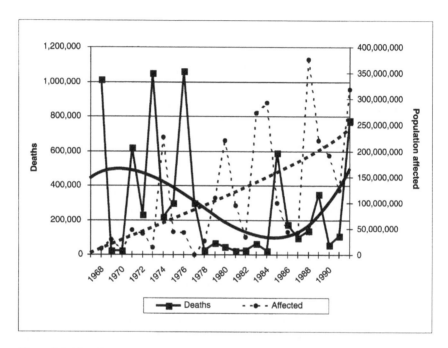

Figure 1.1 Trends in deaths and population affected by weather-related disasters, 1967–91. Heavy solid line shows the trends in deaths and population affected. *Source:* IDNDR (1994)

practicable response strategies (see Box 1.1). While public organisations in other sectors (e.g. the Food and Agriculture Organization, the UK National Rivers Authority) and at the regional level (e.g. the Great Lakes Commission, Local Agenda 21 groups) began to address the issue of climate change, the insurance and reinsurance companies began to organise scientific reviews, participate in the Intergovernmental Panel on Climate Change (IPCC), and publicise the issue.

This volume grew out of that concern. The early concern expressed by the insurance industry was met by an astonishing lack of rigorous research on the impact of climate change. The purpose of this book is to present an overview of

Box 1.1 Perceptions of climate change

Clearly, if the 1990s catastrophes are an early signal of climate change, the insurance industry and disaster specialists should be concerned. Two surveys gauge how seriously climate change is taken.

In the UK, the Society of Fellows Study Group of the Chartered Insurance Institute conducted a survey of the insurance sector, in 1993, to assess the industry's perception of the threat of climate change (Chartered Insurance Institute, 1994). Questionnaires were sent to 3,854 agents and officers in general insurance, life assurance, reinsurance, loss assessors and intermediary companies, with 285 returns. While the low response might indicate a lack of concern about climate change, 84 per cent of those who did respond believed that 'a change in weather patterns is one of the reasons for the increase in weather losses'. Further, in response to the question, 'Accepting that the world's climate is changing, what overall effect do you anticipate this will have on the insurance market in the next 10 years?', almost two-thirds replied 'significant' or 'considerable'. More than three-quarters thought the effects would be significant or considerable over the next 20 years. Specific hazards thought to be altered by climate change are shown in Table 1.2.

Seventy-five per cent of respondents anticipated increased underwriting losses, price increases, and changes in cover over the next 10 years. Two-thirds thought international reinsurance would change and one-third expected withdrawal from certain areas (such as flood hazard zones).[2] Few thought climate change would lead to insolvency or government intervention in the next decade. Almost half the respondents expect losses and prices to stabilise over the next 20 years, but with greater withdrawal (half of the responses) and, possibly, government intervention.

Seventy-eight of the respondents worked in organisations which have 'included strategies to manage the effects of climate change in its planning process'. These include monitoring and participating in debates on climate trends and obtaining relevant data (Table 1.3). Some companies appear to have already altered their insurance practices and adopted practical steps towards reducing the potential effects of climate change. Such measures might include developing differential insurance rates and excesses, demanding hazard surveys before acceptance of cover, applying building design requirements, and seeking alternative reinsurance or risk financing.

In an enquiry among natural disaster experts and policy makers at the IDNDR World Conference in Yokohama, May 1994, Olsthoorn *et al.* (1994) found a strong awareness of climate change among this community as well. Generally, an increase in incidence and intensity of various weather disasters was perceived, but the link to climate change was considered doubtful. Nevertheless, climate change may cause additional danger. The major policy option identified was a reduction in the emission of greenhouse gases.

Table 1.2 Specific hazards thought by insurance agents likely to be affected by climate change

Hazard	Threat				
	L	L–M	M	M–H	H
Drought/subsidence	6	17	38	30	5
Freeze	30	42	18	3	0
Windstorm	1	9	32	44	12
Coastal floods	3	12	28	34	20
Local extremes	21	29	19	22	5
Other	6	2	5	2	0

Source: Chartered Insurance Institute (1994)

Note: Percentages are of the total number of returned questionnaires (285)

Table 1.3 Suggestions by insurance agents for practical steps to mitigate the effects of climate change

Practical steps	Percentage of respondents
Research on weather patterns	74
Studies of vulnerable areas	74
Emphasis on risk management	70
Closer co-operation with the authorities	67
Tax-allowable catastrophe reserves	55
Better education and public relations about natural hazards	45
Improved loss control	36
Restrictions on risk acceptance	35
Government-backed catastrophe pool	27
Other	1
No response	1
Total number of respondents	285
Total number of responses (includes multiple responses)	1,385

Source: Chartered Insurance Institute (1994)

research on climatic hazards and climate change, with a focus on societal responses and insurance in particular. Each chapter focuses on specific regions, hazards and themes, collectively addressing such questions as:

- What are the prospects for changes in future climatic hazards?
- What impacts result from present and future climatic hazards?

- What is the range of effective responses?
- What methodologies are appropriate for understanding climatic risks as they change in the future?

The answers are inevitably peculiar to each region, hazard and author, although we attempt a synthesis below. Perhaps more importantly, this volume illustrates a range of methodologies, and we suggest that further research should expand beyond the conventional frameworks for climate change impact assessment towards more explicit evaluation of adaptive responses and adaptive capacity.

1.2 Hazard, vulnerability and risk – and climate change

Definitions of hazard, vulnerability and risk are essential starting-points for conceptualising the impact of climate change and climatic hazards. The UN Department of Humanitarian Affairs (1992) provides accepted definitions for the concepts of hazard, vulnerability and risk (see also Dovers and Handmer, 1995):

> A *hazard* is a threatening event, or the probability of occurrence of a potentially damaging phenomenon within a given time period and area.

Note the two aspects of this definition. Not only is hazard the chance of an event, but also that the event is potentially damaging. Hazardous weather is distinguished from normal weather by its potential to do damage, and not by its physical or statistical properties. What makes an extreme event a hazard is that the extreme event has the potential to damage some aspect of human welfare or life itself. This depends on vulnerability.

> *Vulnerability* is the degree of loss (from 0 per cent to 100 per cent) resulting from a potentially damaging phenomenon.

Vulnerability depends on human infrastructure and socio-economic conditions. To a large extent, these are shaped by considerations other than natural hazards. However, hazard and vulnerability are related. Societies respond to hazards and disasters by reducing the hazard or reducing vulnerability. Coping mechanisms for natural hazards are determined by perceptions of hazard and vulnerability, preferences and budgets. Coping mechanisms are often classified as protection (dealing with the built environment) and mitigation (socio-economic responses). Examples of protection include water reservoirs, dikes, heating and air-conditioning, wind shields, storm shelters and irrigation. Examples of mitigation include savings, solidarity, mutual and commercial insurance, crop diversification, alternative sources of income, temporary migration, charity, government support and foreign aid.

> *Risk* is the expected losses (of lives, persons injured, property damaged, and economic activity disrupted) due to a particular hazard for a given

area and reference period. Based on mathematical calculations, risk is the product of hazard and vulnerability.

Risk is thus the conjecture of natural hazard and socio-economic vulnerability.

These definitions are primarily concerned with short-term natural hazards, assuming known hazards and present (fixed) vulnerability. Figure 1.2 displays a hypothetical probability density function of a certain weather parameter, such as temperature, rainfall or wind speed. It describes the chance of occurrence of a critical value. For a suitably adapted and adjusted society, extreme events (such as cold or hot spells, droughts and floods, or storms) lurk in the tails of the distribution. The more distant from the central part of the distribution, the more unlikely the event and the higher the damages it may cause. Should climate change, what is presently considered hazardous and extreme weather would occur with a different return period than at present – some events more often, others less often. Vulnerability is also changing. In an optimistic case, vulnerability may be decreasing and the increase in hazardous weather may be offset to some extent. Future damages will be different from predictions based on present hazard–loss relationships.

Clearly, the evolving, long-term nature of climate change, over the course of decades to a century or more, requires an extended, dynamic framework in at least three arenas. First, vulnerable places (structures and populations) are likely to evolve dramatically. For instance, the increasing concentration of population in coastal zones greatly increases the risk of damaging floods, cyclones, tsunamis and coastal erosion, even without changes in the hazard itself. On the other hand, risk management will evolve, partly as a result of changes in technological, economic, social, political and cultural circumstances, and partly in response to perceptions of increasing damages and stresses.

Second, the way hazards are estimated needs to differentiate between the extremes of current weather and changes that can be attributed to the enhanced greenhouse effect. This is not likely to be possible for several decades. Climate and climatic extremes have a large natural variability that masks the relatively small expected changes in climate.

Third, analyses must incorporate the likelihood and consequences of purposeful responses to climate change over the next few decades. For example, a scenario of accelerated policies to control greenhouse gas (GHG) emissions, significant transfer of technology to developing countries, low population growth and moderate economic growth implies a reduction in the lives at risk from natural disasters (Downing *et al.*, 1994).

This difference between static and evolutionary risk requires consideration of the continuum of knowledge between known risks, uncertainty and surprise. A conventional typology (Schneider and Turner, 1994) characterises the continuum as:

• risk, where events, processes or outcomes are known and probabilities are estimated from observed (stationary) data;

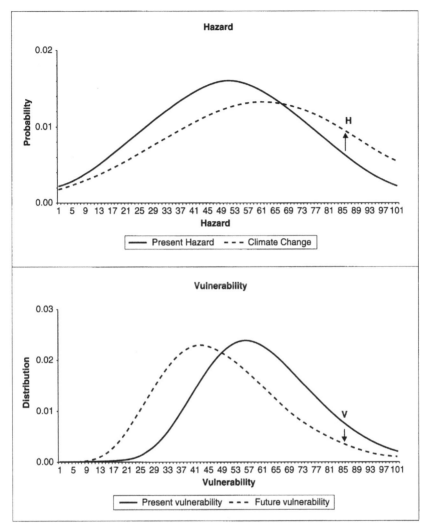

Figure 1.2 A conceptual illustration of normal and extreme weather. Hypothetical distributions display the present (solid line) and future (dashed line) probability of climate change for a given climatic hazard (such as temperature) and for socio-economic vulnerability, using artificial ranges of 0 (low hazard or vulnerability) to 101 (high). The tail of the present distribution of hazard (for example, beyond a critical value of H) increases significantly. However, vulnerability will also change in the future. In this case, vulnerability (for example, supposing a critical value of V) decreases, indicating that the increase in future hazards is offset to some extent by a decrease in vulnerability. Of course, the hazard may decrease in the future and vulnerability may increase; the example is only illustrative

- uncertainty, where events, processes or outcomes are known but their probabilities are not known, or are assigned by subjective estimates;
- surprise, where events, processes or outcomes are not known, or unexpected. 'Surprise is a condition in which perceived reality departs qualitatively from expectations' (Anticipating Global Change Surprise Workshop, after Holling, 1986: 294).

Present extreme events should be considered risks in this typology. However, many of the distributions are uncertain, owing to the availability only of short time series of data and changing natural and social systems. Projections of climate change, for mean conditions, are uncertain, with some elements of surprise, for example in forecasts of GHG emissions, large-scale changes in the ocean circulation and economic sensitivity to climate impacts.

1.3 Scenarios of extreme events

Given the present state of the art in climate modelling, and the uncertainty over future GHG emissions, the unfolding pattern and timing of climate change are still uncertain. Projections of global mean climate change suggest that warming

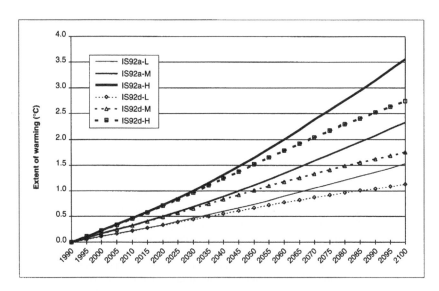

Figure 1.3 Range of uncertainty in projected global mean temperature. Projections are for scenarios of greenhouse gas emissions for the 'business as usual' (the IS92a scenario) and policies that reduce GHG emissions (IS92d). The low, medium and high values correspond to different values for global climate sensitivity (1°C, 2.5°C and 4.5°C)

Source: Data from MAGICC (see Wigley and Raper, 1992)

by 2050 will be approximately 1°C, but diverging between 1.5 and 2.5°C by 2100 for two different emission scenarios (the middle estimates in Figure 1.3). However, the 80 per cent confidence interval for each projection spans a much greater range of estimates. For the business-as-usual scenario of GHG emissions (called the IS92a), warming could be 0.7–1.6°C in 2050 and 1.5–3.6°C in 2100. The scenario with lower GHG emissions (IS92d) results in very similar projected warming in 2050 and about 0.5°C lower global temperatures in 2100.

The most common scenarios of climate change involve two runs of a global climate model (GCM) that simulates the dynamics of the climate system (with various assumptions about atmospheric, oceanic and biotic processes) (see Carter *et al.* 1994). The control run (labelled the $1 \times CO_2$ experiment) seeks to replicate the current climate. The effects of increased GHG concentrations are simulated in a second run. In an equilibrium experiment, the second run includes the equivalent of a doubling of carbon dioxide from pre-industrial times (the $2 \times CO_2$ run) and is continued until the climate stabilises at a new equilibrium. The difference between the two experiments is an estimate of the atmospheric sensitivity to the enhanced greenhouse effect.

Such equilibrium experiments do not provide estimates of the timing of the scenarios of climate change. Transient, or time-dependent, scenarios require specific scenarios of GHG emissions. These can be incorporated directly into GCMs, or run on a simplified model and then linked to regional patterns of climatic changes derived from equilibrium $2 \times CO_2$ experiments. For example, more recent GCM experiments simulate the effect of climate change by incrementally increasing GHG concentrations each year from the present to the future. Both these methods have been employed in the chapters of this volume.

Two observations seem common in the modelling community:

- GCMs that couple the atmospheric and oceanic circulations are beginning to have sufficient resolution and realism to model complex circulation changes, such as the El Niño–Southern Oscillation (ENSO), and to simulate major synoptic features, such as tropical cyclones and mid-latitude depressions.
- These model developments are not yet sufficiently well developed, robust across different models, or analysed in depth to provide a reliable basis for generating scenarios of extreme events (in contrast, for example, with the expected patterns of temperature change).

For example, the Technical Summary of the IPCC Second Assessment Report (IPCC, 1996: 44) did not present scenarios at the regional level, although improvements in GCMs were expected. Regarding changes in variability and extremes, the Summary noted the likely effects of global warming on temperature extremes and drought. However, little agreement between models was reported for mid-latitude storms and tropical cyclones. The effect of warmer sea surface temperatures on ENSO could increase precipitation variability, but robust GCM scenarios are not yet available.

It may be useful to break down the uncertainty in future extreme events into three domains. For events directly related to *heat*, there is some agreement in present understanding of responses to climate change. Almost all models show regional warming, although some locales may diverge markedly from the regional pattern, for example owing to oceanic effects. Present projections for heat-related events tend to at least agree on the sign of the change. For example, with 2°C of warming in Oxford, different statistical assumptions result in a range of a two- to eightfold increase in the number of heatwaves. While this is the current consensus, it must be noted that a major change in oceanic circulation could lead to regional cooling, particularly in northern Europe. Such a change is possible, but difficult to evaluate at present (see Lean, 1996).

For hazards related to long-term accumulations of temperature and precipitation (e.g. *subsidence, drought* and some types of *river floods*), scenarios based on GCMs can provide a reasonable range of results. For example, accumulated soil moisture deficits and drought hazard are likely to follow predictable patterns based on how effective precipitation is in meeting increased thermal demand. Although scenarios can differ quite markedly between GCMs, they provide a realistic range against which analysts could assign probabilistic outcomes.

The most difficult hazards to gauge are complex assemblages of meteorological elements, and ones that are relatively short-lived and local. *Windstorms* and intense *flood* events are typical: to describe their current or future distribution requires realistic representation of extreme pressure gradients and frontal systems. Often these are relatively short-lived events, reaching peak velocities over the course of a day or less. Other examples of local, short-lived hazards are hail, lightning, intense precipitation and tornadoes. While they may be associated with other changes, such as a shift in summer rainfall from large-scale to convective sources, their distribution will be difficult to estimate from current climate models. Present modelling of tropical cyclones is equally problematic. Coupled atmosphere–ocean global climate models are essential for estimating storms that are driven by land–sea interactions.

For any given region, changes in weather hazards may involve essentially new hazards (such as flash floods and tornadoes in the UK), increased frequency, severity and duration of some present hazards (drought may become more common), and decreases in other present hazards (e.g. frost days in most of the world).

In summary, there is considerable divergence in the approaches and scenarios that have been suggested by researchers. Global climate models are not sufficiently well developed to provide a robust scientific consensus for many climatic hazards. This is certainly true for specific locales and probably also the case for broad regions.

1.4 Overview

The chapters in this volume begin with a national overview of climatic hazards. Regional studies of subsidence, drought, floods, hurricanes, cyclones and sea-level rise, windstorms and heatwaves provide insight into specific hazards and places. The volume concludes with chapters on the economics of natural disasters, a simulation of weather insurance and an evaluation of policy responses.

Chapter 2 systematically reviews *climate change and climatic hazards in Australia*, pulling together a wide array of research (Pittock *et al.*, this volume). Natural hazards are common in Australia, particularly drought and heatwaves, tropical cyclones and floods. Climatic variability is high, strongly linked to ENSO events. So far there is little indication as to whether ENSO anomalies or hail occurrence will or will not change in the future. The possible impacts of flooding, tropical cyclones, storm surges and coastal inundation, and temperature extremes and fire are documented. The increased risks are still highly uncertain, although understanding of drought and flood hazards has advanced considerably, and there are promising lines of research on ENSO and tropical cyclone behaviour. The potential for increased costs of climatic hazards is one rationale to support GHG abatement. At the same time, adaptation will be required but will depend on the involvement and willingness of affected communities to change.

Land subsidence, related to shrinkage of clay soils during prolonged drought, is a major issue in the UK, with average losses amounting to £400 million per year in the late 1980s to the mid-1990s (Brignall *et al.*, Chapter 3, this volume). Property insurance in the UK covers subsidence. The hazard exists elsewhere in Europe, but is generally not covered by private insurance. Drought-induced patterns of subsidence are expected to shift with climate change. Management of the risk includes mapping the risk, monitoring sensitive areas (especially in the UK, Spain, France and Germany), altering insurance cover and premiums (as has already happened for affected homes in the UK), and alteration to building designs. While the responses should prove adequate over the next few decades, the competitive nature of the insurance sector has hindered a sector-wide approach.

The risk of *agricultural drought* in Europe is likely to change with climate change (Brignall *et al.*, Chapter 4, this volume). The chapter illustrates analyses at both the site and broad-scale. Two GCM scenarios suggest that the hazard may alleviate somewhat in northern Europe and worsen in southern Europe, but this should be seen as a preliminary conclusion that needs to be verified against scenarios that include potential changes in rainfall variability. For specific crops, especially winter wheat, the combined effects of CO_2 fertilisation and temperature change, with little change in precipitation, would increase mean yields, at least in the short term. Many options exist to cope with drought: land management at the farm level; choice of crop variety and land use; altered planting and harvesting regimes; soil fertility and pest management; and economic adjustments and incentives. However, scarcity of irrigation water, economic competition and spatial development policies are also important, and may either enhance or restrict farm adaptability to drought risks.

Case studies of *riverine floods* include the Thames (upstream from London, UK), the Seine (Île de France) and the Meuse (Limburg, the Netherlands)

(Handmer *et al.*, Chapter 5, this volume). Climate change could well increase precipitation and run-off in northern Europe, but the predicted changes in precipitation are, overall, fairly small. In all three cases, however, a small increase in flood depths could increase damages substantially. In absolute terms, the additional damages are modest, and substantial changes in flood-risk management policies may not be justified by expectations of climate change. However, scenarios of flood hazards are highly uncertain, and managing flood risks often involves prolonged public debate and considerable investment in the case of structural responses. An enhanced capacity for adaptability and flexibility is likely to be an essential policy tool in the face of increasing uncertainty about the future. In turn, this may require fundamental institutional change.

A 1,000-year history of *river floods and flood management in the Dutch Rhine delta* is summarised by Langen and Tol (Chapter 6, this volume). The Dutch experience illustrates the linkages between economic, political and technological development and future climate change. A comprehensive, dynamic look at hazards and vulnerability is the only fruitful way forward to a full understanding and hence proper management of weather-related risks.

Pulwarty (Chapter 7, this volume) reviews *hurricane risk* along the eastern coast of the USA. The intense debate among climatologists regarding climate change and hurricane hazards is reviewed, it being noted that the future remains uncertain. The potential for increased storm damage in the eastern USA has attracted global attention; storms in the past decade have been extraordinarily costly and economic vulnerability has been increasing with population migration to the sunbelt. For example, if historical storms recurred with a 15 per cent increase in wind speed, as might be possible with climate change, the expected damages would more than double. A brief history of hurricane risk management shows the difficulty of adapting local, state and federal policies to mitigate the hazard.

The combination of *sea-level rise and storm surge* is addressed by West and Dowlatabadi (Chapter 8, this volume). The authors identify potential sources of bias in the methods currently used for assessing the economic impacts of sea-level rise on developed coasts. These sources of bias derive both from the perspective of coastal management and from economic theory. In particular, the ability of storms to cause structural damage is emphasised, and future assessments will need to consider whether storm damage is likely to be greater if sea-levels rise. A modelling exercise based on a developing coastal community in the eastern USA, illustrates a novel approach to understanding the dynamics of responses to perceived risk and actual losses. In addition to erosion, storms also cause damages, and investor decisions to build (or rebuild) houses along the coast are modelled according to the risk that is perceived through time. Results suggest that the economic impacts of sea-level rise may be substantially higher than indicated by methods that assume perfect foresight.

A review of *tropical cyclones* in the southwest Pacific documents the hazard and impacts in Fiji (Olsthoorn *et al.*, Chapter 9, this volume). An inhabitant of a rural area in Fiji faces a risk of falling victim to a cyclone that is orders of magnitude greater than the risk of drowning in a flood in the Netherlands. The present return period of a major tropical cyclone is about 10 years, although projections of future risk are inconclusive. The current risk, however, is already a constraint for various economic sectors, and specific disasters consume a significant portion of private and public financial reserves. Tropical cyclones combined with sea-level rise constitute a new risk that warrants additional policy measures.

Wind storms in northwest Europe may increase somewhat in intensity, although large uncertainties remain (Dorland *et al.*, Chapter 10, this volume). For an extreme storm, a 6 per cent increase in wind speed – not unlikely in 50 years' time – could result in a tripling of damage due to an extreme storm and a doubling of annual average damages. At present, windstorms have only temporary effects, with little impact on socio-economic development. Insurance companies would increase premiums and deductibles, while the building sector might adapt building codes.

The clearest projection of climate change in Europe (and elsewhere) is for increased temperatures, which would dramatically increase the frequency and duration of *heatwaves* (Gawith *et al.*, Chapter 11, this volume). The effects of such prolonged periods of high temperature and humidity are mixed. The direct effects on health are apparent, but not economically large, and are covered, if serious, through medical insurance or health care. Northern Europe appears more vulnerable than southern Europe because of a lack of current adaptation to hot weather. Significant costs could accrue through increased cooling, particularly as northern Europe passes a threshold for adoption of air-conditioning in stores, workplaces and homes.

Although the direct and indirect impact of a natural disaster is, by definition, negative, higher-order *economic impacts* are mixed (Tol and Leek, Chapter 12, this volume). Conventional economic indicators, such as GDP or unemployment, usually conceal the real impact. The loss of capital over a longer term can have a positive effect if reinvestment from designated reserves takes place. The true economic impact of weather hazards is revealed in the costs of disaster management. Scarce resources are allocated to actions that are not inherently productive, detracting from general welfare and economic growth. Unfortunately, no reliable or complete estimate of the actual expenditure on disaster management has been found. Should the risk imposed by extreme weather change, the level of adaptation is bound to lag behind the actual change in risk. In addition to the change in opportunity costs (either for the better or for the worse), this induces a transition loss which is invariably for the worse (unless economic decisions are made haphazardly).

Weather insurance is a major form of risk management (Tol, Chapter 13, this volume). At present, property insurance is common for windstorms, but varies from country to country for drought, floods and subsidence. Health insurance (whether private or public) covers the more extreme cases of heatwaves. Insurance protection for climatic hazards typically encompasses public and private

13

measures. The impact of storms and floods (and subsidence in the UK) on the insurance sector can be dramatic, particularly for companies that have misjudged the risk. The insurance sector has a suite of instruments to respond to changing risks. Measures affecting price and cover can be applied rapidly, often to the disadvantage of the insured. Buffer capacity and risk management require more time and imply structural change, but are to the advantage of both insurer and insured. Empirical evidence from the first half of the 1990s suggests that all instruments are applied in response to changing risks. Changes in price and cover are particularly successful (for the insurers). A stylised model of the insurance market suggests that short-term reactions of market variables depend on the speed with which actors acknowledge changes. In the longer term, increases in risk are largely borne by the insured through price increases and cover restrictions.

A *policy analysis* reflects on how scientific information eventually may lead to decisions by policy-makers (Hisschemöller and Olsthoorn, Chapter 14, this volume). It stresses the importance of the interaction between the domains of policy development and science in decision-making. For most climatic hazards, the short-term responses are within the domain of a small set of actors – individual insurance companies setting premiums and cover for windstorms, homeowners in zones subject to subsidence, vulnerable populations exposed to heatwaves. For other hazards, such as droughts, floods and cyclones, responses to changes in risk require concerted action among a dispersed set of private companies, public agencies and vulnerable communities. In such circumstances, timely and effective adaptation may be more difficult.

1.5 Shared conclusions

Given the uncertainty of projections of climate change, particularly for extreme events, and the difficulty of estimating future impacts, the book's conclusions are more general than local, more strategic than specific.

The impact of an individual disaster, and a trend toward increasing impacts, is certain to support the drive to reduce *greenhouse gas emissions.* Years of intense and frequent hurricanes, large floods and prolonged drought stimulate interest in climate change and climate policy. The toll of climatic disasters, if related to climate change, certainly adds to the cost of climate change, also supporting GHG abatement policy in cost–benefit frameworks.

In economic assessments, the largest costs of disasters are associated with the *loss of life.* Sea-level rise, cyclones and floods in Bangladesh, drought and famine in Africa are familiar icons in public presentations. The annual (macro)economic losses from changes in climatic hazards in developed countries are not likely to be large, especially if insurance premiums offset the (insured) losses. However, this may not be the case for the combination of sea-level rise and storm surges, as indicated for the eastern USA, or if cyclone intensity and rainfall extremes increase significantly. The economic value of lives lost remains a contentious issue, caught between a global hegemony and moral economy.

Impacts in *less developed regions* are likely to rise. Certainly economic vulnerability will continue to increase; whether lives at risk are protected in the future remains uncertain. As developing countries become wealthier, commercial insurers may develop new markets. The extent to which real or perceived risks of climate change hinder this global development remains uncertain. The vulnerability of small island states to climate change, as illustrated for Fiji, reinforces their position in international negotiations. However, schemes to create a global mandate for climate change insurance are fraught with difficulties (as noted by Tol and Leek, Chapter 12, this volume). The traditional response has been disaster preparedness and mitigation through official development assistance. Proactive development policies to reduce vulnerability would be a major step forward in adapting to climate change in developing regions, indeed in every country.

For the immediate future, *disaster managers* must intensify their efforts and adapt their policies to new climatic conditions. The best way forward seems to be to focus on robust policies, ones that work reasonably well under a wide variety of weather conditions. Strategies that are effective only for a specific future scenario may not be warranted and could prove costly.

From this review, it appears that the *insurance industry* as a whole will not be threatened with collapse by climate change, at least in the next few decades. Continued monitoring of trends in climatic hazards will be essential to detect significant shifts in risk, and to respond appropriately, not an easy task (see Engeström, 1995, and ISO, 1996, for overviews of risk from the insurance perspective). Individual companies may of course suffer losses and even cease trading. On the other hand, if climatic changes force a sudden change in climatic risks and a cluster of serious disasters around the world, the insurance industry may be hard pressed to survive with its present structure. Since the residual risk from unexpected climate change cannot be estimated and covered, the insurance industry should support GHG emission abatement. Since the 1995 Berlin Conference of the Parties the insurance industry has acted accordingly.

1.6 Future research and impact methodologies

In the course of this research, lessons have been learned in developing suitable methodologies for climate change impact assessment. Three concerns underlie our sense of the requirements for future research on climate change and extreme events.

First, the Second Assessment Report of the IPCC marked a turning point in climate change policy. The scientific consensus (e.g. Houghton *et al.*, 1996) is increasingly seen as an authoritative benchmark that warrants consideration of GHG abatement, as agreed by the Conference of Parties to the UN Framework Convention on Climate Change. The global climate question is no longer 'what if?' but 'what should we do?'. While there is still a need for impact studies in assessments of abatement strategies, the focus has begun to shift to responses.

This shift in focus from impacts to adaptation requires a shift in the substance of climate change impact assessment. Instead of static evaluations of single resources, such as winter wheat, assessments are required of the interaction within and between sectors and of the dynamics of decision-making. Adaptive studies must eschew assumptions that the future will be like the present and look at stakeholders and how they make decisions about risks.

Second, robust scenarios of changes in climatic hazards due to anthropogenic climate change are not likely to be accepted in the next decade (although there may be some exceptions in some regions). As noted above, the issues of temporal and spatial scale and the difficulties of modelling extreme events even in static systems are fundamental challenges for climate modellers. Still, climate information has value, perhaps most promisingly in predicting medium-term episodes of climatic hazards related to El Niño.

Third, addressing climatic hazards is different from studying trends in mean conditions. Decision-making in natural hazards is generally viewed from risk perspectives rather than trend projections. Present hazard, vulnerability and risk are often poorly understood. Small changes in assumptions about the distribution of hazards and vulnerability can lead to large differences in expected losses and recommendations about specific actions.

Studies of strategic and specific responses to changing climatic hazards must go beyond the common IPCC impact assessment methodology (Carter *et al.*, 1994). An alternative paradigm must be based on stakeholder decision-making and a focus on their strategies to utilise climatic resources and adapt to climatic variations. The case studies in this book show that changes in other factors for other reasons than the enhanced greenhouse effect (e.g., per capita income, urbanisation of the watershed) substantially influence vulnerability to and impact of climate change. Autonomous and purposeful changes in disaster management are particularly important. Realistic representations of adaptation is essential.

Acknowledgements

Helpful advice and reviews of this chapter have been provided by Ian Burton, Kees Dorland, Hadi Dowlatabadi, David Favis-Morlock, Matthijs Hisschemöller, Use van der Hul, Huib Jansen, Frank Leek, Angela Liberatore, Ernst Lohman, Barrie Pittock and Peter van der Werff, as well as the authors of the chapters in this volume.

Notes

1 Ninety per cent of the insured catastrophe losses between 1986 and 1994 were due to windstorms (Tucker, 1997).
2 These responses occurred in the USA after Hurricanes Hugo in 1989 and Andrew in 1992.

References

Carter, T.R., Parry, M.L., Nishioka, S. and Harasawa, H. (1994) *Technical Guidelines for Assessing Climate Change Impacts and Adaptations*, Geneva: IPCC, WMO and UNEP.

Chartered Insurance Institute (CII) Society of Fellows Study Group (1994) *The Impact of Changing Weather Patterns on Property Insurance*, London: CII.

Dlugolecki, A., Harrison, P.A., Leggett, J. and Palutikof, J. (1995) 'Implications for insurance and finance', in M.L. Parry and R. Duncan, eds, *Economic Implications of Climate Change in Britain*, London: Earthscan, 87–102.

Doherty, N. (1997) Insurance markets and climate change. *The Geneva Papers on Risk and Insurance* 22 (83): 223–237.

Doornkamp, J. (1993) Clay shrinkage induced subsidence. *Geographical Journal* 159 (2): 196–202.

Dovers, S.R., and Handmer, J.W. (1995) Ignorance, the precautionary principle and sustainability. *Ambio* 24, 2: 92–97.

Downing, T.E., Greener, R.A. and Eyre, N. (1994) *Global Emissions and Impacts*, Report to the International Energy Agency. Oxford: Environmental Change Unit.

Engeström, J. (1995) *Natural Catastrophes: A Reinsurance Perspective*, London: Mercantile & General.

Greenpeace (1994) *The Climate Timebomb*, London: Greenpeace.

Holling, C.S. (1986) 'The resilience of terrestrial ecosystems: Local surprise and global change', in W.C. Clark and R.E. Munn, eds, *Sustainable Development of the Biosphere*, Cambridge: Cambridge University Press.

Houghton, J.T., Meira Filho, L.G., Callander, B.A., Harris, N., Kattenberg, A. and Maskell, K., eds (1996) *Climate Change 1995: The Science of Climate Change*, Cambridge: Cambridge University Press.

IDNDR (1994) *Disasters around the World: A Global and Regional View*, World Conference on Natural Disaster Reduction, Yokohama, Japan, 23–27 May 1994.

Insurance Services Office (ISO) (1996) *Managing Catastrophe Risk*, New York: ISO Inc.

IPCC (1996) 'Technical summary', in J.T. Houghton, L.G. Meira Filho, B.A. Callander, N. Harris, A. Kattenberg and K. Maskell, eds, *Climate Change 1995*, Cambridge: Cambridge University Press.

Lean, G. (1996) Changes in the Gulf Stream may mean even colder winters. *Independent on Sunday* (London), 18 February.

Leggett, J. (1993) *Climate Change and the Insurance Industry: Solidarity among the Risk Community*, London: Greenpeace.

Munich Re (1993) *Winter Storms in Europe, Analysis of 1990 Losses and Future Loss Potential*, Munich: Münchener Ruckversicherungs-Gesellschaft.

Olsthoorn, A.A., van der Werff, P.E. and de Boer, J. (1994) *The Natural Disaster Reduction Community and Climate Change Policy Making*, Amsterdam: Institute for Environmental Studies, Vrije Universiteit.

Schneider, S.H. and Turner, B.L. (1994) *Report of the 1994 Aspen Global Change Institute Summer Session on Surprise and Global Environmental Change*, Aspen: Aspen Global Change Institute.

Simons, P. (1992) Why global warming could take Britain by storm. *New Scientist* (7 November): 35–38.

Tong Jiang, Jiaqi Chen and Junfeng Gao (1997) 1996 flood disaster in China: Brief information, E-mail from the Nanjing Institute of Geography and Limnology, Chinese Academy of Sciences, Nanjing 210008, China.

Tucker, M. (1997) Climate change and the insurance industry: The cost of increased risk and the impetus for action. *Ecological Economics* 22: 85–96.

UN Department of Humanitarian Affairs (1992) *Internationally Agreed Glossary of Basic Terms Related to Disaster Management*, Geneva: UNDHA.

Wigley, T.M.L. and Raper, S.C.B. (1992) Implications for climate and sea level of revised IPCC emissions scenarios. *Nature* 357: 293–300.

Wilson, N.C. (1994) Surge of hurricanes and floods perturbs insurance industry. *Journal of Meteorology* 19 (185): 3–9.

2

CLIMATE CHANGE, CLIMATIC HAZARDS AND POLICY RESPONSES IN AUSTRALIA

A. Barrie Pittock, Robert J. Allan, Kevin J. Hennessy, Kathy L. McInnes, Ramasamy Suppiah, Kevin J. Walsh, Peter H. Whetton, Heather McMaster and Ros Taplin

2.1 Introduction

Australia has been characterised by one of its more famous poets as 'a land of droughts and flooding rains'. This arises largely from the strong influence of the El Niño–Southern Oscillation (ENSO) which produces widespread drought during El Niño years and heavy rains during the opposite La Niña phase (Allan *et al.*, 1996; Philander, 1990; Pittock, 1975). Australia is also subject to visitations from tropical cyclones on most of the west coast, the whole of the north coast, and as far south as Brisbane on the east coast. The most famous recent tropical cyclone was Cyclone Tracy, which destroyed Darwin on Christmas Eve, 1974. Severe thunderstorms, with often damaging hail, are also a widespread feature of the weather, with hail damage second only to that from tropical cyclones in insured losses (Insurance Council of Australia, 1997). Mid-latitude coastal storms can also lead to extensive damage from storm surges and high wave energy.

Table 2.1, from Coates (1996), summarises the number of fatalities in Australia caused by a range of natural hazards. This suggests that, in a relatively low-latitude country such as Australia, summer heatwaves are the dominant cause, followed by tropical cyclones. The latter is rather surprising considering the sparse population of the Australian tropics. It was dominated in the early years by deaths at sea in pearling and fishing fleets, and more recently by growth of coastal settlements. The trend to greater exposure to tropical cyclones is likely to continue. Flood fatalities rank third, dominated in the north by summer events associated with tropical cyclones and in the south by winter storms. Deaths from bushfires are largely confined to dry summers in the Mediterranean climates of the southeastern coastal strip, while lightning strikes from summer thunderstorms are nearly as numerous.

Table 2.1 Fatalities from natural hazards in Australia

Natural hazard	Period covered	Fatalities
Heatwaves	1803–1992	4,287
Tropical cyclones	1827–1989	1,863–2,312
Floods	1803–1994	2,125
Bushfires	1827–1991	678
Lightning	1803–1992	650
Landslides	1803–1994	32

Source: After Coates (1996)

The situation in regard to damages to property is rather different, with heatwaves apparently losing in importance, to be replaced by hail. Some of the largest insurance claims have in fact arisen from hailstorms in urban areas, with extensive damage to cars, roofs and windows. Table 2.2 is a listing of some of the major insurance losses due to climatic hazards in Australia from 1967 to January 1997. It should be noted that flood damage from rising waters is in general not covered by insurance in Australia except for commercial buildings.

Although damages from heatwaves do not appear in Table 2.2, they may be important at present in regard to crop and livestock losses and disruption to infrastructure (e.g. rail derailments). Evidence for this has been brought together in an unpublished thesis by Kylie Andrews at Macquarie University (Andrews, 1994). She cites a single day in January 1990 when a maximum temperature of 47°C caused Australian $12 million damage to grapevines at Mildura. Global warming could well lead to a much greater impact from heatwaves.

Unfortunately, the climate-related hazards listed here are all phenomena which are difficult to model reliably, especially in global climate models, so their future behaviour under enhanced greenhouse conditions is a major uncertainty. This chapter reviews some of the efforts to understand and model these phenomena, in Australia and elsewhere, and to draw inferences about the future. It also touches on how Australians might best cope with any changes brought about by global warming.

2.2 Regional climate change: The changing context

Possible changes in climatic hazards in the Australian region are of course closely related to the regional manifestations of the global change in average climate. CSIRO has derived estimates of the latter from a range of global climate models (GCMs), generally following the procedures and ranges of uncertainties adopted by the Intergovernmental Panel on Climate Change (IPCC) (Houghton *et al.*, 1990, 1992, 1996).

The procedure adopted by CSIRO (CSIRO, 1992, 1996) has thus been to use time-varying (transient) estimates of global average warming, combined with

Table 2.2 Major climatic catastrophe losses in Australia from 1967 to January 1997

Date	Event	Insurance loss	
		Original dollars ($M)	Dec 1996 dollars ($M)
Feb. 1967	Bushfires, Hobart, Tas.	14	101
June 1967	Rain and hail, Brisbane, Qld	~5	36
Jan. 1970	Tropical cyclone (TC), Ada, Qld	12	79
Aug. 1970	Flooding, Tas.	~5	31
Feb. 1971	Flooding, Vic.	>2	12
Dec. 1971	TC Althea, Qld	>25	147
Feb. 1972	TC Daisy, Qld	2	12
March 1973	TC Madge, N. Aus.	30	150
Jan./Feb. 1974	TC Wanda and floods, Qld	68	328
March 1974	TC Zoe, NSW and Qld	>2	12
Apr. 1974	Flooding, Sydney, NSW	20	98
May 1974	Flooding, Vic.	>4	20
May 1974	Wind and hail, Sydney, NSW	20	98
Dec. 1974	TC Tracy, Darwin, NT	200	837
March 1975	Flooding, Sydney, NSW	15	63
Dec. 1975	Cyclone Joan, WA	20	74
Jan. 1976	Hailstorm, Toowoomba, Qld	12	49
Feb. 1976	Cyclone Beth, Bundaberg, Qld	3	12
Nov. 1976	Hailstorm, NSW	40	131
Dec. 1976	TC Ted, Qld	15	49
Jan. 1977	Thunderstorms, NSW	15	49
Feb. 1977	Storm, Tongala, Echuca, Vic.	4	13
Feb. 1977	Fires, Western District, Vic.	9	30
March 1977	Floods, NSW	5–7	23
Feb. 1978	Storms, Sydney, Newcastle and Wollongong, NSW	10–15	44
March 1978	Storms, north coast, NSW	3–5	15
April 1978	TC Alby, WA	13	39
June 1978	Wind and floods, Sydney, NSW	5–7	21
March 1979	TC Hazel, WA	15	41
Nov. 1979	Hailstorm, SA	10	24
Feb. 1980	TC Dean, Pilbara, WA	20	49
Feb. 1980	Bushfires, Adelaide Hills, SA	13	34
Dec. 1980	TC Brisbane, Qld	7.5	17
Dec. 1980	Storm, Brighton, Qld	15	36
Feb. 1981	Floods, Dalby, and storms, Qld	20	49
Nov. 1982	Storm, Melbourne district, Vic.	8–10	19
Feb. 1983	Bushfires (Ash Wednesday) (single event), Vic.	138	255
Feb. 1983	Bushfires (Ash Wednesday) (single event), SA	38	69
March 1984	Cyclone Kathy, NT	5	12

Table 2.2 Continued

Date	Event	Insurance loss	
		Original dollars ($M)	Dec 1996 dollars ($M)
Nov. 1984	Floods, NSW	80	132
Sept. 1984–Feb. 1985	Bushfires, NSW	25	45
Jan. 1985	Hailstorms, Brisbane, Qld	180	299
Sept. 1985	Hailstorms, Melbourne, Vic.	10	17
Jan. 1986	Hailstorm, Orange, NSW	25	41
Jan. 1986	TC Winifred, Cairns to Ingham, Qld	40	65
Aug. 1986	Storms and floods, Sydney, NSW	35	53
Oct. 1986	Hailstorm, western suburbs, Sydney, NSW	104	161
Dec. 1986	Storm, Adelaide, SA	10	15
Feb. 1987	Bushfires, southern Tas.	7	12
Nov./Dec. 1987	Rain and floods, Melbourne, Vic.	8	12
April 1988	Floods, Alice Springs, NT	10	14
April 1988	Floods, Sydney, NSW	25	36
May 1988	TC Herbie, Carnarvon to Denham, WA (including ship, *Korea Star*)	20	30
Sept. 1988	Storms, widespread, WA	8	12
Nov. 1988	Rainstorms, Vic.	11	15
Feb. 1989	Rainstorms, Melbourne, Vic.	17	24
April 1989	TC Aivu, Qld	26	35
Nov. 1989	Hailstorm, Ballarat, Vic.	17–20	25
Feb. 1990	Hailstorm, Dubbo, NSW	9	12
Feb. 1990	TC Nancy, Qld and NSW	33	42
Feb. 1990	Hailstorm, Sydney, NSW	10	12
March 1990	Hail, Sydney, NSW	319	384
April 1990	Floods, south Qld and western NSW	30	38
Aug. 1990	Storms, Sydney, NSW	12	15
Dec./Jan. 1990/91	Flood and wind, Qld (from TC Joy)	Coal loss: 32 Cyclone: 30	75
Dec./Jan. 1990/91	Bushfires, Vic.	10	12
Jan. 1991	Storms, Sydney, NSW		226
Jan. 1991	Hail, Orbost, Vic.		12
Jan. 1991	Hail, Adelaide, SA (revised 1993)		30
June 1991	Storms and floods, southeast, NSW		15
Oct. 1991	Bushfires, NSW, Central Coast		12
Dec. 1991	Floods and water damage, Melbourne and Ballarat, Vic.		24
Feb. 1992	Storms, Sydney, NSW		118
Sept./Oct. 1993	Flood, Benalla/Shepparton, Vic.		12
Dec. 1993	Storms, Melbourne, Vic.	10	12
Jan. 1994	Bushfires, NSW	48	56
May 1994	Windstorms, Perth, WA	24	39

Table 2.2 Continued

Date	Event	Insurance loss	
		Original dollars ($M)	Dec 1996 dollars ($M)
Nov. 1994	Windstorms, NSW	14	15
Nov. 1994	Windstorms, NSW	13	14
March 1995	Cyclone Bobby, WA		11
Nov./Dec. 1995	Hailstorms, southeast Qld		40
Dec. 1995	Storms, Hunter Valley, NSW		10
Jan./Feb. 1996	Storms, Sydney, NSW		14
April 1996	Cyclone Olivia, WA		10
May 1996	Floods, southeast corner Qld		31
Aug. 1996	Storm, Sydney, NSW		10
Sept. 1996	Hail, Armidale/Tamworth, NSW		150
Nov. 1996	Floods, Coffs Harbour, NSW		35
Nov. 1996	Hail, Tamworth, NSW		10
Dec. 1996	Hail, Singleton, NSW		50
Jan. 1997	Bushfires, Dandenong Hills, Vic		10

Source: Based on Insurance Council of Australia (1997)

estimates of the pattern of local temperature and precipitation change per degree global warming. The range of uncertainty was derived by using the IPCC ranges for greenhouse gas emission scenarios (highest is IS92e, and lowest IS92c) and global climate sensitivities (highest is 4.5°C warming for an equilibrium doubling of CO_2, lowest 1.5°C). These were combined with the second highest and second lowest estimates (at each grid point) for local change per degree global warming from five GCMs.

Australia is situated in an oceanic hemisphere, and at a relatively low latitude. It is also in the southern hemisphere, where the cooling effect of sulphate aerosols (microscopic particles in the atmosphere formed from sulphur emissions) is much smaller than in the industrialised areas of the northern hemisphere. Thus it is necessary to allow for the greater warming in the southern hemisphere, relative to that in the northern, in global warming estimates (CSIRO, 1996).

CSIRO's earlier estimates of future climates (CSIRO, 1992) were based on the results of five slab-ocean GCMs (models with a simple representation of the oceans). New estimates based on five coupled-ocean GCMs, in which the deep ocean circulation and ocean currents are explicitly represented, give significantly different results, particularly in summer (Whetton *et al.*, 1996). This is largely due to the coupled models sequestering heat into the deep waters of the Southern Ocean, leading to reduced surface warming at high southern latitudes, and the suppression of the Australian monsoon by relatively cool surface waters moving equatorward in the eastern Indian Ocean.

In brief summary, the results of these assessments are as follows. Northern coastal areas of Australia may warm at about the same rate as the global average (by 0.9 to 1.3 times the global average), while southern coastal areas have a greater range of possibilities (0.8 to 1.6 times). Inland Australia may warm by about or more than the average amount (1.0 to 1.8 times). Using the IPCC ranges for global averages, these are readily turned into local warming ranges at any time up to 2100.

Seasonal rainfall changes over Australia are expected to differ between summer and winter, with both the slab and coupled models suggesting winter rainfall decreases in the range 0 to 10 per cent per degree of global warming over most of southern Australia. In the extreme south of mainland Australia, changes in winter could be in the range +5 to −5 per cent, while over Tasmania they could be 0 to +10 per cent per degree of global warming.

The situation is more complex in summer, with the slab-ocean models indicating increases of the order of 0 to 10 per cent per degree of global warming in monsoon rainfall in the north and west, while the coupled models indicate a similar or slightly larger-magnitude decrease over the whole country. Changes in the southwest and southeast in the slab models are in the range +5 to −5 per cent per degree of global warming.

In any assessment of future changes in the frequency, location and magnitude of climatic hazards, the uncertainties in the changes in average climate are clearly an important limiting factor. Thus continuing improvements in the reliability and detail of estimates of changes in mean climate, from global models, are essential to a better understanding of possible changes in climatic hazards. With this caveat in mind, the following sections will discuss what is known about how to estimate changes in hazards.

2.3 ENSO in a warmer world

ENSO events are complex interactions between the atmosphere and the tropical oceans which have significant effects in many mid- and low-latitude countries. They are one of the principal causes of climatic hazards in Australia. El Niño years are characterised by widespread drought in northern and eastern Australia, while La Niña years bring heavy rains, with frequent and widespread flooding (Allan et al., 1996; McBride and Nicholls, 1983; Suppiah and Hennessy, 1996). Moreover, ENSO affects the location of the South Pacific Convergence Zone (SPCZ) (Vincent, 1994), which lies further east in El Niño years than in La Niña years. Similarly, tropical cyclones occur further east in the tropical Pacific in El Niño years than in La Niña years (Evans and Allan, 1992; Basher and Zheng, 1995). Local mean sea-level around the Australian coast also varies by tens of centimetres during the ENSO cycle (Wyrtki, 1985; Pariwono et al., 1986; Bryant, 1988).

While a variety of models exist which attempt to reproduce ENSO behaviour, until recently there have been relatively few studies of possible changes to ENSO. In 1993, a workshop was held (Bureau of Meteorology Research Centre, 1993) which concluded that ENSO was likely to continue relatively unchanged into the future, with a possible increase in the intensity of associated droughts and floods. This brief review is mainly concerned with the results of model and observational studies since then.

Unfortunately, the reliable instrumental record of ENSO events is little more than a century long (Allan *et al.*, 1996), while the historical and palaeoclimatic record is fragmentary. Solow (1995) has analysed the historical ENSO record and found no statistically significant trend in the frequency of El Niño events over the period 1525–1987. Latif *et al.* (1995) analysed atmospheric changes over recent decades and concluded that the extended El Niño in the early 1990s is not a result of greenhouse warming, but rather a manifestation of natural variability. This is supported by a reconstruction of the Southern Oscillation by Allan and D'Arrigo (1998) which suggests that there have been several protracted El Niño and La Niña sequences in the record. However, Trenberth and Hoar (1996) lean towards an enhanced greenhouse explanation, on the basis of statistical modelling of the 100–year Darwin mean sea-level pressure (MSLP) record. See discussion by Kerr (1994).

2.3.1 Simplified and limited area models

Basic models attempt to capture the important physics of the coupled atmosphere–ocean system in a simplified form. Although tightly constrained, such models can now provide relatively good representations of ENSO features. This is partly borne out by their performance when used in predictive mode (e.g. Chen *et al.*, 1995), although none of them did well in the early 1990s (Kerr, 1993). In a recent study involving one such model, Kleeman *et al.* (1996) report that the imposition of steadily increasing background temperature leads to an increase in the frequency of ENSO-like oscillations. Similar results have also been found using the Scripps hybrid model (R. Kleeman, private communication).

Relatively high-resolution models have been developed which include explicit physics but have an ocean domain which is confined to the tropical Pacific Ocean. The capabilities of these types of models are indicated by recent results of the CSIRO ENSO General Circulation Model (CEGCM).

In Figure 2.1, the ocean model is forced by observed climatological mean winds. The upper plot is a north–south section across the Equator of the top 300 m of the eastern Pacific, showing the east–west velocity. The lower plot is an east–west cross-section along the Equator showing ocean temperatures. These plots show that the model realistically simulates the dynamic and thermal structure of the tropical Pacific, including an equatorial undercurrent at the observed depth, and a 'tight' thermocline.

Simulations of interannual variability using the CSIRO CEGCM model are also quite realistic, with irregular ENSO events occurring at intervals of 4 or 5 years. Sea surface temperature anomalies of about the observed magnitude occur in the eastern Pacific and propagate westward, as actually occurs. During modelled El Niños westerly wind bursts occur in the west and extend towards the

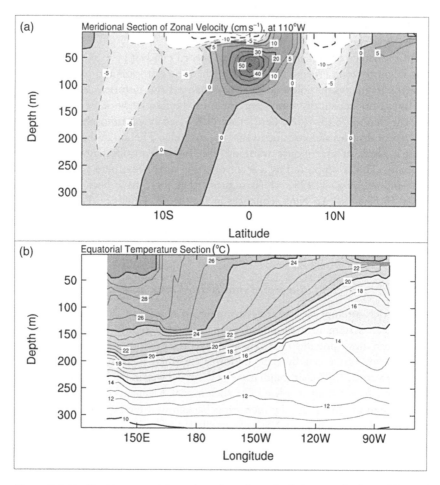

Figure 2.1 Pacific Ocean model results using climatological mean forcings: (a) zonal velocity versus depth at longitude 110° W; (b) temperature versus depth along the Equator

dateline, generating Kelvin waves and the eventual onset of the La Niña phase of the oscillation. This work is presently being extended using a global atmospheric model coupled to the CEGCM to investigate the effect of global warming.

There are reports that the European Centre for Medium-Range Weather Forecasting and Max Planck Institute models, when coupled to atmospheric models, yield increases in the amplitude and frequency of El Niño-type oscillations under warmer background conditions (Stockdale, private communication).

Work is progressing on coupling the DAR high-resolution ocean model to the CSIRO nine-level atmospheric GCM (AGCM) for use as an ENSO forecast tool. These models can provide some indications of the sensitivity of the climate system to changes in the background state but, in the long term, computationally more expensive global atmosphere–ocean models will be needed to address fully the implications of greenhouse warming.

2.3.2 Global models

Meehl *et al.* (1993) analysed experiments with an atmospheric general circulation model coupled separately to both a global ocean model and a slab-ocean model. Simulated warm anomalies in the eastern Pacific for a $1 \times CO_2$ climate simulation were interpreted as ENSO-like warm events. Neither the frequency nor the strength of these anomalies altered significantly in a $2 \times CO_2$ climate simulation. Although anomalously wet and dry regions bordering the Pacific were similar in pattern in both simulations, the anomalies were intensified. This apparent intensification of the hydrological cycle was attributed to the higher mean sea surface temperatures (SSTs) that, near the rising branch of the Walker circulation, led to greater evaporation, low-level convergence and therefore precipitation. Their finding that dry areas tended to be drier and wet areas to be wetter was related to an overall strengthening of the Walker circulation. Smith *et al.* (1997) describe the results of a similar study that makes use of both observed and slab-ocean model equilibrium SST. They find little evidence of a significant change to the intensity of the hydrological cycle during ENSO events under possible future greenhouse conditions. Their results indicated that increased stability of the tropical atmosphere under $2 \times CO_2$ conditions acts to dampen any rainfall intensification associated with the higher SSTs. A similar increase in stability is also reported by Boer (1993) in an analysis of a $2 \times CO_2$ climate simulation.

Tett (1995) performed an experiment with a United Kingdom Meteorological Office (UKMO) coupled model in which CO_2 was increased at a rate of 1 per cent (compounded) per year for 70 years. These results also did not reveal any significant changes to the interannual variability of either SSTs or precipitation over the Pacific Ocean. Knutson and Manabe (1994, 1995) have also analysed the results of a transient coupled-model experiment with quadrupled CO_2 values. They noted that the amplitude of ENSO-like SST anomalies decreased slightly. ENSO-like precipitation anomalies were enhanced, but the associated surface wind anomalies were not greatly affected. They also noted several mechanisms which acted to nullify intensification of associated atmospheric anomalies.

Coupled-model results obtained by CSIRO reveal ENSO-like warm and cold events under $1 \times CO_2$ conditions although the amplitude is much less than observed (Gordon and O'Farrell, 1997). Under enhanced (up to $3 \times CO_2$) greenhouse conditions, these events were also found to continue but with no major changes in either amplitude or frequency. This is indicated in Figure 2.2, which shows the simulated temperature anomalies for the eastern equatorial Pacific. Interdecadal variability of the ENSO-like signals are greater than changes due to enhanced greenhouse warming.

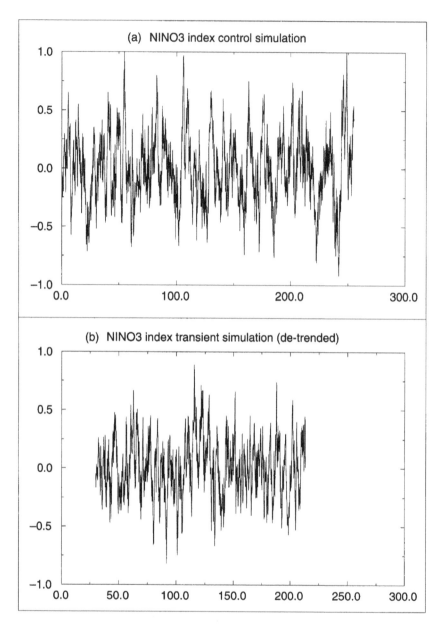

Figure 2.2 Simulated sea surface temperature (SST) anomalies °C in the eastern equatorial Pacific (region NINO3) from the CSIRO coupled model experiments: (a) anomalies simulated under $1 \times CO_2$ conditions; (b) anomalies simulated under gradually increasing (1 per cent compounded) CO_2 conditions, with the trend in SST removed. Note the period of decreased amplitude of the year-to-year variations between years 150 and 180.

Although the global coupled models can exhibit ENSO-like oscillations, the coarse resolution of the models limits their ability to resolve all the underlying processes. A recent exception is the model developed by Tokioka *et al.* (1995), in which the resolution of the ocean model is enhanced near the tropics. They also found that CO_2-induced warming had no definite effect on ENSO-like oscillations in their model, which suggests that resolution may not be important for this type of problem.

2.3.3 *Conclusions regarding ENSO*

At present the weight of evidence (from observations and global GCM studies) does not point to significant greenhouse-induced changes to the amplitude or frequency of ENSO SST anomalies in the future. Neither is there yet any conclusive evidence that atmospheric anomalies will be more intense. It is thus arguable that any future changes in average ENSO behaviour are likely to be less than the interdecadal variability seen in the historical record.

This does not mean that drought and flood frequencies and intensities in Australia will not change. The same length of time without rain, but with higher temperatures, would lead to more rapid drying of the soil, and thus longer and more intense drought. Conversely, any change in rainfall intensities, even without changes in the number of rainy days, could alter flood frequencies and intensities. Moreover, while ENSO is the dominant cause of flood and drought events in Australia, it accounts for at most about 50 per cent of the variance of rainfall in the most affected regions and seasons.

2.4 Extreme rainfall, run-off and floods

Extreme rainfall events cause significant damage to agriculture, ecology and infrastructure, disruption to human activities, injury and loss of life (Coates, 1996). For example, as a result of the flood that affected Sydney in August 1986, $100 million worth of damage occurred and six lives were lost (Lynch, 1987). Joy (1991) has estimated that the annual cost of flooding to Australia is $380 million. A change in the probability of extreme rainfall would have important implications for engineering, the insurance industry and other sectors that, at least until recently, have assumed that the climate is constant. Increases in heavy rainfall may lead to an increase in the frequency of flood events, landslides, soil erosion, accumulation of silt in dams, inundation of lowland areas, aquifer recharge, and possibly salinisation due to rising water tables.

Analyses of observational records during the past century (Karl *et al.*, 1995; Suppiah and Hennessy, 1996) indicate a tendency for increasing heavy rainfall events in the USA and in the Australian tropics over the past 80 years, but little change in the former Soviet Union and China. In particular, at the wettest tropical Australian stations, the top 10 per cent of daily rainfall intensity has increased by 20 per cent, and the top 5 per cent of daily rainfall intensity has increased by 10 per cent. Trends in the intensity and frequency of heavy rainfall events show decadal-scale fluctuations, suggesting a possible link with variability due to the El Niño–Southern Oscillation (ENSO). However, after removing the ENSO influence from heavy rainfall trends, increasing trends remain. This implies that the increases are not solely due to the interannual variability of ENSO.

Suppiah and Hennessy (in press) have extended the analysis of heavy rainfall to 125 stations covering most of Australia during the summer half-year (November to April) and winter half-year (May to October). In the summer half-year, there is a general increasing trend in the top 10 per cent of daily rainfall, with twenty-one stations having statistically significant increases. During the winter half-year, there is also an increasing trend except for significant decreases in southwest Western Australia. This seasonal dependency strongly dominates annual trends in the top 10 per cent of rainfall intensity shown in Figure 2.3.

Figure 2.3 Locations of stations showing an increasing (+) or decreasing (−) trend in the magnitude of the annual top 10 per cent of daily rainfall intensity over the period 1910–90. Bold positive and negative symbols indicate trends statistically significant at the 95 per cent confidence level.

GCMs from the United Kingdom Meteorological Office (UKHI), CSIRO and the United States National Center for Atmospheric Research (NCAR) simulate increases in heavy rainfall intensity almost globally under enhanced greenhouse conditions (Gordon *et al.*, 1992; Gregory and Mitchell, 1995; Fowler and Hennessy, 1995; Henderson-Sellers *et al.*, submitted). At middle latitudes, these models simulate a decrease in light rainfall intensity and number of rain days, and a disproportionately large increase in the intensity of heavy rainfall. This is shown in Figure 2.4. The heavy rainfall threshold of 12.8 mm/day is small in the real world, but is large in the models because simulated rainfall represents an areal average over at least $100,000$ km^2. To overcome the difference between real and simulated magnitudes of heavy rainfall, the increase in simulated rainfall intensity can be expressed as a change in the return period of an event with a given magnitude. For a doubling of CO_2, heavy-rainfall return periods are reduced by a factor of 2 to 5 (i.e. they become two to five times more frequent). Moreover, the CSIRO nine-level GCM indicates that synoptic systems producing heavy summer rainfall over central Australia become more intense under doubled CO_2 conditions (Suppiah, 1994).

Bates *et al.* (1996) investigated the potential impacts of climate change on runoff in six Australian catchments. They used a stochastic weather generator to simulate 1,000-year daily weather sequences from present and doubled CO_2

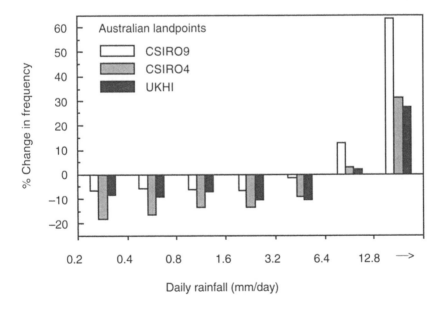

Figure 2.4 Percentage of change in the frequency of simulated daily rainfall (mm day^{-1}) by CSIRO nine-level, CSIRO four-level and UK Meteorological Office high resolution GCMs averaged over the Australian continent

climates simulated by the CSIRO9 GCM. The weather sequences were used to drive two different rainfall run-off models, previously calibrated with historical data. For most catchments, the CSIRO9 GCM simulated sizeable increases in summer rainfall. Both hydrological models gave generally large increases in median monthly run-off in five of the six catchments during the high run-off periods of the annual cycle, and notable increases in the annual maximum monthly run-off. The latter were related to increases in extreme rainfall events, and indicate an increase in flood frequency and magnitude. While the magnitude of these increases varied between catchments and hydrological models, typical reductions in return period of heavy flow events for a CO_2 doubling were by a factor of 2 to 5.

An urban flood damage study was carried out for the Hawkesbury–Nepean and Queanbeyan regions of eastern Australia by Minnery and Smith (1996) using the ANUFLOOD damage model. They used GCM-based scenarios for the year 2070, in which flood events with return periods of 1 in 100 years or more could occur four times more often (e.g. a flood event with a 1 in 200 year return period could become a 1 in 50 year event), and events with return periods of less than 1 in 100 years could occur three times more often.

Highly preliminary estimates suggest that the combined residential and commercial direct damage cost simulated by ANUFLOOD increases by factors of about 5 to 10 in the two regions. There are significant increases in the simulated number of residences with over-floor flooding and the number liable to collapse. Since methodologies exist to assess flood risks to urban areas (Smith, 1991), these could be applied to flood-prone areas to assess future vulnerability to climate change together with existing information on trends in rainfall intensity. Moreover, climate change needs to be included in the formulation of management policies for flood-prone areas at the regional and national level.

A study of impacts on run-off and soil moisture by Schreider *et al.* (1996), for a number of major Victorian tributaries of the Murray River, used GCM-based 'most wet' and 'most dry' scenarios for 2030 and 2070 in the IHACRES unit hydrograph model. The scenarios were based on results from five slab-ocean GCMs. For 2030, the most dry scenario had a mean warming of 2°C, no change in summer half-year rainfall, but a 10 per cent decrease in the winter half-year. This resulted in about a 35 per cent decrease in mean run-off, more low-flow days, and more frequent dry soil. The corresponding most wet scenario had a mean warming of 1.5°C, a 20 per cent increase in summer half-year rainfall, and a 10 per cent increase in the winter half-year. This resulted in little change in mean flow, but a large increase in the frequency of heavy-flow (flood) events. The results for daily flow are summarised, for both 2030 and 2070, in Figure 2.5, and for soil moisture in Figure 2.6.

Using generally similar scenarios, in a more elaborate hydrological model (MODHYDROLOG), Chiew *et al.* (1995) found annual run-off changed by +50 per cent to −50 per cent in catchments widely distributed around Australia

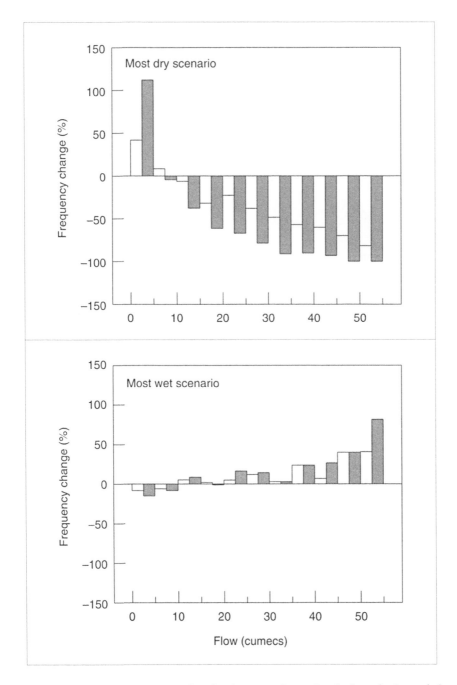

Figure 2.5 Changes in daily river flow for the Upper Ovens river basin under 'most dry'
and 'most wet' scenarios for the years 2030 (unshaded) and 2070 (shaded).
Source: After Schreider *et al.* (1996)

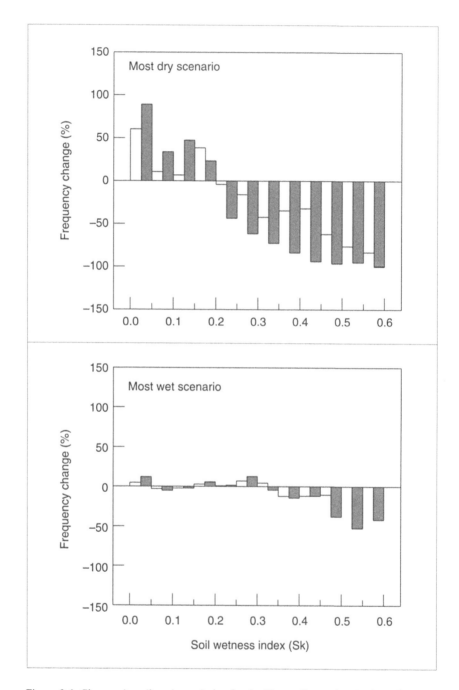

Figure 2.6 Changes in soil moisture index for the Upper Ovens river basin under 'most dry' and 'most wet' scenarios for the years 2030 and 2070
Source: After Schreider *et al.* (1996)

by the year 2030. Their hydrological model showed less sensitivity to temperature increases than the IHACRES model used by Schreider *et al.* (1996).

If the studies by Bates *et al.* (1996), Chiew *et al.* (1995) and Schreider *et al.* (1996) were to be repeated using the more recent coupled-ocean GCM results (Whetton *et al.*, 1996; CSIRO, 1996), the simulated reductions in summer rainfall, despite reduced warming, would probably lead in general to reduced average run-off and greater aridity, relative to the present.

2.5 Tropical cyclones

The frequency, intensity and location of tropical cyclones in the Australian and southwest Pacific regions are of vital importance. In Australia, tropical cyclones have been responsible for some 2,000 fatalities since 1827 (Coates, 1996), with many of the earlier deaths occurring at sea among fishing and pearling vessels. However, the rapid growth in population in northern Australia, especially on the Queensland coast, is increasing the exposure to this hazard on land, especially from storm surges, high winds and flood damage.

Despite widespread fears of greater numbers and intensities of cyclones, the potential effects on cyclones of climate change are highly uncertain (Lighthill *et al.*, 1994; Emanuel, 1995; Broccoli *et al.*, 1995). There are several reasons why this is so. At present, tropical cyclones cannot be simulated faithfully in climate models, mostly because of the coarse horizontal resolution of the models compared to the scales of the important processes occurring in observed tropical cyclones. For example, the maximum wind speed of an observed cyclone typically occurs inside a radius 100 km from the centre of the storm, which is less than the horizontal resolution of most climate models. Additionally, in several regions of the globe, the numbers and geographical distribution of tropical cyclones are strongly influenced by ENSO (Evans and Allan, 1992; Basher and Zheng, 1995). As discussed above, the state of ENSO in a changed climate may not alter dramatically, but nevertheless, subtle changes in atmospheric circulation and underlying sea surface temperatures (SSTs) could affect tropical cyclone frequency and tracks.

Despite current uncertainties, significant advances have been made in recent years towards understanding the effects of climate change on tropical cyclones. While the representation of tropical cyclones in climate models is crude in several respects, recent experiments have demonstrated improvements in the simulation of the observed climatology. Bengtsson *et al.* (1995) showed that a GCM at about 125 km resolution could simulate the observed geographical distribution and seasonal variation of tropical cyclones. Similar ability was shown in the work of Tsutsui and Kasahara (1996), using a coarser-resolution GCM. Hirakuchi and Giorgi (1995) briefly examined the performance of a regional climate model in simulating tropical cyclones in the northwest Pacific region. They found that the intensities of the systems generated in their model were substantially less than observed. A preliminary simulation of the $2 \times CO_2$ climatology of tropical cyclones has been performed by Bengtsson *et al.* (1996) using a slab-ocean GCM, but with the many current uncertainties it would be premature to draw conclusions from this study.

The CSIRO regional climate model DARLAM (McGregor, 1987) can simulate tropical cyclone-like vortices when run at a horizontal resolution of 125 km. Figure 2.7 shows a sequence of mean sea-level pressure fields generated by DARLAM driven by January climatological SSTs (Walsh and McGregor, 1995). The maps depict the southward progress of a tropical low-pressure system

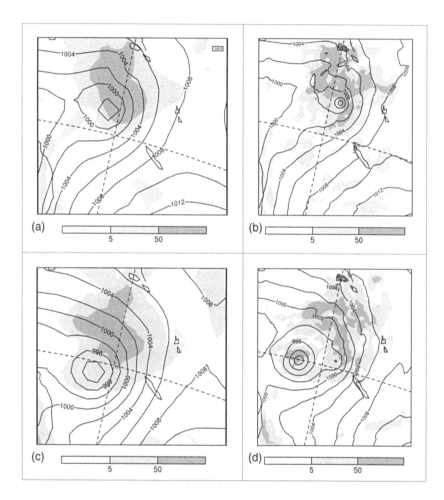

Figure 2.7 Tropical cyclone-like vortices simulated in the CSIRO regional climate model (DARLAM). Daily sequence of mean sea-level pressure (MSLP) and precipitation in the Coral Sea region for (a and c) 125 km horizontal resolution and (b and d) 30 km horizontal resolution. Contour interval for MSLP is 2 h Pa. Shading for precipitation is indicated in key, in mm day^{-1}

in the Coral Sea. On the left is a simulation at 125 km resolution, and on the right results for the same system when the model is doubly nested at 30 km resolution. An increase in intensity with resolution is clear. These systems have several of the observed characteristics of a tropical cyclone: a warm core, a low-level wind maximum and an associated area of heavy precipitation. The vertical composite structure of these systems (not shown) also has some similarity to composites of observed tropical cyclones (e.g. Frank, 1977). Work is now in progress to determine the climatology of these cyclone-like vortices in DARLAM and to compare it to reality (Walsh and Watterson, 1997). If this control climatology is satisfactory, simulations with DARLAM nested in the CSIRO GCM under doubled-CO_2 conditions will be assessed for changes in cyclone behaviour.

Another approach has involved the use of parameters derived from theoretical arguments regarding the relationship of tropical cyclones to large-scale atmospheric fields, either observed or simulated. These parameters are evaluated from GCM output to obtain an indication of the changes in their values under $2 \times CO_2$ conditions. Both numbers and intensities of tropical cyclones have been examined in this fashion. Numbers have been evaluated using the seasonal genesis parameter (SGP) of Gray (1975). Ryan *et al.* (1992) used the SGP to assess the effect of changes in atmospheric dynamics and thermodynamics on tropical cyclone numbers as a result of global warming. Large changes in numbers under $2 \times CO_2$ conditions were simulated, but these were dominated by changes in the thermodynamic terms associated with SSTs.

As the observed correlation between SST and tropical cyclone numbers is poor, this result points to the need to improve the genesis parameter. Watterson *et al.* (1995) went on to use Gray's SGP to evaluate the ability of a climate model to simulate observed interannual variations in cyclone numbers. They found that the SGP could do so only in certain regions, which implied deficiencies in either the GCM or the SGP, or both.

Theoretical maximum intensities of tropical cyclones have been formulated by Emanuel (1988, 1991) and Holland (1997). These estimates of maximum potential intensity (MPI) in a given region are based on factors such as the temperature structure of the atmosphere. Simulated changes in maximum intensities under enhanced greenhouse conditions using the Holland formulation were assessed by Tonkin *et al.* (1996). They found that over the North Pacific there was a modest increase in MPI and no significant change in regions of cyclone formation. At present, it must be emphasised that the simulated changes in intensities under $2 \times CO_2$ conditions are smaller than the errors in the MPI simulated under the current climate. Thus predictions using such techniques currently suffer from serious uncertainties.

Climate change may also influence the preferred paths of tropical cyclones, as the motion of tropical cyclones is affected by the large-scale circulation patterns (Harr and Elsberry, 1995). This can in principle be investigated either from the climatological relationship between tropical cyclone paths and the large-scale circulation, and then simulated changes in the circulation caused by climate change, or more directly by analysing the paths of simulated tropical cyclones under present and enhanced greenhouse conditions in high-resolution GCMs or regional models.

In summary, while the estimation of the effects of climate change on tropical cyclones is being actively investigated, no firm conclusions can yet be reached. Methodologies are being developed for studying their genesis, paths and, to a limited extent, their intensities. In each case the credibility of the results is critically dependent on the veracity of the large-scale climate simulations from the climate models.

2.6 Storm surges and inundation

Inundation of low-lying coastal terrain due to rising sea level is a serious outcome of the enhanced greenhouse effect. However, the temporary sea-level fluctuations which occur as a result of severe storms are potentially at least as destructive to the coastal zone. A combination of strong winds and low atmospheric pressure can lead to the generation of oceanic storm surges which can cause flooding over low-lying coastal regions, and can enable the destructive forces of wind-generated waves to penetrate further inland. Elevated sea levels can also combine with high rainfall to impede drainage and prolong upstream flooding.

The issue of coastal inundation on both long and short time scales is highly site specific. Significant variations in local sea-level rise may occur as a result of changes in ocean circulation and local atmospheric pressure associated with global warming and climate change (Pittock *et al.*, 1995). For sea-level fluctuations due to storm surges, the impact of climate change on the frequency, intensity and movement of severe storms must be considered. Local coastal geomorphology and bathymetry are also vital factors (Pittock and Flather, 1993).

In tropical Australia, storm surges up to several metres high are often caused by intense low pressures and strong onshore winds associated with tropical cyclones (Hubbert *et al.*, 1991). In mid-latitudes, extra-tropical cyclones and associated cold fronts can generate storm surges. However, since such storms are usually of larger spatial scale and less intense in terms of central pressure, it is often sustained winds parallel to the coast which are responsible for the generation of surges. For example, along the southern coast of Australia, storm surges typically of about 1 m are generated by westerly winds associated with wintertime cold fronts (McInnes and Hubbert, 1994; Pittock *et al.*, 1996).

These surges affect not only the open coastline, but also embayments such as Port Phillip, where the city of Melbourne is located. Figure 2.8 shows an example of the observed and modelled sea-level heights which occurred during a storm surge event at the northern end of Port Phillip Bay. The predicted normal tidal levels are shown also and illustrate how the sea-level increase due to the storm surge, in this type of event, is comparable to the worst estimates of mean sea-level rise by late in the twenty-first century.

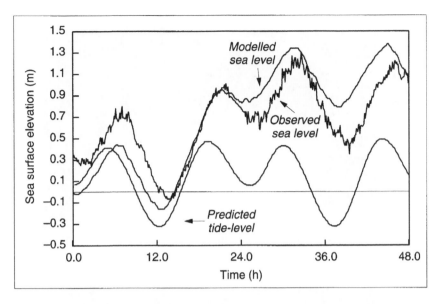

Figure 2.8 Observed and modelled sea levels at St Kilda (on Port Phillip Bay, Victoria) over a 60-hour interval for a storm surge event, November 1994. The predicted tide level is also shown. All sea levels are referenced to the Australian Height Datum.

While local estimates of changes to severe storm climatologies and long-term sea-level rise are not yet available, it is possible to estimate the potential impact of storm surges in a given region using high-resolution storm surge inundation models under different plausible scenarios. Figure 2.9 illustrates this for the storm surge shown in Figure 2.8. The inundation model has been run on a 30 m grid (McInnes and Hubbert, 1996) under observed conditions (Figure 2.9a) and shows inundation confined mainly to the beach front. When the model was rerun with the imposition of an 80 cm mean sea-level rise and a 10 per cent increase in the wind strength (to simulate the possible effects of a stronger storm) (Figure 2.9b), the areas of inundation increased in the region shown from 0.9 km^2 to 2.9 km^2, and included some low-lying upstream and residential areas.

While coastal erosion and deposition can be accelerated during storm surge events, even slight shifts in the strength or direction of the prevailing wind conditions can have dramatic effects on beach orientation (Cowell and Thom, 1994). This is brought about by changes in the wave energy impinging on a particular region of coastline which can alter the net transport of sediment along the beach. Shifts in prevailing wind regimes and local sea-level variations associated with the ENSO phenomenon can cause coastal erosion (Bryant, 1983). In the context of climate change, reliable estimates of the long-term changes in

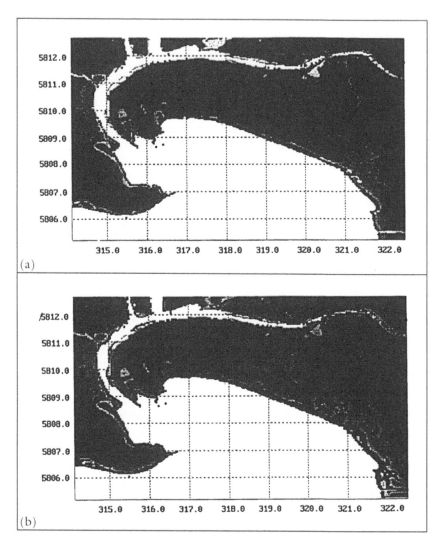

Figure 2.9 The modelled inundation at Port Phillip Bay for (a) an observed present-day storm and (b) the same storm with a superimposed background sea-level rise of 80 cm and 10 per cent increase in wind speed. This is a worst-case scenario for the year 2070

Source: McInnes and Hubbert (1996)

wind regime affecting coastal areas, and of local sea-level variations, will be needed to address this issue.

Impacts on particular coasts due to climate change and sea-level rise will not be determined until more reliable scenarios are available for regional variations in mean sea-level, changes in ENSO, and changes in tropical and mid-latitude storms. Specific calculations will be needed for each locality to account for the local topography and bathymetry, and possible changes in beach alignment and transport of sand or coral debris. As the inundation results for Port Phillip Bay illustrate, such detailed studies are possible once scenarios are available.

2.7 Hail

Hail is the occasional product of a thunderstorm. The possibility of an increase in the frequency and/or severity of thunderstorms with global warming arises from theoretical arguments for an increase in the strength of the hydrological cycle.

Hail falls frequently in some part of every continent except Antarctica. The highest frequencies of hailstorms world-wide tend to be in the continental interiors of mid-latitudes. Frequencies drop off rapidly towards the poles, where moisture and vigorous cloud development are limited. In the tropics, thunderstorms are frequent but there are few hailstorms. Hailstorms that do occur in the tropics, occur mostly at high elevations (Gokhale, 1975).

2.7.1 Conditions necessary for hailstorm formation

The fundamental mechanisms that generate a hailstorm are the same in any region of the world (Admirant et al., 1985). Since hail is the occasional product of a thunderstorm, the pre-existing conditions necessary for hailstorm development are the same as those necessary for thunderstorms generally: low-level moisture, vertical instability and a lifting mechanism such as a cold front or a mountain range.

A condition that has been associated with hailstorms in particular is the height of the freezing level. Pappas (1962) suggested a technique of hail forecasting based on the height of the freezing-level and the ratio of cloud depth below the freezing level to the cloud's estimated vertical development. Steiner (1989) found that in New Zealand, a mean freezing level height below about 3,400 m was necessary for hail occurrence. He also found that the mean annual freezing level at a weather station was highly correlated with the average annual hail day frequency at that station. A lower than usual height of the freezing level may well account for occasional reports of hailstorms in the tropics (Neumann, 1965).

Miller (1972) related hailstorm occurrence and the size of hailstones to the height of the wet-bulb freezing level. The wet-bulb freezing level is lower than the freezing level, its height being a measure of dryness at mid-levels as well as temperature. Miller found that large hail in the United States mostly fell when the wet-bulb freezing level was between 7,000 and 11,000 feet (2,100–3,400 m) above the surface.

Severe thunderstorms which sometimes produce long hailswaths and large hailstones driven by strong winds have been associated with dry mid-level air, moderate to strong wind shear (variation in horizontal wind with height) and strong upper tropospheric winds (Gokhale, 1975; Barnes and Newton, 1986). The climatology of particularly severe hailstorms has been studied in a few regions. La Dochy (1985), for example, examined synoptic conditions in Manitoba, Canada, on 50 days when hailstorms caused the most crop damage claims in a 10-year period. All these hailstorms were frontal, and were associated with a polar jet stream (upper tropospheric wind) and distinctive synoptic conditions of instability and wind shears. Willemse (1995) found the main features favouring the largest hailstorms in the Mittelland of Switzerland to be airflow from the southwest (warm, moist air), frontal disturbances (particularly cold fronts), dry air around the 700 hPa level and a freezing level between 3,000 and 4,500 m above sea level.

2.7.2 *Hailstorms in Australia*

In Australia, hailstorms are the second most costly weather phenomenon after tropical cyclones (Insurance Council of Australia, 1997). The greatest insurance payout in the past 30 years was for damage caused by Cyclone Tracy in 1974, and the next most costly event was a Sydney hailstorm in March 1990 which caused an insurance loss of $384 million (1996 value). Insurance payouts for hailstorm damage have been highest in New South Wales. Crop insurance records show the northern tablelands of New South Wales to be the most hail-prone region of the state.

Hailstorms are most frequent and most severe in late spring and summer. In New South Wales, the frequency of all hailstorms (McMaster, 1997) and the frequency of the most severe hailstorms (hailstone diameter >2 cm) peaks in November (Griffiths *et al.*, 1993).

2.7.3 *Hailstorm prediction using GCMs*

Severe hailstorms are not represented in GCMs because they are small-scale features requiring specific synoptic conditions. GCMs are still a long way from being able to represent such conditions accurately on a daily basis and on a regional scale. It may be possible using nested models at very high resolution, or future higher-resolution GCMs, but this would require enormous computing power to establish a good climatology.

Alternatively, GCMs can be used to predict changes in thermodynamic conditions related to hail damage on a broader scale and on a seasonal rather than a daily basis. McMaster (1997) has used this approach in relation to an insured winter cereal crop loss record for New South Wales. The period of the year during which the crop was most susceptible to hail damage corresponded well with the season when hailstorms were most frequent and most severe. A seasonal index combining measures of low-level moisture and vertical instability was found to be highly correlated with crop losses within a 200 km radius of each of two weather stations ($r=0.68$, chance probability <0.01 for both stations). Because the topography around the stations was fairly uniform, it was not considered to have an important influence on hailstorm formation. Output from three slab-ocean GCMs was then used to predict changes in the index under enhanced greenhouse conditions. All GCMs predicted no significant change in the seasonal index. Increases in low-level (850 hPa) temperature and moisture were predicted by all three models for both locations, but the increased probability of hailstorms predicted by these near-surface variables was offset by equivalent increases in temperature higher in the troposphere (at 500 hPa). The three GCMs also predicted that freezing levels would rise. If freezing levels rise above historical levels, this could result in less frequent hailstorms or smaller hailstones at lower latitudes.

The damage caused by hailstorms also depends on the speed of wind driving the hailstones. Conditions favouring thunderstorms with strong downdrafts (i.e. medium to strong wind shear and dry mid-level air) were not included in McMaster's model. However, from observations at both weather stations in New South Wales, McMaster found a recent increase in the humidity of mid-level air to correspond with a decline in hail damage to crops in the surrounding areas. This decline occurred without a corresponding decline in the seasonal index or in the frequency of thunderstorms.

2.8 Temperature extremes

Day-to-day temperature variability is important because short periods of extreme temperature may adversely affect human health, crops and animals, and the viability of coral reefs in tropical waters. Indeed, surprising as it may seem, among all natural hazards, heatwaves have historically been the major cause of fatalities in Australia, with an estimated 4,287 deaths directly attributable to heatwaves during the period 1803–1992 (Coates, 1996). Most of those who died were elderly, and there is a decreasing trend.

It should be noted, however, that while a general warming trend may increase the frequency of heatwaves in summer, it may reduce periods of extreme cold in winter. Indeed, a study of the relationship between deaths and climate for five Australian cities indicates that the increase in deaths due to warmer summers, under enhanced greenhouse climate, would be almost offset by a decrease in winter deaths (Guest et al., in preparation a, b).

There is little consistency between GCMs as to the direction of change in daily temperature variability simulated in enhanced greenhouse experiments. While numerous studies have analysed greenhouse-gas-induced changes in variability in selected regions (Wilson and Mitchell, 1987; Rind et al., 1989; Mearns et al., 1990, 1995; Fowler et al., 1992), only the UKMO study (Cao et al., 1992) has considered changes on a global scale. It is generally concluded that there is little change in variability and diurnal range (the difference between daily maxima and minima), even though increases in the mean temperature are

Table 2.3 Effect of a warming of 1.5 or 2°C on the number of summer days over 35°C, and of winter days below 0°C, assuming no change in variability.

No. of summer days > 35°C			No. of winter days < 0°C		
City	*Present*	*+2°C*	*City*	*Present*	*+2°C*
Melbourne	8	13	Ballarat	9	2
Sydney	2	4	Wangaratta	18	5
Perth	15	23	Orange	24	7*
Canberra	4	10	Griffith	17	6*
Rockhampton	10	19*	Roma	11	5*

* Warming of 1.5°C, not 2°C

significant. A small change in variability and a significant increase in mean temperature imply a decrease in the frequency of extremely low temperatures and an increase in the frequency of extremely high temperatures under enhanced greenhouse conditions.

To test the sensitivity of extreme temperatures to an increase in the mean, a warming of 0.5–2.0°C with no change in variability was imposed on observed temperatures at sites in southeastern Australia (Hennessy and Pittock, 1995). This is a scenario for the year 2030. For a 0.5°C warming, there was little change in the frequency of days above 35 or 40°C, but the frequency of days below 0°C decreased by 20–30 per cent. However, the 2°C warming increased the frequency of days over 35 or 40°C by 50–100 per cent and reduced the number of days below 0°C by 50–75 per cent. Results for a number of Australian cities are given in Table 2.3. The probability of summer heatwaves (defined as at least five days at or above 35°C increases by 20–30 per cent for a 0.5°C warming and almost doubles for a 2°C warming (Figure 2.10).

An analysis of observed trends in intraseasonal (1–30 day) temperature variability over the past 30–80 years (Karl *et al.*, 1995) reveals that there has been a decrease in the northern hemisphere, but mixed trends over Australia. Analysis of Australian observations for the period 1951–1992 by Plummer *et al.* (1995) indicates a small decrease (of about 0.1°C per decade) in the diurnal temperature range over most of the country, except for a small region of western Queensland, where the decrease exceeds 0.3°C per decade, and in eastern Victoria, where there is a small increase. These patterns are probably influenced by local effects such as changes in land use, cloud cover or soil moisture.

Over the southwest Pacific, Salinger (1995) found that warming trends in

Figure 2.10 (opposite) Probability (per cent) of at least five consecutive summer (December, January, February) days with maximum temperatures greater than or equal to 35°C for (a) the present climate, (b) the low scenario for 2030 and (c) the high scenario for 2030

(a) Pr (5 days over 35°C) (present climate)

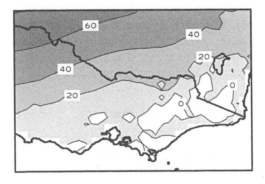

(b) Pr (5 days over 35°C) (low scenario for 2030)

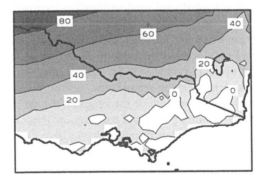

(c) Pr (5 days over 35°C) (high scenario for 2030)

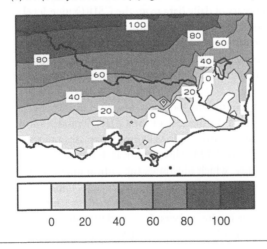

recent decades vary with location relative to the SPCZ, with a general tendency for warmings in both mean daily maximum and mean daily minimum temperatures, and little change in the diurnal temperature range. The situation in New Zealand is more complicated, owing to the interaction with topography (Folland and Salinger, 1995). In general, in coastal and maritime regions, land-based temperatures will reflect trends in SSTs in the region. However, in more arid continental areas, there may be a greater increase in extremely hot conditions, associated with larger increases in mean temperatures.

2.9 Fire

Historically, 'bushfires' (i.e. wildfires) in Australia have lagged behind heatwaves, tropical cyclones and floods as a cause of death, but there is an increasing trend. The southeast coastal strip of Australia is among the most fire-prone in the world, owing to its quasi-Mediterranean climate (mild, wet winters and long, hot summers) and highly flammable, fire-adapted vegetation. Increasing population, often housed in well-vegetated semi-rural settings, has increased the exposure to this hazard.

Fire danger is estimated in Australia using the forest fire danger rating systems and behaviour models developed by McArthur (1967, 1973), Peet (1965), and Sneeuwjagt and Peet (1985). Fuel loading is a critical factor, with fire spread also dependent on temperature, relative humidity, wind speed and fuel moisture. Other factors include the length of the ignition line, which is influenced by multiple ignition points and coalescence. These models appear to work reasonably well in experimental fires under moderate conditions, but are unreliable for predicting the behaviour of wildfires, especially under extreme conditions (Cheney and Gould, 1996).

Beer and Williams (1995) applied the McArthur fire danger index to 3 years of daily data for $1\times$ and $2\times CO_2$ conditions provided by the CSIRO four-level GCM, and to 30 years of daily data from the CSIRO nine-level slab-ocean GCM. This required estimates of air temperature, relative humidity and wind speed, plus a drought index calculated using daily rainfall and temperature. Comparison with fire danger indices calculated from observed data at a single location indicated that the models tended to underestimate humidity and thus overestimate the fire danger.

Under $2\times CO_2$ conditions, both GCMs gave an increase in fire danger over large parts of Australia. In the case of the CSIRO9 model, most of the fire-prone southern and eastern coastal regions showed increases in fire danger in excess of 10 per cent (see Figure 2.11). This result was dominated by modelled changes to the relative humidity, which illustrates the critical importance of correct simulation of the hydrological cycle in climate change impact studies. Widespread decreases in relative humidity under enhanced greenhouse conditions were regarded as plausible by the authors. It should be noted, however, that this study did not take account of changes in fuel loading which might occur owing to

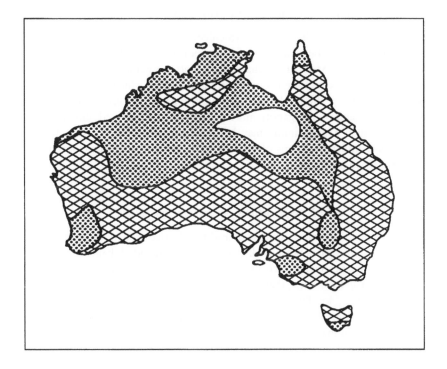

Figure 2.11 The difference between $2 \times CO_2$ and $1 \times CO_2$ average annual accumulated fire danger, as calculated from CSIRO9 experiments. Stippled regions indicate an increase in fire danger of 0–10 per cent and cross-hatched regions indicate increases of > 10 percent

Source: Beer and Williams (1995)

changes in vegetation and litter brought about by climatic change or the physiological effects of higher ambient CO_2 concentrations. Nor did it take into account the fact that persistently higher fire frequency would change the vegetation type.

Lightning is generally considered to be a major cause of wildfires in the United States. Price and Rind (1994) have applied a method relating the depths of convective clouds to lightning frequencies, to obtain estimates from a GCM of changes in fire frequency and area burned in a $2 \times CO_2$ climate. Globally they found an increase in lightning frequency and consequent fires, with the largest increases in tropical ecosystems, where they say few natural fires occur today. However, lightning is not generally considered to be a major cause of fires in settled areas of Australia, and it is not clear that the Price and Rind methodology is applicable to wildfires. Nevertheless, it is of interest to ask whether the importance in Australia of lightning-induced fires might increase with global warming (T. Beer, private communication).

2.10 Responses and policy implications

Broadly speaking, the policy response to enhanced greenhouse-induced climate change will inevitably be a mixture of adaptation to unavoidable changes, and action to reduce the extent of the changes. The latter is commonly referred to as mitigation. To be effective for a country such as Australia, which contributes only a small percentage of the total global greenhouse gas emissions (although high emissions per capita), mitigation must be part of a concerted international programme. Such a programme may be implemented within the near future via a protocol under the terms of the United Nations Framework Convention on Climate Change (FCCC).

2.10.1 Implications for mitigation policy

Mitigation policies are those which will lead to effective reductions in the emissions of greenhouse gases both locally within Australia, and globally. In so far as these policies are perceived as costly, in terms of reductions in the use of convenient and cheap fossil fuels, they will be agreed upon only if the perceived costs of climate change, locally or globally, are greater. Such a decision involves complex socio-political arguments and considerations, including value judgements regarding human life and well-being, as well as other non-market 'goods' such as biological diversity, amenity, and the survival of threatened island states and cultures. Economic discounting of future costs, broadly based on the argument that costs in the future can be met or avoided by investments and economic growth in the present, is particularly important and controversial (Bruce *et al.*, 1996). Part of this complex equation is undoubtedly the potential cost of changes in the incidence and severity of natural hazards.

It is at least arguable that the greatest costs of climate change will arise through changes in the incidence of natural hazards. A vital point here is that future risk from extreme events such as tropical cyclones, storm surges, floods or droughts is a product of hazard by vulnerability. Moreover, vulnerability is likely to increase with economic and population growth, unless measures are taken to reduce vulnerability by risk-aversive planning; that is, by adaptation policies.

It is vital, then, in the calculus of developing national and global mitigation policies, to better evaluate the potential costs of changes in extreme events due to climate change. This will require integrated impact assessments which take account of likely changes in the hazard *per se*, changes in exposure to the potential hazards, and the capacity of communities to prepare for, respond to, and recover from potential impacts of hazards. In so far as adaptation may be a response to such assessments, the process of assessment will be an iterative one, with greater awareness leading to greater adaptation, and thus to less exposure and reduced cost.

At present, the likely changes in the risk of extreme events are highly uncertain. Probably for this reason, there have been few attempts to quantify the potential costs of changing hazard magnitudes and frequencies, and regrettably little attention paid to adaptation policies, beyond those for present disaster management, which might reduce the costs.

The 'state of the art' is summarised in Chapter 6 of Bruce *et al.* (1996). Based largely on the work of economists such as Nordhaus, Cline, Fankhauser and Tol, estimates have been made of the global costs of climate change. Monetary estimates have been made for agriculture (but largely ignoring extreme events), coastal protection (but almost exclusively mean sea-level rise, not extreme events), forestry (ignoring changes in fire frequency) and hurricane damage. The last has been largely based on hypothesised increases in tropical cyclone frequency and intensity which have little support in the recent scientific literature. Assessments of the costs of changes in drought frequency have been partial at best, and almost nothing has been attempted on the costs of changes in flood frequency and magnitude, despite a high probability of increases. This last factor may be particularly important in developed countries, where there are high infrastructure investments at or above prescribed historical return period (usually 100-year) flood levels (Minnery and Smith, 1996).

Global cost–benefit estimates have included estimates of the number of additional lives at risk from summer heatwaves (of the order of 130,000 in a $2 \times CO_2$ climate), assuming that these far outweigh reductions in deaths associated with cold in winter. Moreover, such cost–benefit estimates have almost completely ignored estimates of the numbers at risk of hunger, which are variously estimated to increase by some tens to hundreds of millions (Rosenzweig and Parry, 1994).

The upshot, as far as national and global mitigation policy is concerned, is that the potential cost of changes in extreme events is at present very poorly estimated, although such costs may well dominate future estimates of the total cost of climate change. Until this unsatisfactory situation is rectified, the economic rationale for adopting, or failing to adopt, any particular mitigation target is very shaky.

2.10.2 *Adaptation policy*

Even on optimistic assumptions of low greenhouse gas emissions and low climate sensitivity, global average warming (relative to the 1990s) is expected to be at least 0.8°C by the end of the twenty-first century (Houghton *et al.*, 1996). In Australia, local warmings are likely to be in the range 0.8 to 1.8 times the global average (CSIRO, 1996). Thus some adaptation to climate change will be necessary.

The official Australian response is outlined in the National Greenhouse Response Strategy (NGRS, 1992), adopted by the Commonwealth (i.e. federal), state and territory governments in December 1992, and scheduled for revision in 1998. The broad elements of the strategy which relate to adaptation include:

- research into impacts, vulnerability assessments of the natural, built and human environment;
- incorporation of possible effects of climate change into planning;
- specific application to disaster planning;
- ensuring that natural resource management takes climate change into account.

Most systems affected by short-term climatic variations are already evolving and changing in response to climatic fluctuations and other pressures such as economics, land use and technological change. Where these changes are adaptive to potential climate change they may be termed 'autonomous adaptations'. These include changing crops or farming practices; building water storage, irrigation systems or coastal defences; flood zoning; insurance, or emergency relief schemes; and building fire-fighting capacity.

Other planned adaptations will consist either of building greater resilience or short-term adaptive capacity to cope with changed but unknown risks, or of anticipating the specific nature of changing hazards so that measures can be taken to avoid or minimise their impact. Adaptations should also attempt to maximise potential benefits associated with climate change, such as those possible with reduced frost frequency, or increased run-off.

Anthropogenic climate change may lead to unexperienced climatic conditions, including unprecedented magnitudes or frequency of climatic hazards. Thus past experience may not be a sufficient guide to the best adaptation techniques. Possible non-linearities and thresholds mean that simple extrapolation from an ability to cope with existing climate variability may not be sufficient.

There have been a number of reviews of policy responses to climate change in Australia (e.g. Taplin, 1994; Carruthers, 1996), and reports dealing with integrated impact assessment and adaptation methods (Braaf et al., 1995; Henderson-Sellers and Braaf, 1996; Hennessy and Pittock, 1996a, b). Henderson-Sellers (1996) has discussed adaptation with particular reference to Oceania, while Kay et al. (1996) review planning for climate change and sea-level rise in the coastal zones of Australia and New Zealand. Options for adapting agriculture to climate change are discussed by Gifford et al. (1996) and Pittock (1996), while the Tropical Cyclone and Coastal Impacts Program (Falls et al., 1996) is attempting to address the problem of the increasing vulnerability of the rapidly growing Queensland coastal communities to tropical cyclones.

In the case of floods, tropical cyclones, storm surges and inundation, and fire, it is clear that particular locations can be identified where exposure to these hazards is more likely now (e.g. Davidson and Dargie, 1996). Further development in these areas will increase the exposure, so that a clear adaptive response is to plan developments either to avoid such exposure, or to 'harden' the development so as to withstand the extreme events when they occur. To a lesser degree the same is possible with drought and hail, at least through improving the resilience of the enterprise through diversification or other institutional measures.

Adaptive responses will thus include:

- avoiding increased exposure, through planning and zoning;
- engineering development to withstand the stresses from extreme events;
- setting up or strengthening social mechanisms to cope with the stresses through cost-sharing, emergency preparedness, or other measures;
- capitalising on any reductions in risk.

In many cases the degree of present exposure is not well understood owing to the shortness or other inadequacy of historical records, particularly in areas of recent or planned future development. Lack of implementation of appropriate zoning and emergency planning measures to cope with existing hazards is also common, and is particularly exacerbated in Australia by the division of responsibility between federal, state and local government authorities (May, 1996; May et al., 1996; Smith, 1996).

In so far as these measures need to go beyond those necessary to adapt better to the present climatic variability, the critical issue is better information on the likely change in exposure due to climate change. Two key uncertainties must be recognised. One concerns the rate and precise nature of the large-scale climate changes, an uncertainty due both to scientific and to socio-economic problems. The other is that the usefulness of climatic change scenarios is limited by our inability to obtain details at the spatial and temporal scales most relevant to planning authorities and decision-makers.

The actual implementation of desirable adaptation measures is a matter both of perception and of cost–benefit analysis. The long time perspective of climate change, compared with that of financial return on investments, favours adaptation measures which relate to long-term planning and infrastructure investments, and those which will have short-term benefits under the present climate. These will include measures which will have other, non-climate-related advantages, such as those appropriate for ecologically sustainable development or the preservation of scenic or other values. Appropriate zoning of coastal and riparian environments may well qualify under such considerations.

Priority-setting and decisions on rates of investment in adaptation are dependent on quantitative assessments and cost–benefit analyses. Involvement of stakeholders in assessing potential impacts and adaptation strategies is a key element in developing measures which may be implemented (Cohen, 1995; Henderson-Sellers and Braaf, 1996).

2.11 Conclusions

It is clear from the above brief review of climatic change and climatic hazards in Australia that the uncertainties are still quite large. It is not yet possible to give definitive advice on the likely changes in any of the hazards reviewed. The specific mechanisms determining particular types of extreme events, and the broader-scale regional climate change scenarios within which these mechanisms operate, are in the main both poorly determined.

Nevertheless, considerable progress is being made in several areas, notably in regard to the ENSO and related droughts, possible changes in the locations and numbers of tropical cyclones, extreme temperatures, and high-rainfall and high run-off events. Tropical cyclone intensities, fire danger and hail are perhaps more difficult to quantify.

The first requirement to improve this situation is a better understanding of the factors governing the magnitude and frequency of extreme events under present climatic conditions. Even when that is accomplished, however, the accuracy of any scenario of changes in extreme events will be totally dependent on the reliability of the scenario for regional climate change. This has been well illustrated above in the cases of tropical cyclones, hail and fire. Further improvements in the modelling of global and regional climates are thus essential.

In order to develop realistic policy responses, a holistic approach to climate change impacts, adaptation and mitigation assessment is needed, including the involvement of stakeholders. This applies particularly to developing policies in regard to the potential impacts of extreme events on major infrastructure and development in Australia.

Assessments of the exposure to present and possible future climatic hazards should be conducted prior to or during the planning phase of major infrastructure, particularly in riverine and coastal environments. Hazard reduction is especially cost-effective in new developments, where engineering standards and siting can be readily modified before large investments are made.

Successful adaptation will depend on the support, involvement and willingness of the community to change. Thus successful adaptation will occur only through a partnership-based approach which requires that the whole community be educated, informed and consulted in regard to potential climate change and its effects on extreme events. In the case of hazards involving risk to life and property, the legal responsibility of 'due care' may turn out to be an onerous one, and should provide a strong incentive for decision-makers to keep abreast of the best possible estimates of future changes in risk due to extreme events.

As part of the adaptation process, targeted education programmes that aim to increase awareness of climatic risks associated with particular hazards, the possibility of increased exposure to them, and strategies for coping should be developed and implemented. This may improve investment decisions and ensure that the market is adequately informed.

Acknowledgements

The authors thank other staff in the CSIRO Divisions of Atmospheric Research and Oceanography for their contributions to this work, and in particular Drs S.P. O'Farrell, I.N. Smith and S.G. Wilson in relation to ENSO. Funding for this work came in part from the Australian Department of the Environment, Sport and Territories, through the CSIRO Climate Change Research Program, and the governments of New South Wales, Queensland, the Northern Territory, Western Australia and Victoria. Support from CSIRO and Macquarie University is also acknowledged.

References

Admirant, P., Goyer, G.G., Wojtiw, L., Carte, E.A., Roos, D., and Lozowki, E.P. (1985) A comparative study of hailstorms in Switzerland, Canada and South Africa. *Journal of Climate* 5: 35–51.

Allan, R.J., and D'Arrigo, R.D. (1998) 'Persistent' ENSO sequences: How unusual was the recent El Niño? *The Holocene* (accepted).

Allan, R.J., Lindesay, J.A., and Parker, D.E. (1996) *El Niño Southern Oscillation and Climatic Variability*, Melbourne: CSIRO publishing.

Andrews, K. (1994) The consequences of heatwaves in Australia. BSc (Hons.) dissertation, Sydney: School of Earth Sciences, Macquarie University.

Barnes, S.L., and Newton, C.W. (1986) 'Thunderstorms in the synoptic setting', in E. Kessler, ed., *Thunderstorm Morphology and Dynamics*, 2nd edition, Norman: University of Oklahoma Press.

Basher, R.E., and Zheng, X. (1995) Tropical cyclones in the southwest Pacific: Spatial patterns and relationships to the Southern Oscillation and sea surface temperature. *Journal of Climate* 8: 1249–1260.

Bates, B.C., Jakeman, A.J., Charles, S.P., Sumner, N.R., and Fleming, P.M. (1996) 'Impact of climate change on Australia's surface water resources', in W.J. Bouma, G.I. Pearman and M.R. Manning, eds, *Greenhouse: Coping with Climate Change*, Melbourne: CSIRO Publishing, 235–247.

Beer, T., and Williams, A. (1995) Estimating Australian forest fire danger under conditions of doubled carbon dioxide concentrations. *Climatic Change* 29: 169–188.

Bengtsson, L., Botzet, M., and Esch, M. (1995) Hurricane-type vortices in a general circulation model. *Tellus* 47A: 175–196.

Bengtsson, L., Botzet, M., and Esch, M. (1996) Will greenhouse gas-induced warming over the next 50 years lead to higher frequency and intensity of hurricanes? *Tellus* 48A: 57–73.

Boer, G.J. (1993) Climate change and the regulation of the surface moisture and energy budgets. *Climate Dynamics* 8: 225–239.

Braaf, R., Taplin, R., Henderson-Sellers, A., Fagan, R., Curson, P., and Blong, R. (1995) *A Study of Adaptation Responses to Climate Change for Australia*, For the Department of Environment, Sport and Territories, Government of Australia. Macquarie: Climatic Impacts Centre, Macquarie University.

Broccoli, A.J., Manabe, S., Mitchell, J.F.B., and Bengtsson, L. (1995) Comments on 'Global climate change and tropical cyclones', Part II. *Bulletin of the American Meteorology Society* 76: 2243–2245.

Bruce, J.P., Lee, H., and Haites, E.F. (1996) *Climate Change 1995: Economic and Social Dimensions of Climate Change*, Cambridge, UK: Cambridge University Press.

Bryant, E. (1983) Regional sea level, Southern Oscillation and beach change, New South Wales, Australia. *Nature* 305: 213–216.

Bryant, E. (1988) 'Sea-level variability and its impact within the greenhouse scenario', in

53

G.I. Pearman, ed., *Greenhouse: Planning for Climate Change*, East Melbourne and Leiden: CSIRO and E.J. Brill, 135–146.

Bureau of Meteorology Research Centre (1993) *Climate Change and the El Niño–Southern Oscillation: Proceedings of a Workshop Held 31 May and 4 June 1993*, Melbourne.

Cao, H.X., Mitchell, J.F.B., and Lavery, J.R. (1992) Simulated diurnal range and variability of surface temperature in a global climate model for present and doubled CO_2 climates. *Journal of Climate* 5: 920–943.

Carruthers, I. (1996) 'Changing course to cope with climate change: Directions in national response', in W.J. Bouma, G.I. Pearman and M. Manning, eds, *Greenhouse: Coping with Climate Change*, Melbourne: CSIRO Publishing, 662–669.

Chen, D., Zebiak, S.E., Busalacchi, A.J., and Cane, M.A. (1995) An improved procedure for El Niño forecasting: Implications for predictability. *Science* 269: 1699–1702.

Cheney, N.P., and Gould, J.S. (1996) 'Development of fire behaviour models for high-intensity forest fires' in *Proceedings of NDR96 Conference on Natural Disaster Reduction*, Institute of Engineers Australia, Barton, ACT, 165–170.

Chiew, F.H.S., Whetton, P.H., McMahon, T.A., and Pittock, A.B. (1995) Simulation of the impacts of climate change on run-off and soil moisture in Australian catchments. *Journal of Hydrology* 167: 121–147.

Coates, L. (1996) 'An overview of fatalities from some natural hazards in Australia', in *Proceedings of NDR96 Conference on Natural Disaster Reduction*, Institute of Engineers Australia, Barton, ACT, 49–54.

Cohen, S.J. (1995) Broadening the climate change debate. *Ecodecision* 17: 34–37.

Cowell, P.J., and Thom, B.G. (1994). 'Coastal impacts of climate change: modelling procedures for use in local government', in *Proceedings of 1st National Coastal Management Conference, Coast to Coast '94*, Hobart, Tasmania, 43–50.

CSIRO (1992) *Climate Change Scenarios for the Australian Region*, Aspendale: Climate Impact Group, CSIRO Division of Atmospheric Research.

CSIRO (1996) *Climate Change Scenarios for the Australian Region*, Aspendale: Climate Impact Group, CSIRO Division of Atmospheric Research.

Davidson, J., and Dargie, S. (1996) 'Improving our knowledge of the cyclone hazard in Queensland', in *Proceedings of NDR96 Conference on Natural Disaster Reduction*, at Institution of Engineers Australia, Barton, ACT, 347–352.

Emanuel, K.A. (1988) The maximum intensity of hurricanes. *Journal of Atmospheric Science* 45: 1141–1155.

Emanuel, K.A. (1991) The theory of hurricanes. *Annual Review Fluid Mechanics* 23: 179–196.

Emanuel, K.A. (1995) Comments on 'Global climate change and tropical cyclones'. Part I. *Bulletin of the American Meteorology Society* 76: 2241–2243.

Evans, J.L., and Allan, R.J. (1992) El Niño–Southern Oscillation modification to the structure of the monsoon and tropical cyclone activity in the Australasian region. *International Journal of Climatology* 12: 611–623.

Falls, R., Angus, D., Barr, J., Holland, G., and Henderson-Sellers, A. (1996) 'Managing tropical cyclone risk in Queensland', in *Proceedings of NDR96 Conference on Natural Disaster Reduction*, Institution of Engineers Australia, Barton, ACT, 353–358.

Folland, C.K., and Salinger, M.J. (1995) Surface temperature trends and variations in New Zealand and the surrounding ocean. *International Journal of Climatology* 15: 1195–1218.

Fowler, A.M., and Hennessy, K.J. (1995) Potential impacts of global warming on the frequency and magnitude of heavy precipitation. *Natural Hazards* 11: 283–303.

Fowler, A.M., Wang, Y.P., Pittock, A.B., and Mitchell, C.D. (1992) *Regional Impact of the Enhanced Greenhouse Effect on New South Wales. Annual Report to NSW EPA*, Aspendale: CSIRO Division of Atmospheric Research.

Frank, W.M. (1977) The structure and energetics of the tropical cyclone. 1. Storm structure. *Monthly Weather Review* 105: 1119–1135.

Gifford, R.M., Campbell, B.D., and Howden, S.M. (1996) 'Options for adapting agriculture to climate change: Australian and New Zealand examples', in W.J. Bouma, G.I. Pearman and M. Manning, eds, *Greenhouse: Coping with Climate Change*, Melbourne: CSIRO Publishing, 399–416.

Gokhale, N.R. (1975) *Hailstorms and Hailstone Growth*, Albany: State University of New York Press.

Gordon, H.B., and O'Farrell, S. (1997) Transient climate change in the CSIRO coupled model with dynamic sea ice. *Monthly Weather Review* 125: 875–907.

Gordon, H.B., Whetton, P.H., Pittock, A.B., Fowler, A.M., and Haylock, M.R. (1992) Simulated changes in daily rainfall intensity due to the enhanced greenhouse effect: implications for extreme rainfall events. *Climate Dynamics* 8: 83–102.

Gray, W.M. (1975) *Tropical Cyclone Genesis*, Paper no. 234. Fort Collins, CO: Department of Atmospheric Science, Colorado State University.

Gregory, J.M., and Mitchell, J.F.B. (1995) Simulation of daily variability of surface temperature and precipitation over Europe in the current and $2 \times CO_2$ climates using the UKMO climate model. *Quarterly Journal of the Royal Meteorological Society* 121: 1451–1476.

Griffiths, D.J., Colquhoun, J.R., Batt, K.L., and Casinader, T.R. (1993) Severe thunderstorms in New South Wales: Climatology and means of assessing the impact of climate change. *Climate Change* 25: 369–388.

Guest, C.S., Willson, K., Woodward. A.J., Hennessey, K.J., Kalkstein, L.S., Skinner, C. and McMichael, A.J. (in preparation a) Climate and mortality in Australia, I: Retrospective study, 1979–1990. *International Journal of Epidemiology*.

Guest, C.S., Willson, K., Woodward, A.J., Hennessey, K.J., Kalkstein, L.S., Skinner, C., and McMichael, A.J. (in preparation b) Climate and mortality in Australia, II. Predictions of health impacts of global warming. *International Journal of Epidemiology*.

Harr, P.A., and Elsberry, R.L. (1995) Large-scale circulation variability over the tropical western North Pacific, Part 1: Spatial patterns and tropical cyclone characteristics. *Monthly Weather Review* 123: 1225–1246.

Henderson-Sellers, A. (1996) 'Adapting to climatic change: Its future role in Oceania', in W.J. Bouma, G.I. Pearman and M. Manning, eds, *Greenhouse: Coping with Climate Change*, Melbourne: CSIRO Publishing, 349–376.

Henderson-Sellers, A., and Braaf, R. (1996) 'Developing new perspectives on climate change, impacts assessment and response', in T.W. Giambelluca and A. Henderson-Sellers, eds, *Climate Change: Developing Southern Hemisphere Perspectives*, Chichester: John Wiley, 449–466.

Henderson-Sellers, A., Hoekstra, J., Kothavala, Z., Holbrook, N., Hansen, A.-M., Balachova, O., and McGuffie, K. (submitted) Assessing simulations of daily variability by global climate models for present and greenhouse climates. *Climatic Change*.

Hennessy, K.J., and Pittock, A.B. (1995) Greenhouse warming and threshold temperature events in Victoria, Australia. *International Journal of Climatology* 15: 591–612.

Hennessy, K.J., and Pittock, A.B. (1996a) *Climate Impacts Assessment Workshop Report*, Aspendale: CSIRO Division of Atmospheric Research.

Hennessy, K.J., and Pittock, A.B. (1996b) *Climate Impacts Assessment Workshop Abstracts*, Aspendale: CSIRO Division of Atmospheric Research.

Hirakuchi, H., and Giorgi, F. (1995) Multi-year present day and $2 \times CO_2$ simulations of monsoon-dominated climate over eastern Asia and Japan with a regional climate model nested in a general circulation model. *Journal of Geophysical Research* 100: 21105-21126.

Holland, G.J. (1997) The maximum potential intensity of tropical cyclones. *Journal of Atmospheric Science* 54: 2519-2541.

Houghton, J.T., Jenkins, G.J., and Ephraums, J.J., eds (1990) *Climate Change: The IPCC Scientific Assessment*, Cambridge: Cambridge University Press.

Houghton, J.T., Callander, B.A., and Varney, S.K., eds (1992) *Climate Change 1992: The Supplementary Report to the IPCC Scientific Assessment*, Cambridge: Cambridge University Press.

Houghton, J.T., Meira Filho, L.G., Callander, B.A., Harris, N., Kattenberg, A., and Maskell, K., eds (1996) *Climate Change 1995: The Science of Climate Change*, Cambridge: Cambridge University Press.

Hubbert, G.D., Holland, G.J., Leslie, L.M., and Manton, M.J. (1991) A real-time system for forecasting tropical cyclone storm surges. *Weather and Forecast* 6: 86-97.

Insurance Council of Australia (1997) *Major Disasters since June 1967 Revised to 30 June 1995*, Melbourne: Insurance Council of Australia Ltd.

Joy, C.S. (1991) 'The cost of natural disasters in Australia' in *Workshop on Climate Change Impacts and Adaptation: Severe Weather Events*, Climatic Impacts Centre, Macquarie University, NSW, 13-15 May, 1991, 9.

Karl, T.R., Knight, R.W., and Plummer, N. (1995) Trends in high-frequency climate variability in the twentieth century. *Nature* 377: 217-220.

Kay, R., Kirkland, A., and Stewart, I. (1996) 'Planning for future climate change and sea-level rise induced coastal change in Australia and New Zealand', in W.J. Bouma, G.I. Pearman and M. Manning, eds, *Greenhouse: Coping with Climate Change*, Melbourne: CSIRO Publishing, 377-398.

Kerr, R.A. (1993) El Niño metamorphosis throws forecasters. *Science* 262: 656-657.

Kerr, R.A. (1994) Did the tropical Pacific drive the world's warming? *Science* 266: 544-545.

Kleeman, R., Colman, R.A., and Power, S.B. (1996) A recent change in the mean state of the Pacific Ocean: Observational evidence, atmospheric response and implications for coupled modelling. *Journal of Geophysical Research (Oceans)* 101: 20483-20499.

Knutson, T.R., and Manabe, S. (1994) Impact of increasing CO_2 on simulated ENSO-like phenomena. *Geophysical Research Letters* 21: 2295-2298.

Knutson, T.R., and Manabe, S. (1995) Time-mean response over the tropical Pacific to increased CO_2 in a coupled ocean-atmosphere model. *Journal of Climate* 8: 2181-2199.

La Dochy, S. (1985) Climatic characteristics of hailstorms in agricultural Manitoba, Canada. *Geographical Perspectives* 55: 15-25.

Latif, M., Kleeman, R., and Eckert, C. (1995) *Greenhouse Warming, Decadal Variability or El Niño? An Attempt to Understand the Anomalous 1990s*. Report no. 175, Hamburg: Max-Planck-Institut für Meteorologie.

Lighthill, J., Holland, G., Gray, W., Landsea, C., Craig, G., Evans, J., Kurihara, Y., and Guard, C. (1994) Global climate change and tropical cyclones. *Bulletin of the American Meteorological Society* 75: 2147-2157.

Lynch, A.H. (1987) Australian East Coast cyclones, III: Case study of the storm of August 1986. *Australian Meteorology Magazine* 35: 163–170.

McArthur, A.G. (1967) *Fire Behaviour in Eucalypt Forest.* Leaflet no. 107, Canberra: Commonwealth Australia Forestry and Timber Bureau.

McArthur, A.G. (1973) *Forest Fire Danger Meter Mk 5,* Canberra: Commonwealth Australia Forestry and Timber Bureau.

McBride, J.L., and Nicholls, N. (1983) Seasonal relationship between Australian rainfall and the Southern Oscillation. *Monthly Weather Review* 111: 1998–2004.

McGregor, J.L. (1987) 'Accuracy and initialization of a two-time-level split (semi-Lagrangian model): Short- and medium-range numerical weather prediction', in T. Matsuno, ed., Special volume of *Journal of Meteorological Society of Japan*: 233–246.

McInnes, K.L., and Hubbert, G.D. (1994) 'The impact of east coast lows on storm surges: Implications for climate change', in *Climate Impact Assessment Methods for Asia and the Pacific,* Canberra: Australian International Development Assistance Bureau, 97–103.

McInnes, K.L., and Hubbert, G.D. (1996) *Extreme Events and the Impact of Climate Change on Victoria's Coastline,* Melbourne: EPA (Victoria) and Melbourne Water.

McMaster, H.J. (1997) Climate and hail losses to winter cereal crops in New South Wales, PhD dissertation, Macquarie University, Sydney.

May, P.J. (1996) 'Addressing natural hazards: Challenges and lessons for public policy', in *Proceedings of MOR 96 Conference on Natural Disaster Reduction,* Canberra, 121–129.

May, P.J., Burby, R.J., Handmer, J.W., and Ingle Smith, D. (1996) 'Hazard management and governance: Evaluating intergovernmental approaches', in *Proceedings of NDR 96 Conference on Natural Disaster Reduction,* Institution of Engineers, Barton, ACT, 69–75.

Mearns, L.O., Schneider, S.H., Thompson, S.L., and McDaniel, L.C. (1990) Analysis of climate variability in general circulation models: Comparison with observations and changes in variability in $2 \times CO_2$ experiments. *Journal of Geophysical Research* 95: 20469–20490.

Mearns, L.O., Giorgi, F., McDaniel, L., and Shields-Brodeur, C. (1995) Analysis of variability and diurnal range of daily temperature in a nested regional climate model: Comparison with observations and doubled CO_2 results. *Climate Dynamics* 11: 193–209.

Meehl, G.A., Branstator, G.W., and Washington, W.M. (1993) Tropical Pacific inter-annual variability and CO_2 climate change. *Journal of Climate* 6: 42–63.

Miller, R.C. (1972) *Notes on Analysis and Severe Storm Forecasting Procedures of the Air Force Global Weather Central.* Technical Report 200. Air Weather Service, USAF.

Minnery, J.R., and Smith, F.I. (1996) 'Climate change, flooding and urban infrastructure', in W.J. Bouma, G.I. Pearman and M. Manning, eds, *Greenhouse: Coping with Climate Change,* Melbourne: CSIRO Publishing, 235–247.

National Greenhouse Response Strategy (NGRS) (1992) *National Greenhouse Response Strategy,* Canberra: Australian Government Publishing Service.

Neumann, C.J. (1965) Mesoanalysis of a severe Florida hailstorm. *Journal of Applied Meteorology* 4: 161–171.

Pappas, J.J. (1962) A simple yes–no hail forecasting technique. *Journal of Applied Meteorology* 1: 353–354.

Pariwono, J.I., Nye, A.T., and Lennon, G.W. (1986) Long period variations of sea level in Australasia. *Geophysical Journal of the Royal Astronomical Society* 87: 43–54.

Peet, G.B. (1965) *A Fire Danger Rating and Controlled Burning Guide for Northern*

Jarrah Forest of Western Australia. Bulletin no. 74. Forestry Department of Western Australia.

Philander, S.G.H. (1990) *El Niño, La Niña and the Southern Oscillation,* New York: Academic Press.

Pittock, A.B. (1975) Climatic change and the patterns of variation in Australian rainfall. *Search* 6: 498–504.

Pittock, A.B. (1996) 'Adapting agriculture to climate change: A challenge for the 21st century', *Proceedings of the Second Australian Conference on Agricultural Meteorology,* University of Queensland, 1–4 October, 1996, 28–34.

Pittock, A.B., and Flather, R.A. (1993) 'Severe tropical storms and storm surges', in R.A. Warrick, E.M. Barrow and T.M.L. Wigley, eds, *Climate and Sea Level Change: Observations, Projections and Implications,* Cambridge: Cambridge University Press, 392–394.

Pittock, A.B., Dix, M.R., Hennessy, K.J., Katzfey, J.J., McInnes, K.L., O'Farrell, S.P.O., Smith, I.N., Suppiah, R., Walsh, K.J., Whetton, P.H., Wilson, S.G., Jackett, D.R., and McDougall, T.J. (1995) Progress towards climate change scenarios for the southwest Pacific. *Weather and Climate* 15: 21–46.

Pittock, A.B., Walsh, K., and McInnes, K. (1996) Tropical cyclones and coastal inundation under enhanced greenhouse conditions. *Water, Air and Soil Pollution* 92: 159–169.

Plummer, N., Lin, Z., and Torok, S. (1995) Trends in the diurnal temperature range over Australia since 1951. *Atmospheric Research* 237: 79–86.

Price, C., and Rind, D. (1994) The impact of a $2 \times CO_2$ climate on lightning-caused fires. *Journal of Climate* 7: 1484–1494.

Rind, D., Goldberg, R., and Ruedy, R. (1989) Change in climate variability in the 21st century. *Climatic Change* 14: 5–37.

Rosenzweig, C., and Parry, M.L. (1994) Potential impact of climate change on world food supply. *Nature* 367: 133–138.

Ryan, B.F., Watterson, I.G., and Evans, J.L. (1992) Tropical cyclone frequencies inferred from Gray's yearly genesis parameter: Validation of GCM tropical climates. *Geophysical Research Letters* 19: 1831–1834.

Salinger, M.J. (1995) Southwest Pacific temperatures: Trends in maximum and minimum temperatures. *Atmospheric Research* 37: 87–99.

Schreider, S.Y., Jakeman, A.J., Pittock, A.B., and Whetton, P.H. (1996) Estimation of possible climate change impacts on water availability, extreme flow events and soil moisture in the Goulburn and Ovens basins, Victoria. *Climatic Change* 34: 513–546.

Smith, D.I. (1991) 'Extreme floods and dam failure inundation: Implications for loss assessment', in N.R. Britton and J. Oliver, eds, *Natural and Technological Hazards: Implications for the Insurance Industry,* Armidale, NSW: University of New England, 149–166.

Smith, D.I. (1996) 'Flooding in Australia: progress to the present and possibilities for the future', in *Proceedings of NDR96 Conference on Natural Disaster Reduction,* Institution of Engineers, Barton, ACT, 11–22.

Smith, I.N., Dix, M., and Allan, R.J. (1997) The effect of greenhouse SSTs on ENSO simulations with an AGCM. *Journal of Climate* 10: 342–352.

Sneeuwjagt, R.J., and Peet, G.B. (1985) *Forest Fire Behaviour Tables for Western Australia,* Department of Conservation and Land Management, Australia.

Solow, A.R. (1995) Testing for change in the frequency of El Niño events. *Journal of Climate* 8: 2563–2566.

Steiner, J.T. (1989) New Zealand hailstorms. *New Zealand Journal of Geology and Geophysics* 32: 279–291.

Suppiah, R. (1994) Synoptic aspects of wet and dry conditions in central Australia: Observations, and GCM simulations for $1 \times CO_2$ and $2 \times CO_2$ conditions. *Climate Dynamics* 10: 395–405.

Suppiah, R., and Hennessy, K.J. (1996) Trends in the intensity and frequency of heavy rainfall in tropical Australia and links with the Southern Oscillation. *Australian Meteorological Magazine* 45: 1–17.

Suppiah, R., and Hennessy, K.J. (in press) Trends in total rainfall, heavy rain events and dry days in Australia, 1910–1990. *International Journal of Climatology*.

Taplin, R. (1994) Greenhouse: An overview of Australian policy and practice. *Australian Journal of Environmental Management* 1, 3: 142–155.

Tett, S.F.B. (1995) Simulation of El Niño/Southern Oscillation like variability in a global AOGCM and its response to CO_2 increase. *Journal of Climate* 8: 1473–1502.

Tokioka, T., Noda, A., Kitoh, A., Nikaidou, Y., Nakagawa, S., Motoi, T., and Yukimoto, S. (1995) A transient CO_2 experiment with the MRI CGCM: Quick report. *Journal of the Meteorological Society of Japan* 73: 817–826.

Tonkin, H., Holland, G., Landsea, C., and Li, S. (1996) 'Tropical cyclones and climate change: A preliminary assessment', in W. Howe and A. Henderson-Sellers, eds, *Assessing Climate Change Results for the Model Evaluation Consortium for Climate Assessment*, New York: Gordon & Breach.

Trenberth, K.E., and Hoar, T.J. (1996) The 1990–1995 El Niño–Southern Oscillation event: Longest on record. *Geophysical Research Letters* 23: 57–60.

Tsutsui, J.-I., and Kasahara, A. (1996) Simulated tropical cyclones using the National Center for Atmospheric Research community climate model. *Journal of Geophysical Research* 101: 15013–15032.

Vincent, D.G. (1994) The South Pacific Convergence Zone (SPCZ): A review. *Monthly Weather Review* 122: 1949–1970.

Walsh, K., and McGregor, J.L. (1995) January and July climate simulations over the Australian region using a limited area model. *Journal of Climate* 8: 2387–2403.

Walsh, K., and Watterson, I.G. (1997) Tropical cyclone-like vortices in a limited area model: Comparison with observed climatology. *Journal of Climate* 10: 2240–2259.

Watterson, I.G., Evans, J.L., and Ryan, B.F. (1995) Seasonal and interannual variability of tropical cyclogenesis: Diagnostics from large-scale fields. *Journal of Climate* 8: 3052–3066.

Whetton, P., England, M., O'Farrell, S., Watterson, I., and Pittock, A.B. (1996) Global comparison of the regional rainfall results of enhanced greenhouse coupled and mixed layer ocean experiments: Implications for climate change scenario development. *Climatic Change* 33: 497–519.

Willemse, S. (1995) A statistical analysis and climatological interpretation of hailstorms in Switzerland. Doctor of Natural Sciences dissertation, Swiss Institute.

Wilson, C.A., and Mitchell, J.F.B. (1987) Simulated climate and CO_2-induced climate change over western Europe. *Climatic Change* 10: 11–42.

Wyrtki, K. (1985) Sea level fluctuations in the Pacific during the 1982–83 El Niño. *Geophysical Research Letters* 12: 125–128.

3

ASSESSING THE POTENTIAL EFFECTS OF CLIMATE CHANGE ON CLAY SHRINKAGE-INDUCED LAND SUBSIDENCE

A. Paul Brignall, Megan J. Gawith, John L. Orr and Paula A. Harrison

3.1 Introduction

In 1971 domestic insurance cover in England was extended to include damage to household dwellings from subsidence (DoE, 1991). A few years later, England experienced its worst drought on record with the hot, dry summers of 1975 and 1976, separated by an unusually dry winter (Doornkamp *et al.*, 1980; Doornkamp, 1993; an overview of summer drought from 1961 to 1990 is presented by Brignall *et al.*, 1996a). Soil moisture loss and consequent shrinkage and subsidence in clay-rich soils resulted in numerous insurance claims: one subsidence claim was lodged per 1,000 dwellings in England and Wales for this 18-month period at a cost of £100 million to the insurance industry (Doornkamp, 1993). By 1979, subsidence claims comprised 10 per cent of the cost of all household insurance claims. Such claims have continued to mount in recent years, with 133 per cent more claims being submitted in the first quarter of 1996 than for the same period the previous year (*Property Weekly*, 1996). Figure 3.1 illustrates the dramatic increase in the cost of domestic subsidence claims in England. Subsidence now follows windstorms as the second most expensive 'natural hazard' affecting the insurance industry in the UK (Chartered Insurance Institute Society of Fellows Study Group, 1994).

Land subsidence or ground movement can be divided into two principal categories: deep-seated downward movement of the ground and soil or ground surface movements (Table 3.1) (Alexander, 1993). The latter category includes clay shrinkage and swelling in response to changes in soil moisture content, and may be influenced by climate. The former category includes the extraction of minerals and fluids, and is independent of climate. While both have important consequences for building stability, only the former climate-related causes will be

60

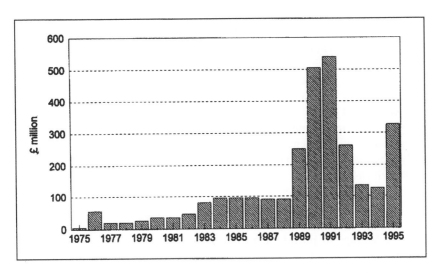

Figure 3.1 Cost of domestic insurance claims in England, 1975–95. Note that data have not been adjusted to take account of inflation.
Source: Association of British Insurers (1996)

Table 3.1 Types of ground movement

Deep-seated causes Not affected by climate change	Soil/ground surface causes May be affected by climate change
Extraction of minerals (e.g. coal, salt) Extraction of fluids (e.g. water) Extraction of hydrocarbons (e.g. oil, gas) Collapse of shallow mine workings	Clay shrinkage/swelling Absorption of water by trees Landslip following period of concentrated rainfall Sinkholes in karst terrain Development of voids in sand

considered in this chapter, specifically the property of clay-rich soils to shrink as soil moisture decreases, thereby damaging building foundations and causing subsidence.

There is growing concern that climate change may increase the cost of subsidence to the insurance industry. Warm, dry conditions have a direct influence on the number of subsidence claims submitted. Climate change projections suggest that a future climate may be warmer and in some places drier than at present. Such conditions would lead to an increase in soil moisture deficits and hence increase the hazard of clay shrinkage-induced subsidence in some places. A number of studies have been undertaken to determine subsidence hazard in a future climate on the basis of assessments of climate and soils data at various scales (DoE, 1991; Doornkamp, 1993; Harrison and Gawith, 1994; Palutikof, 1997).

The present study aims to develop a more reliable index of subsidence hazard.[1] The index could be used to predict potential claims and to plan preventive responses. Previous indices of subsidence hazard are first examined and a revised index is then developed for Yorkshire, based on recorded clay-related subsidence claims are examined. Claims are compared with available data on soils, geology and the outputs from a soil water balance model to establish a relationship between ground and climatic conditions. The index is then applied at the European scale to examine the distribution of potential subsidence hazard in Europe both under present conditions and possible future climates. Limitations to the approach are identified, and implications of findings for the insurance industry discussed.

3.2 Indices of subsidence hazard

The first attempt to provide a preliminary estimate of long-term subsidence hazard in the UK was undertaken by Clark (1980) following the 1975–6 drought. Clark used the 1956–75 average end-August potential soil moisture deficit to predict 'expected' claims in an average year. Southeast England was shown to experience high hazard and northern England and Scotland medium hazard of subsidence.

A UK government study (DoE, 1991) argued that a critical threshold of precipitation could be identified below which clay shrinkage and building subsidence were likely to occur. This threshold was based on the relationship between observed annual damage to buildings and accumulated rainfall for December to August from 1976 to 1989 in the UK. Those areas receiving less than 500 mm were considered to be susceptible to subsidence. It was argued that the potential impacts of climate change on the frequency and distribution of subsidence claims in England and Wales could be determined from examining rainfall distribution.

Clay shrinkage-induced subsidence can occur only where damaging clays (i.e. those particularly prone to shrinking and swelling) are present. Doornkamp (1993) identified differential hazard categories based on the location of clay-rich soils in the UK. These results revealed that soils in central England contain a high proportion of damaging clays likely to lead to subsidence problems. Soils in southern and eastern England contain a significant proportion of potentially damaging clays. A similar approach was adopted in the Geo-Hazard Susceptibility Package for the Insurance Industry (GHASP) (BGS, 1993), where geological maps are used to inform insurers of their exposure to risk to different types of subsidence hazard. Being a commercial product, GHASP's distribution of subsidence risk is not available for comparison with other indices.

The behaviour of clayey soils is dependent on soil water content: clays with an intermediate water content are plastic in consistency, but become dry and brittle with water loss (Gerrard, 1988). Subsidence potential will be realised only when long periods of reduced soil moisture result in a critical reduction of the water content of clay-rich soils, causing shrinkage. The location of clay-rich soils alone will not necessarily identify subsidence risk. Information on soil moisture conditions, which influence the behaviour of clay-rich soils, must also be considered. A risk-mapping package developed by the Soil Survey and Land Research Centre, Cranfield University, called VENTECH (Stephen Hallet, 1997, personal communication) uses a 100 m coverage of soils and climate data in the UK to identify areas of subsidence risk. Although it is a sophisticated approach, it is data intensive and its application is limited to the UK. As with GHASP, the package is a commercial product and results are not available for comparison.

Harrison and Gawith (1994) derived a less data-intensive index to assess subsidence hazard in Europe. Climatic data were integrated with information concerning the distribution of clay-rich soils. A soil hazard index was derived on the basis of the proportional clay content in the first 4 m of the soil profile. A clay content of 18 per cent distinguished low- from medium-hazard categories, and a 35 per cent clay content distinguished medium- from high-hazard areas. This was overlain with a drought index based on atmospheric water availability. Those areas with a negative annual atmospheric water balance and a high clay content were considered to present subsidence hazard.

This index was applied at the European scale to identify regions of Europe potentially susceptible to drought-induced subsidence as a function of both climatic and edaphic factors. Southern Spain was identified as an area of high hazard and central England an area of medium hazard (Harrison and Gawith, 1994). Output from three general circulation models (GCMs) was utilised to provide a range of possible future climates to assess the potential impacts of climate change on subsidence hazard across Europe.

3.3 Yorkshire case study

A case study of Yorkshire in northern England provides a detailed assessment of a hazard index based on observed claims data. Such data are generally confidential and not readily available. However, subsidence claims data were provided by a major insurance company for Yorkshire, General Accident (GA). The main objectives of this case study were to verify and refine a subsidence index.

3.3.1 Data and methods

Subsidence claims from GA for the whole UK included 12,400 claims over the period 1983–94 providing details on location, by postcode, cost of claim and approximate cause. A major problem with these data was that the cause of subsidence had not been rigorously recorded, with most causes being classified as 'other'. An additional set of checked records was obtained for analysis. The company had evaluated subsidence claims for Yorkshire from the period 1988–94, and from this set of 204 records, 88 claims were extracted which were explicitly clay related. Figure 3.2 shows the total number of policies by postcode district, compared with the total number of subsidence claims and clay-related claims for Yorkshire. Subsidence-related claims represent, on average, 6 per cent of the company's profile. Coincident peaks occur in the total subsidence claims and the clay-related claims over time (Figure 3.2), indicating that the clay-related claims are not an insignificant component of the overall dataset. This may be partially due to the misclassification of claims, as some subsidence claims may be clay related even though they are not recorded as such.

Data on subsoil characteristics of the eighty-eight clay-related subsidence claims were obtained from surface geology maps of the Yorkshire area. Subsoil characteristics are a more important determinant of subsidence hazard than surface soil type. Surface soils are often removed during construction, and building foundations are set in underlying soils. Geological maps were available for most of the area. Drift types associated with most claims are shown in Table 3.2.

These data confirmed that most of the claims for which detailed locations were available were based on a substrate containing clayey, or at least mixed, material. No particular type of clay consistently contained the recorded claim as the definitions are based on the provenance of material rather than the clay content. Data available from these sources do not provide any additional insights concerning subsidence hazard as all the subsoil types could be interpreted as providing some degree of clay-related subsidence hazard.

Previous broad-scale assessments of subsidence hazard relied on the use of simple climatic parameters to determine drought. While these can be reliable indicators, they do not contribute to understanding climatic controls on subsidence. An index based on a soil water balance provides a more process-based approach to assessment. The soil water balance model is defined as a layered 'bucket' type model, in which the balance between total monthly precipitation (P), total monthly potential evapotranspiration (PET) and the available water-holding capacity of the soil (AWC) is computed. The AWC is divided into three layers: a top, middle and lower layer representing 40, 40 and 20 per cent of the total AWC respectively. In each month a soil water deficit accrues if PET > P. Water is abstracted from the top layer at the full rate of PET, from the second layer at half the PET rate and from the third layer at one-quarter of the PET rate. When the profile is empty, no more extraction occurs. Once P > PET, the profile fills until the soil profile is full, or at field capacity. Any additional precipitation is then run-off. The model was initialised assuming that the soil profile was at field capacity. It was then run for five iterations using the average 1961–90 climatology to determine more realistic start values for soil water content in the different regions of Europe. The model was run continuously throughout the 34-year temperature and precipitation series (1961–94). For each winter, the length of field capacity between July and June was recorded; for each summer, the maximum soil water deficit between January and December was noted.

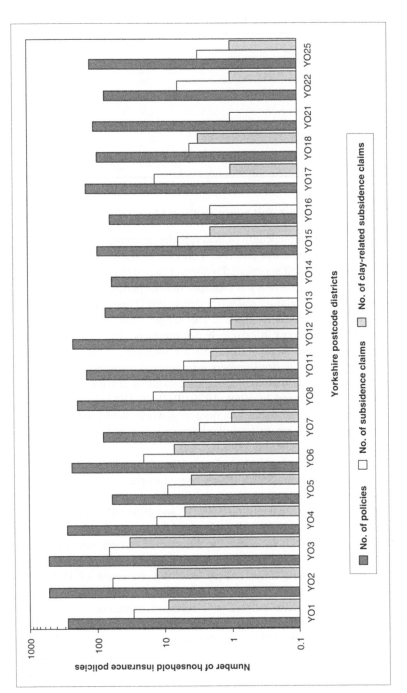

Figure 3.2 Household insurance policies and subsidence claims per postcode district for Yorkshire, 1988–94. Note that the scale is logarithmic

Table 3.2 Subsoil characteristics of clay-related insurance claims in Yorkshire

Subsoil type	Number of claims
Alluvium	22
Sand and gravel	20
Boulder clay	14
River alluvium	9
Warp and lacustrine clay	9
Solid geology only available	7
Missing	7
Total	88

3.3.2 Soil water and subsidence

The relationship between the length of field capacity, maximum soil water deficit and the incidence of subsidence claims in Yorkshire is illustrated in Figure 3.3. Field capacity, when the soil is saturated is always reached over winter in this area (that is, there are no years when soil moisture deficits persist over winter). The second axis shows the maximum soil water deficit (SWD) as a percentage of the available water-holding capacity (AWC). This ranges from 30 per cent in wet summers to over 80 per cent in extremely dry summers. The two components give an indication of two factors relating to soil moisture: winter wetness and summer dryness. Extremely dry summers were defined as occurring where the summer deficit was greater than the mean plus 1 standard deviation (i.e. >81 per cent in York). This occurred in 1969, 1970, 1975–6, 1983–4 and 1989–91. Extremely dry winters were defined as occurring where field capacity was less than or equal to the mean minus 1 standard deviation (i.e. ≤4 months in York). This occurred in 1961, 1976 and 1990.

The second axis also shows the clay-related claims for Yorkshire from 1988 to 1994. The peak in claims in 1991 is closely related to the extremely dry year of 1990. Unfortunately, no claims records were available prior to 1988, but a peak in claims might have been expected in 1976/77 and 1984/85. These two years are identified as being characteristically dry years from the 1961–90 baseline, but differ from the period 1989–91 in that the duration of dryness was only 1 year. An interesting feature of this relationship, which is difficult to explore, is that the number of reported claims drops sharply after the dry event of 1990. This suggests that there is a finite number of houses at hazard of subsidence, and, once these households have suffered and reported subsidence damage, the potential number of additional claims will not continue to increase even if dry conditions persist. The years 1988 and 1989 were very dry, with large summer deficits and relatively wet winters. An equal number of claims was reported in 1989 and 1990, with the maximum number occurring in 1991, more than a year after the peak of the dry event. Claims in 1992 were almost as numerous as they were at the peak of the event, dropping to almost none in 1993 and 1994, when

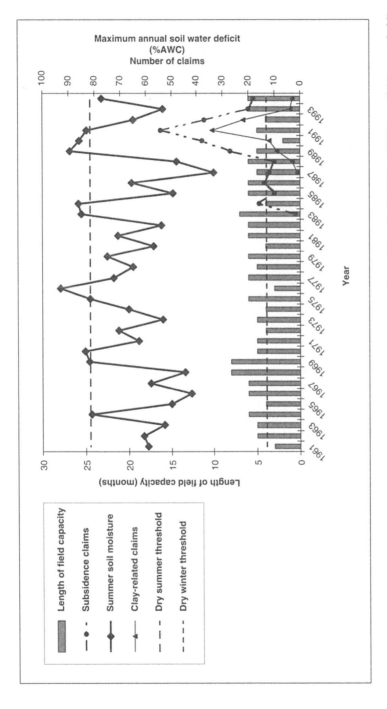

Figure 3.3 Relationship between maximum soil water deficit, field capacity and subsidence claims for Yorkshire. AWC = available water-holding capacity of the soil

SWD was low. Examining records for 1995 should indicate whether this phenomenon is real.

The same analysis was applied to north London (Figure 3.4) using all subsidence claims data from 1982 to 1994 to test the index on a longer time series. The assumption is made that, as in Yorkshire, peaks in the data were caused by clay-related claims. A peak in subsidence claims occurs in 1984 and another in 1989–92. The latter appears to be related to the extended extremely dry winters and summers which occurred at the time. Similarities in the trends displayed in London and Yorkshire indicate that the increase in claims in these years may well be due to clay-related subsidence. The slight peak in 1984 appears, however, to be unrelated to extreme conditions. Rather, this peak may be caused by a rapid rate of change in soil moisture conditions during the previous season. The unusually wet winter of 1982/83 was followed by a dry summer in 1983. The rate of soil desiccation between these seasons may have caused instability and subsidence.

3.4 European subsidence hazard

The subsidence hazard in Europe was evaluated, at a generalised level, on the basis of soil and climatic conditions. The thresholds of water stress identified in Yorkshire were used, although this procedure should be revised using locally calibrated thresholds from national studies. The intention here is to assess the relative impact of climate change, rather than to predict local subsidence risk everywhere in Europe.

The climate database consisted of a mean 1961–90 baseline climatology of monthly variables at a 0.5 degree longitude by 0.5 degree latitude resolution for Europe, and a 34-year (1961–94) time series of temperature and precipitation values for the same region (Hulme *et al.*, 1995). The temperature datasets were used to calculate *PET* using the Thornthwaite method (Thornthwaite and Mather, 1955).

In order to apply the index to the European scale, soils information was extracted from the Global Soil Particle Size Properties dataset of Webb *et al.* (1992), which contains data on a number of variables for characteristic soil classes within 1 degree grid cells across the globe. The depth and percentage clay content arrays over fifteen soil horizons were extracted using the method outlined by Harrison and Gawith (1994).

Using the relationship between SWD and claims potential defined by the case study, the full results of the soil water balance (SWB) model were analysed. Each 0.5 degree cell was assessed on an annual basis from 1961 to 1994 to determine whether the 'dry summer' threshold of mean SWD plus 1 standard deviation and the 'dry winter' threshold of length of field capacity less than or equal to the mean minus 1 standard deviation had been reached. Where both conditions were met the cell was considered to experience a subsidence hazard for that year.

For most of Europe, at least one year experiences subsidence hazard. Areas of Spain and Italy and southern parts of the former Soviet Union record 'no hazard'

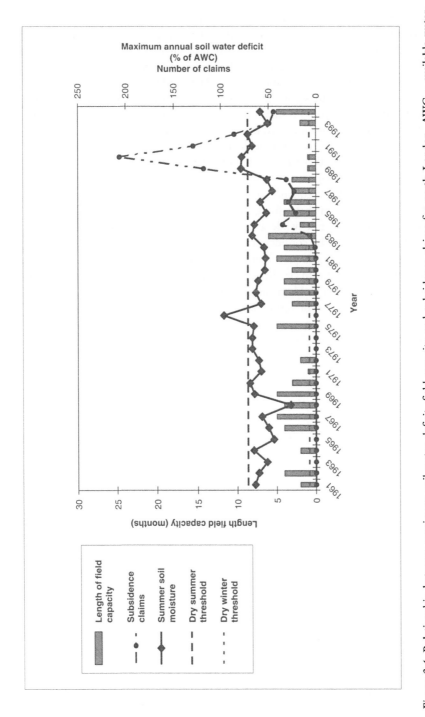

Figure 3.4 Relationship between maximum soil water deficit, field capacity and subsidence claims for north London. AWC = available water-holding capacity of the soil

because these areas are consistently dry. It is therefore inappropriate to apply the meteorological index in these regions. Areas with the greatest hazard are located in southern Spain, a result that is consistent with findings by Carcedo *et al.* (1986). Hazardous areas in northern Finland and Sweden may be caused by irregularities in the water balance model. The model does not accommodate snow, so winter rainfall is considered to be run-off, which is not necessarily the case. Under these conditions summer moisture deficits become too high because snowmelt waters are not included in the model calculation of soil water. For the British Isles most areas record some degree of hazard. The most hazardous areas are in central and southern England, particularly East Anglia. This is consistent with results from other studies of subsidence hazard (e.g. Doornkamp, 1993).

It is unclear what frequency of meteorological hazard can be considered critical. It would seem that a frequency of less than 10 per cent presents negligible meteorological risk, while a frequency of more than 20 per cent in the period studied might be considered a high meteorological risk.

To establish a more complete picture of subsidence hazard the meteorological hazard was combined with clay hazard (Figure 3.5). The patterns of hazard agree well with those shown by Harrison and Gawith (1994), identifying southern and eastern England, parts of Spain, France and Germany, and parts of eastern Europe as experiencing hazardous conditions. According to the present assessment, central Britain and Ireland as well as eastern Scotland are likely to experi-

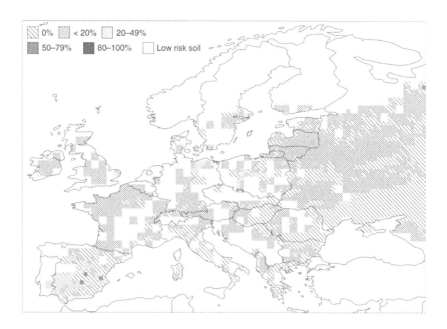

Figure 3.5 Percentage of years from 1961 to 1990 with a meteorological subsidence hazard found in areas of medium and high clay subsidence hazard

ence subsidence hazards in less than 20 per cent of years. East Anglia is at higher (20–49 per cent) risk. Areas at greatest risk in Europe are southwest Spain, northern Germany and parts of Italy and Poland. Large areas of northern France, southern Germany and the former Soviet Union experience less than 20 per cent subsidence risk through the 34-year period 1961–94.

The meteorological subsidence hazard index highlights the extreme dryness of 1976 and 1990. Of the two events, 1976 was the more severe, with a maximum SWD of >160 mm being recorded over most of England, France, Benelux, northern Germany, Poland and Finland. Coupled with this, winter field capacity was reached in less than 3 months over much of the area. In most of England a moisture deficit was carried into the next season. The second event, in 1990, was less severe than that of 1976 but was part of a sequence of three years (1989–91) where unusually dry conditions persisted over England, northern France and Germany. In southern regions, these years were not significantly drier than usual.

Figure 3.6 and 3.7 show the extent of subsidence hazard for these two years. Hazard assessments are based on the amount of clay within those soils where a meteorological hazard occurred. For 1976 (Figure 3.6), the area of hazard extended over west and northwest Europe, with an area of greater hazard occurring in central and southern Spain. The 1990 assessment, by comparison, shows a smaller area of hazard in England, Wales and Spain, a larger area in France and an area of greater hazard in the Balkans (Figure 3.7).

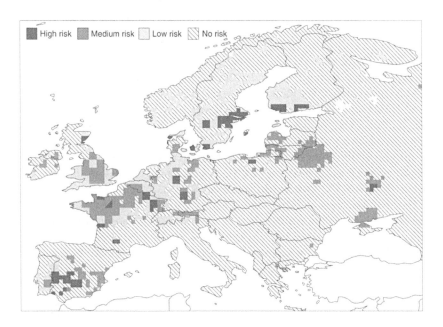

Figure 3.6 Meteorological subsidence hazard for 1976

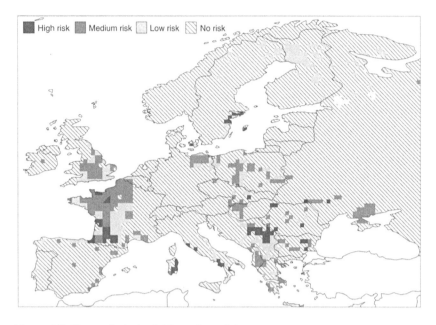

Figure 3.7 Meteorological subsidence hazard for 1990

3.5 Subsidence hazard in the future: The UKHI and GFDL scenarios

Two scenarios of climate change were applied to the 1961–94 time series to assess potential subsidence risk under altered climatic conditions. These were the UK Meteorological Office High Resolution (UKHI) scenario and the Geophysical Fluid Dynamics Laboratory model (GFDL) transient scenario, corresponding to the 2060s. UKHI model results were scaled to the IPCC IS92a emissions scenario, which equates to an increase in global average temperature of 1.37°C by the year 2050 (Hulme *et al.*, 1994). The GFDL scenario is based on a mean global temperature increase of 1.8°C. Average temperatures in Europe are expected to be 1.9°C higher than at present.

The two scenarios generate very different patterns of subsidence risk as a function of their differing projections of temperature and precipitation change. The principal factor controlling changes in the pattern of subsidence risk under the UKHI scenario is precipitation change. Areas south of 48° N generally become drier under this scenario; those north of this latitude become wetter. The subsidence risk for northern and western Europe remains the same or declines (e.g., in northeast Germany) (see Figure 3.8). Risk in France increases, in some areas by 10 per cent. South of 48° N risk generally increases, affecting areas such as Spain, the Balkans and the Ukraine. Effects of increased tempera-

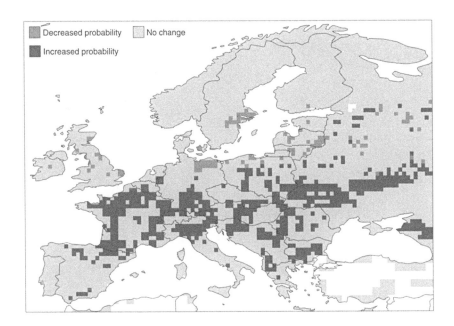

Figure 3.8 Meteorological subsidence hazard for the UK Meteorological Office High
Resolution Model (UKHI), 2050

ture in the British Isles are completely offset by increased precipitation, and
subsidence risk there is projected to decline or remain unchanged.

The GFDL scenario yields a different distribution of subsidence risk from that
described above. Under this scenario, temperatures increase in most of Europe
by 2–3°C in winter and summer, while precipitation changes little. In summer,
northern Europe and Scandinavia experience slightly wetter and cooler conditions than at present. Subsidence risk in those countries remains unchanged
(Figure 3.9). Effects of increased summer temperatures in most of central Europe
are accentuated by the small decreases in precipitation. Consequently a marked
increase in subsidence risk occurs throughout most parts of Europe between
45° N and 60° N. The zone of increased subsidence risk stretches further north
under the GFDL projection than it does under the UKHI scenario. Unlike the
UKHI scenario, only one cell (in northern England) is projected to experience
reduced risk. This pattern is not surprising, given that the GFDL scenario
predicts a mean temperature increase for Europe that is 0.5°C greater than the
UKHI projection.

The extent to which the UKHI and GFDL projections of subsidence risk differ
is illustrated in Figure 3.10. There is broad agreement between the scenarios in
those areas north of 60° N and south of 45° N. However, between these lines of
latitude the GFDL scenario predicts a greater probability of subsidence risk than
does the UHKI scenario, except in the Balkans. The differing results obtained

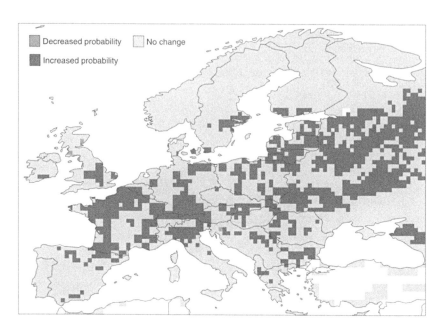

Figure 3.9 Meteorological subsidence hazard for the Geophysical Fluid Dynamics Laboratory model (GFDL), 2060s

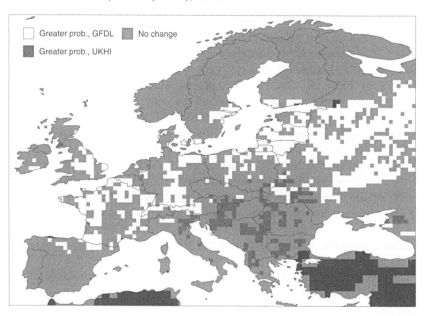

Figure 3.10 Difference in meteorological subsidence hazard between the UKHI 2050 and GFDL 2060s climate change scenarios

from these two scenarios highlight the difficulties associated with assessing potential changes in subsidence risk in a future climate. Results suggest that more detailed assessment of the distribution of subsidence risk in the future will be influenced by the extent to which climate change scenarios capture local variations in temperature and precipitation change.

3.6 Discussion

It has been shown that the soil water balance model can identify extreme years. These can, in turn, be correlated with the occurrence of subsidence claims. This index needs to be validated with data from other areas and for longer periods before it can be accepted as representative. It does, however, include some of the important factors affecting clay-related subsidence. The combination of wet/dry winters and summers is important for establishing patterns of clay shrinkage/swelling. The soil water balance model realistically describes the condition of the soil in response to antecedent and prevailing moisture conditions. It is considerably more robust than the initial precipitation threshold suggested by the DoE (1991).

For areas such as Yorkshire, a single meteorological event can cause a significant number of subsidence claims, but, using a broad-scale assessment, the region might be classified as experiencing a low or medium hazard. The difference in the scale of assessment is important. The broad-scale assessment provides a guide to areas where subsidence hazard is greatest. Application of a fixed set of climatic parameters to define hazard over the whole of Europe may, however, be inappropriate. The results for Spain and southern parts of the former Soviet Union suggest that there is no hazard. In the model, these regions have moisture deficits that often persist over winter. The presence of continually dry conditions prevents a useful broad-scale assessment being made, as the localised soil water balance becomes more important.

Assessment of subsidence hazard is complex, requiring consideration of a number of climatic and edaphic factors. The approaches adopted here have been partially successful but have also highlighted a number of problems.

The Yorkshire study revealed difficulties in obtaining useful data with which to develop and validate hazard assessments. Data from insurance companies are commercially valuable and often confidential. They are also difficult to use, because the information has not necessarily been recorded to meet the needs of a researcher. The most common cause of subsidence is classed as 'other' simply because the cause may have been incorrectly identified. Another difficulty in interpreting claims data for the whole of the UK is that companies tend to have a regional bias so can supply only localised information. Further, companies may not know where their policies are located until a claim is made, as many household policies are associated with mortgages. Such policies are administered by mortgage companies, which act as agents between insurers and policy-holders. Data from insurance companies may therefore not reflect claims held with mortgage companies.

The time period over which details of claims are available is too short to detect trends. The analysis presented here includes data from the period 1983–94, and thus excludes the most significant UK droughts in recent time, namely those of 1976 and 1995. Analysis of a longer time series would help identify of meteorological factors associated with subsidence.

The application of a single index over a large area may be inappropriate. The meteorological hazard index is dependent on the maximum soil water deficit and field capacity having a normal distribution for each location. This is clearly not the case in the dry regions of southern Europe, and in the wetter fringes of western Europe.

Interpretation could have been enhanced through analysis of the housing stock. It is likely that there is only a finite set of properties which are likely to subside. Once subsidence has occurred and remedial action been taken, the recurrence of subsidence at the same site should be associated with a more extreme climatic event. No records on the age and building quality of subsided properties currently exist. Analysis of the subsidence potential of housing stock cannot, therefore, be undertaken.

The socio-economic controls on the number of claims require examination. The personal choice of householders to have a policy covering subsidence, and then to claim on that policy following subsidence, will be affected by both the socio-economic status of the householder and the insurance policy. The large numbers of subsidence claims in the 1980s and 1990s (Figure 3.1) has affected the cost of policies, which may in turn reduce the number of claims made.

3.7 Conclusions and implications for the insurance industry

Subsidence hazard is predictable. Insurance companies can assess current exposure of properties to subsidence on the basis of soils and historical claims data. The insurance industry has already responded to an increase in the number of claims in such a way as to reduce possible future losses. The most common response has been to modify the terms of policy cover and insurance premiums in order to control total exposure (Chartered Insurance Institute, 1994). This has occurred in those areas which suffered high claims following the 1975–6 and 1988–92 droughts, where premiums have been increased or variable premiums determined by perceived hazard introduced (Doornkamp, 1995). The insurance industry is thus resilient to current subsidence hazard.

Subsidence is likely to remain a hazard in the future. Although existing response mechanisms may still be relevant, additional measures should be taken to limit exposure. One such measure might be to develop detailed profiles on current, and estimated future, exposure. Potential losses may be reduced by insisting on improved building standards in high-hazard areas. Exposure in these areas may also be contained by fixing an upper limit on claims to a figure acceptable to the insurer (Dlugolecki et al., 1995).

Implementation of such strategies will require an understanding of how subsidence hazard may alter as a result of climate change. It is essential that an appropriate index for assessing subsidence hazard be developed if the insurance industry is to respond to altered frequency and distribution of hazard in a future climate. Greater certainty in scenarios of climate change is also required if more accurate assessments of future distribution of hazard are to be made. Inaccurate assessment could result in high exposure, or too few policies. While this study has improved existing indices, further validation is required before the approach may be generally applicable. Further refinement of the approach may enable identification of hazardous areas and implementation of strategies to reduce future losses.

Acknowledgements

This chapter is a revised version of a report to the European Commission by Brignall et al. (1996b). We would like to acknowledge the assistance of General Accident in providing the subsidence claims data, the Ordnance Survey for providing the co-ordinate locations for some of the claims, and the Climatic Research Unit for providing the climate data and scenarios of climate change. Professor Doornkamp's helpful comments on an earlier draft of the chapter are gratefully acknowledged.

Note

1 Smith's (1992) definition of risk is used in this chapter. Risk is taken to mean 'the probability of hazard occurrence', whereas hazard is 'the potential threat to humans and their welfare'.

References

Alexander, D. (1993) *Natural Disasters*, London: UCL Press.
Association of British Insurers (1996) *Catalogue of Statistics*, London: ABI.
Brignall, A.P., Downing, T.E., Favis-Mortlock, D., Harrison, P.A. and Orr, J.L. (1996a) 'Climate change and agricultural drought in Europe: Site, regional and national effects', in T.E. Downing, A.A. Olsthoorn, and R.S.J., Tol, eds, *Climate Change and Extreme Events: Altered Risk, Socio-economic Impacts and Policy Responses*, Amsterdam: Institute for Environmental Studies, Vrije Universiteit.
Brignall, A.P., Gawith, M.J., Orr, J.L., and Harrison, P.A. (1996b) 'Towards an index for assessing the potential effects of climate change on clay shrinkage induced land subsidence', in T.E. Downing, A.A. Olsthoorn, and R.S.J. Tol, eds, *Climate Change and Extreme Events: Altered Risk, Socio-economic Impacts and Policy Responses*, Amsterdam: Institute for Environmental Studies, Vrije Universiteit.
British Geological Survey (BGS) (1993) *GHASP: The Geo-Hazard Susceptibility Package. The Informed Approach to Loss Reduction in UK Domestic Underwriting*, Nottingham: BGS.

Carcedo, F.J.A., Gijon, M.F., Mazo, C.O., and Rodriguez, J.L.S. (1986) *Mapa Previsor de Riesgos por Expansividad de Arcillas en España a Escala 1 : 1000000 in*, Madrid: Instituto Geológico y Minero de España.

Chartered Insurance Institute (CII) Society of Fellows Study Group (1994) *The Impact of Changing Weather Patterns on Property Insurance*, London: CII.

Clark, M.J. (1980) 'Property damage by foundation failure', in J.C. Doornkamp, K.J. Gregory and A.S. Burns, eds, *Atlas of Drought in Britain, 1975–6*, London: Institute of British Geographers.

Department of the Environment (DoE) (1991) *The Potential Effects of Climate Change in the United Kingdom*, London: HMSO.

Dlugolecki, A., Harrison, P.A., Leggett, J., and Palutikof, J. (1995) 'Implications for insurance and finance', in M.L. Parry and R. Duncan, eds, *Economic Implications of Climate Change in Britain*, London: Earthscan, 87–102.

Doornkamp, J. (1993) Clay shrinkage induced subsidence. *Geographical Journal* 159, 2: 196–202.

Doornkamp, J. (1995) 'Legislation, policy and insurance aspects of landslip and "subsidence" in Great Britain', in D.F.M. McGregor and D.A. Thompson, eds, *Geomorphology and Land Management in a Changing Environment*, Chichester: John Wiley, 37–49.

Doornkamp, J.C., Gregory, K.J., and Burn, A.S., eds (1980) *Atlas of Drought in Britain, 1975–76*, London: Institute of British Geographers.

Gerrard, A.J. (1988) *Rocks and Landforms*, London: Unwin Hyman.

Harrison, P.A. and Gawith, M.J. (1994) 'Potential Effects of Climate Change on Shrinkage Induced Land Subsidence in Europe', ed. T.E. Downing, D.T. Favis-Mortlock and M.J. Gawith (eds), Vol. 7, *Climate Change and Extreme Events: Scenarios of Altered Hazards for Further Research*, Oxford: Environmental Change Unit, University of Oxford.

Hulme, M., Conway, D., Brown, O., and Barrow, E.M. (1994) *A 1961–90 Baseline Climatology and Future Climate Change Scenarios for Great Britain and Europe: Part III, Climate Change Scenarios for Great Britain and Europe*, Norwich: Climatic Research Unit.

Hulme, M., Conway, D., Jones, P.D., Jiang, T.O., Barrow, E.M., and Turney, C. (1995) Construction of a 1961–1990 European climatology for climate change modelling and impact applications. *International Journal of Climatology* 15: 1333–1363.

Palutikof, J.P. (1997) 'Construction and buildings', in J.P. Palutikof, S. Subak and M.D. Agnew, eds, *Economic Impacts of the Hot Summer and Unusually Warm Year of 1995*, Norwich: University of East Anglia.

Property Weekly (1996) 'Insurance costs could climb 10 per cent'. *Property Weekly* 406 (19 September): 1.

Smith, K. (1992) *Environmental Hazards: Assessing Risk and Reducing Disaster*, London: Routledge.

Thornthwaite, C.W. and Mather, J.R. (1995) The water balance. *Climatology* 8: 1–104.

Webb, R.S., Rosenzweig, C.E., and Levine, E.R. (1992) *Global Data Set of Soil Particle Size Properties: Digital Raster Data on a 1-Degree Geographic (lat./long.) 180–360 Grid*, New York: NASA Goddard Institute of Space Studies.

4

AGRICULTURAL DROUGHT IN EUROPE: SITE, REGIONAL AND NATIONAL EFFECTS OF CLIMATE CHANGE

A. Paul Brignall, Thomas E. Downing, David Favis-Mortlock,
Paula A. Harrison and John L. Orr

4.1 Introduction

World-wide, drought is a major hazard. Nearly 1.5 billion people were affected by it in the past thirty years, at least as recorded in the major databases of the CRED and USAID (IFRCRCS 1994). Drought is a pervasive natural hazard. All regions of the world and all societies experience climatic variations that include shortages of water. Water is essential to all economies, notwithstanding wide variations in its use. And climate change probably presents a greater threat to economies and development through the impacts on drought than through the marginal trends in average temperatures and precipitation.

Yet vulnerability to drought and drought impacts and responses differ enormously (Table 4.1). In subsistence economies, a shortage of rainfall can be a matter of survival, threatening food scarcity and famine, and the health effects of water shortages and poor water quality. Typical responses to the risk of drought include income diversification and emergency relief – seeking to reduce the risk over the long term and cope with the consequences of specific episodes. In contrast, among industrialised economies the effects are largely economic (e.g., increased water costs) and environmental (e.g., changes in valued landscapes). Response strategies are more varied, relying on anticipatory planning, strategic investment and insurance (both private and through public acceptance of some of the costs). Of course, most of the developing countries are mixed economies. Drought seldom triggers a famine, unless confounding factors such as civil strife are implicated. And responses include a wide range of strategies, but often targeted toward different vulnerable groups. Reality, of course, is more complex than a simple typology allows; the point is to set the scope of the drought hazard and not to generalise unduly about the world.

Table 4.1 Range of drought vulnerability, impacts and responses

Socio-economic vulnerability		Impacts	Responses	
Industrialised	Low	Water/energy Economic Irrigation Landscape	Demand management Contingency planning Water capacity Insurance	Risk management
Transitional Subsistence		Food security Water/health	Income diversification Emergency relief Water capacity	
	High	Survival		Risk avoidance

Below two examples of differing drought situations are reported. Bangladesh may come closest to the subsistence pole of drought vulnerability. Chile is a typical transitional economy, with strong market and subsistence sectors. The major focus of the chapter, however, is the sensitivity of European agriculture to drought and climate change, incorporating the direct effects of increased atmospheric CO_2. A crop-specific field-scale model is compared with a broad-scale, spatial assessment. A methodological aim is to investigate the utility of a multi-scale approach in addressing policy questions. The analysis is essentially methodological – with a limited investigation of scenarios of future climates. The potential for adaptive responses is described and implications for policy are noted.

4.1.1 Drought in Bangladesh

Three recent studies provide a review of vulnerability and potential climate change in Bangladesh. This review draws primarily upon Karim (1996), but see also Huq *et al.* (1996) and papers in Warrick and Ahmad (1996). Bangladesh is one of the poorest countries in the world, subject to recurrent floods, cyclones and drought. Poverty is endemic: over half the rural population is functionally landless (with less than 0.2 ha of land). Arable land is only 0.1 ha per person. Poverty is a condition of vulnerability – both a state of deprivation and structural exposure to hazards. Over half the rural population live below the food poverty line, consuming less than the basic diet required for healthy and active lives (Hossain, 1992). Seasonal food deprivation is common. Two-thirds of households face two lean months at a minimum. Half of the landless and marginal and small landowners expect a minimum of four months where food consumption is less than adequate. September to mid-November, mid-March to mid-April and mid-June to mid-July are typical lean months, coinciding with periods of risk from natural hazards. Over 80 per cent of farmers are subject to natural hazards, half experience health crises and a third insecurity in their livelihoods (Rahman, 1992). This structural vulnerability means that natural disasters can impoverish households and communities.

The drought hazard has been well mapped in Bangladesh (Table 4.2). About 90 per cent of the rainfall occurs in June to September, during the monsoon. Because of variability in rainfall, droughts of different intensities occur during the main season (when irrigated rice is transplanted), during the dry season (where residual soil moisture is relied upon to grow wheat, potato, rice and vegetables) and during the less reliable transitional season. In all seasons, drought reduces crop yields below the potential, by over 50 per cent in severe cases.

Two scenarios explored the sensitivity of the drought hazard to climate change (Table 4.2). The moderate scenario assumes +2°C, an 8–10 per cent increase in precipitation and a 12–18 per cent increase in evaporation. The severe scenario reflects warmer conditions: +4°C, 15–20 per cent wetter and a 22–33 per cent increase in evaporation. In the dry and transitional seasons, the area subject to severe drought expands two- to threefold, depending on the scenario. Production of high-yielding rice could be reduced by 30 per cent as a result of lower yields and a decrease in the area suitable for rice.

Although drought is a perennial hazard, much can be done to cope with its effects. Within the agricultural system, improved planting methods, altered timing of operations and improved tillage, mulching and inter-cropping are all currently possible. Supplementary irrigation may require additional investment, as would growing non-conventional food crops (such as yams and certain types of beans) and agro-processing industries. More fundamental are restructuring of rural land use, empowerment of vulnerable communities, and sustainable development.

Climate change, even in Bangladesh, need not threaten catastrophe. However, current development trends are not encouraging in Bangladesh. In many developing countries agricultural development has not reduced vulnerability to a significant degree. Efforts to address climate change must be matched with concerted efforts to improve food security (see Downing, 1996).

Table 4.2 Drought hazard in Bangladesh during the dry and transitional seasons

Drought class	Current area (km²)	Yield reduction (%)		Area affected by climate change (km²)	
		Wheat	Rice (B.Aus)	Moderate scenario	Severe scenario
Very severe	3,639	60–70	>40	8,636	12,220
Severe	8,581	50–60	30–40	10,874	15,303
Moderate	18,161	40–50	20–30	15,694	10,780
Less moderate	14,571	30–40	10–20	9,747	19,814
Slight	43,367	<30	<10	43,367	30,203

Source: Karim (1996)

4.1.2 Drought in Chile

The Coquimbo region of the Norte Chico is a semi-arid strip of northern Chile crossed by three major river basins that drain the Andes cordillera (see Downing *et al.*, 1994, and Gwynne *et al.*, 1994, for further details of this case study). The regional economy is dominated by irrigated agriculture in the valley bottoms. In contrast, the interfluvial hills are dominated by rain-fed, traditional agriculture. The transitional Norte Chico region is one of the most sensitive areas in South America, owing to the:

- High fragility of its arid and semi-arid ecosystems. The arid interfluvial area covers 1,858,855 ha (over half of the Norte Chico).
- Precarious reliance on rain-fed grassland production (20–2,500 kg ha^{-1} yr^{-1}), with a high variability, and high pressures from grazing.
- Low and unreliable yields for crop production. Wheat yields, in the rainfall system, are as low as 1,800 kg ha^{-1} in good years, with no yields in dry years.
- Dependence of domestic energy use (95 per cent) on fuelwood.
- High degree of desertification and poverty in non-irrigated areas, especially at the *comunidad* level (annual per capita income about US$300). More than 90 per cent of the land area is used as grassland, with nearly half showing a high degree of desertification.
- High intensity of land and water use in irrigated areas, approaching the saturation level in some valleys.
- Sensitivity of water resources to snowfall and accumulation in the Andes, although water availability for irrigation is regulated by a system of dams in the Limarí valley.
- Dependence of agricultural income, at a regional level, on irrigated areas, with production dominated by one species (table grapes). The Norte Chico is one of the few places in the world where table grapes can be grown and exported to lucrative markets in the USA and Europe in time for the Christmas season, with high profits. Irrigated agriculture and food packing are important for regional labour markets and urban settlements.
- Communal land tenure that favours an 'irresponsible' use of natural resources. Within the non-irrigated area, two land-holding systems coexist: the individual property or *haciendas* and the communal holding or *comunidad*, a system that originated during the colonial period. The *comunidades* represent 25 per cent of the total surface of the region, totalling 1,000,000 ha.

The area cultivated in table grapes, for export, has expanded dramatically in the past two decades, spurred on by the export orientation of economic policy since the mid-1970s. The rapid expansion of land under table grapes, climate change and drought highlight an emerging water crisis. Water use and table-grape production are increasing rapidly, while soil salinity also increases. Irrigation is already highly efficient, limiting the scope to reduce losses. Some reservoir development is possible, but expensive. Prolonged drought and climate change could shift the critical juncture of water supply and demand.

The effect of the drought on land resources depends on the number of successive dry years and the amount and spatial distribution of precipitation in the antecedent wet year. Two or more years of drought in this region eliminate surface run-off and lower the water table. Then springs become dry and inhabitants must emigrate from the region if they have no other source of water. If drought affects the whole region, rain-fed cereal production is reduced. As the drought continues, grazing is abandoned, since transhumance to the higher Andes mountains is not practicable. Finally, irrigated crops are gradually abandoned, starting with areas with access only to gravity-fed systems (fed by furrows), and ultimately to plantations with furrows and water pumped from rivers and wells.

If drought lasts 4 years, the areas cultivated with table grapes also might be abandoned. At the beginning of 1991, with the prospects of a fourth year of drought, water authorities planned to abandon all crops other than table grapes and try to save the plantations by keeping the plants alive but not allowing them to produce grapes (consequently losing the 1992 harvest). Fortunately, in May the rains started again, with an abnormally wet season.

The effects of the recent drought were aggravated in the region, not only by the magnitude and duration of the drought, but also because it affected most of central Chile. As a result, it was not possible to transfer livestock southward, as happened in 1969 during a similar drought. In fact, the impacts of the 1988–90 drought were less severe than in 1969. Over the past two decades a complex interconnected irrigation scheme has been developed, including water reservoirs and lower-consumption, drip irrigation technology. Drought in this region is closely related to the El Niño–Southern Oscillation, and drought prediction could lead to major improvements in water management.

The prospects for climate change in Chile are difficult to gauge. The narrow strip between the coast and the Andes is not represented well in general circulation models (GCMs). Much depends on changes in local fog, which reduces radiation, and depth and duration of the snow pack in the Andes.

Two analyses show the results for a synthetic scenario of climate change: a +3°C warming and a 25 per cent decrease in precipitation. Grassland production simulated for one site, Salamanca, would peak earlier in the season. Carrying capacity would be reduced from over 1 to less than 0.5 animal units per hectare, with a strong shift in the probability of essentially barren pasture. Coping with the decrease in average resources and increased risk of failure of grazing resources would require time to adjust stocking rates.

Farm-level effects of climate change and drought were simulated for a typical *comunidad*. Small farmers cultivate wheat and raise goats. Because of the drought risk to rain-fed agriculture, technology and investment are low. Mean wheat yields are 400 kg ha^{-1}. With the synthetic scenario of climate change, nil or negative farm income from wheat becomes common, occurring in a quarter of the simulated years as opposed to 5 per cent at present. For a typical community, gross farm income might decrease by half on average. More frequent droughts would require external assistance, or place greater pressures on off-farm incomes and urban migration.

As everywhere, much can be done to cope with drought and to ameliorate the effects of climate change. However, in the semi-arid regions of Chile, already marginal in some places in terms of rain-fed agriculture or in the balance between water supply and use, climate change and drought risk could trigger large-scale changes in the economy.

These case studies illustrate two diverging situations regarding drought and climate change. The competition between agriculture and other sectors for water is a common theme, as it is increasingly, in Europe. The focus of the remainder of the chapter is an assessment of agricultural drought in Europe.

4.2 Agricultural sensitivity to drought

4.2.1 Definition of European drought

Indices of drought derived from meteorological or hydrological variables all have in common the notion of an accumulated deficit; that is, drought is a prolonged shortage of water (Palmer, 1965; Marsh and Lees, 1985; Parry *et al.*, 1988; Rind *et al.*, 1990; Salinger *et al.*, 1990; Katz and Brown, 1992). A simple method for assessing drought is to take the minimum precipitation for rain-fed crops to achieve reasonable yields. For example, the minimum summer (June, July, August) precipitation required for rain-fed grain maize in mid-latitudes is approximately 150 mm; addition of irrigation increases yields (Shaw, 1977; Bignon, 1990). Such thresholds vary by crop type, between sites and by the intensity and timing of precipitation in a year.

Precipitation thresholds are based on experience and observation for recent conditions, and may not fully account for possible future conditions (Kenny and Harrison, 1992). A warmer climate in the future could lead to higher levels of water use by crops through increased evapotranspiration. For example, global warming of 1°C results in an annual temperature increase of about 1.5°C in Europe and an increase in annual potential evapotranspiration (PET) of about 10 per cent (some 200 mm).[1] Hence, even with no change in precipitation in a future, warmer climate, there could be significant impacts on the susceptibility of agriculture to drought in the low-rainfall areas of Europe. Thus precipitation alone is not a satisfactory indicator of crop water requirements when assessing the impacts of climate change. Rather, the combined effects of changes in PET and precipitation on agricultural drought must be considered.

Rind *et al.* (1990) defined drought as the excess of the atmosphere's demand for moisture over its supply. The supply of moisture refers to precipitation, while PET is related to the atmospheric demand (AD) for water:

$$AD = \sum_{i=b}^{e} (P_i - PET_i)$$

where: P_i = precipitation in month i;

     PET_i = potential evapotranspiration in month i;

     b = the beginning of the moisture-sensitive phase of crop development;

     e = the end of the moisture-sensitive phase.

This index was used by Harrison (1994) to evaluate the risk of drought to European agriculture.

While water balance approaches to defining drought capture changes in the physical supply of water, drought is the deficit between supply and demand. At the plant/physical level, crop-specific water requirement indices gauge demand. However, people are more concerned with the effects of drought on economic output, behaviour or other resources. For instance, drought might be gauged by its effects on yield, the value of crops, farm income or consumer water shortages.

The economic impacts of drought can be defined at various scales. Table 4.3 shows 19 years of yields and gross margins for four crops in the eastern UK. Yields and margins are affected by a variety of weather effects, including drought, along with agronomic and economic changes. Years of low productivity are marked, based on departures of 1 standard deviation below the linear trend. The trend towards higher yields is significant for all four crops, although there is no clear trend in gross margins. Across southern England, the drought year 1975/76 stands out for its effect on winter wheat yields, with lower gross margins the following year. Drought does not affect all crops to the same extent, and yield losses may not signal comparable decreases in gross margins.

At the national scale, notable droughts can be taken as those that are designated major droughts in public documents. For southern England, Hungary and southern Spain, notable droughts are shown in Figure 4.1. These drought episodes correspond to significant decreases in water availability, as measured by the broad-scale index elaborated below.

4.2.2 Exposure to agricultural drought

Exposure to agricultural drought spans many scales, from individual plants (or indeed cells) to fields, farms, landscapes and human institutions. This chapter primarily reports on results at the site/farm and regional/sector level. Droughts limited to individual plants or fields are rarely of wide societal concern. On the other hand, regional droughts also affect water companies, consumers and industrial production if water supplies or deliveries are insufficient to meet demand.

Agricultural practices in any region are generally adapted to expected climatic

Table 4.3 Yield and gross margins for selected crops in the eastern UK

Year	Winter wheat Yield	S	GM	S	Spring barley Yield	S	GM	S	Winter barley Yield	S	GM	S	Oil seed rape Yield	S	GM	S
74/75	5.26		1,411		3.72		982		4.45		981		2.44		1,405	
75/76	4.46		1,142		3.29		815		3.79	−	815	−	1.95	−	707	−
76/77	3.92	−	1,013	−	3.47		939		4.11	−	1,139	+	1.86	−	866	−
77/78	5.22		1,086		4.20		798		4.50		981		2.66		1,251	
78/79	5.65		1,380		4.13		911		4.76		1,043		2.27		955	
79/80	5.46		1,208		3.90		858		4.78		1,043		2.77		1,304	
80/81	6.42		1,414	+	4.42		909		5.31	+	1,007		3.21		1,499	+
81/82	6.29		1,290		4.22		883		4.92		891		2.41		1,008	
82/83	6.45		1,317		5.21	+	1,181	+	5.75	+	1,146	+	3.27	+	1,545	+
83/84	6.87		1,423	+	4.23		978		5.73	+	1,143	+	2.65		1,192	
84/85	8.36		1,504	+	5.30	+	1,006		6.11	+	1,120	+	3.40	+	1,477	+
85/86	6.52		1,030		4.46		746	−	5.33		797	−	3.13		1,237	
86/87	7.17		1,202		5.10		1,013		5.58		873		2.99		1,273	
87/88	5.85	−	825	−	4.35		873		5.73		897		3.48	+	1,150	
88/89	6.46	−	957	−	3.88	−	596	−	5.43	−	833	−	2.77		863	−
89/90	7.59		1,096		4.36	−	814		6.15		1,033		2.86		1,140	
90/91	7.65		1,147		4.58		733	−	5.95		901		3.08		1,154	
91/92	7.78		1,145		5.61	+	946		6.36		925		3.06		889	
92/93	7.21		1,137		5.41		1,013	+	6.56		998		2.78	−	919	
Regression:																
Slope	0.17		−10.64		0.08		−2.92		0.13		−4.51		0.05		−5.15	
R^2	0.64		0.11		0.47		0.02		0.82		0.05		0.37		0.01	

Source: Data from Murphy (1994)

Significance is:
+   over 1 standard deviation above trend;
−   over 1 standard deviation below trend;
GM is gross margin;
Slope corresponds to the annual trend in yield (t ha^{-1} yr^{-1}) or gross margin (US$ per year);
R^2 is unadjusted

conditions and some tolerance for unusual weather. For example, producers in drought-prone regions of Bangladesh may adopt risk-adverse strategies in order to ensure a minimum yield in, say, three-quarters of the years. Outside this range of tolerable weather, the ability of agriculture to adjust without stress and damage declines as conditions become progressively more extreme (Salinger *et al.*, 1990).

Drought directly affects agriculture by reducing the number of days available for plant growth, leading to decreased yields (Salinger *et al.*, 1990). If insufficient water is available for crops they reduce their growth rate by closing their stomata to limit water loss. When the stomata are closed, carbon dioxide is not absorbed, photosynthesis is decreased and eventually the crop wilts. If this condition lasts sufficiently long, the crop cannot recover (Rind *et al.*, 1990). Higher temperatures during drought increase atmospheric demand for water and may also affect

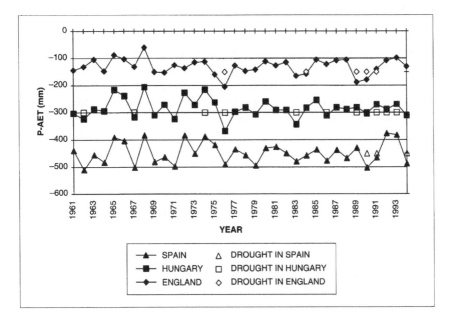

Figure 4.1 Drought index (precipitation – actual evapotranspiration (P – AET)) for selected regions showing selected significant drought episodes. Notable droughts, mentioned in national reviews, are shown by open symbols. Note that this record may not be complete, especially for Spain

the biochemistry of plants. For example, above 40°C proteins begin to break down and yields are decreased. The effect of drought on water supplies for irrigation compounds the direct effects on plants.

However, possible negative effects of increased PET on the water available to a crop must be balanced against possible improved water-use efficiency from elevated concentrations of atmospheric carbon dioxide. It has been known since 1804 that increased concentrations of atmospheric CO_2 can enhance plant growth (Kimball *et al.*, 1993). In many crops increased concentrations of CO_2 cause a higher exchange resistance between leaf and ambient air to gases such as CO_2 and water vapour; this results in lower water losses by transpiration and higher water-use efficiency (Kimball and Idso, 1983; Kimball *et al.*, 1993; Wolf, 1994). These responses – commonly referred to as the 'direct effects' of enhanced CO_2 in contrast to the 'indirect effects' of the changed climate itself – vary greatly between crop types. C_3 crops, such as wheat, are generally more responsive to elevated CO_2 concentrations than C_4 crops, such as maize. However, this is still an area of active research. Interactions between temperature and enhanced CO_2 are not yet fully understood, and crop responses to increased CO_2 in the (relatively) competitive ecosystem of a field may be very different from those encountered during the isolation of pot experiments (Kirschbaum *et al.*, 1996; Porter *et al.*, 1991; Kimball *et al.*, 1993).

The effects of changes in mean climate (e.g., Kenny *et al.*, 1993) interact with changes in climatic variability. Katz and Brown (1992) and Semenov and Porter (1992, 1995) showed that the frequency of extreme climatological events, such as drought, is relatively more dependent on changes in the variability of weather rather than on mean changes in mean climatic conditions. Similarly, Mearns *et al.* (1992) investigated the effects of changes in extreme high-temperature events on wheat production in the USA using a crop simulation model and perturbed historical climate data.

4.2.3 Scenarios of climate change

Two scenarios of climate change from the UK Meteorological Office General Circulation Model (GCM) high-resolution (UKHI) experiment were developed (Mitchell *et al.*, 1990). The UKHI experiments were scaled to global-average temperature projections, assuming no intervention in greenhouse gas emissions (the IS92a scenario), for 2025, based on a low estimate of global climate sensitivity, and 2050 with a high estimate of climate sensitivity (as outlined by Viner and Hulme, 1992a, b).

At Woodingdean (used in the site-scale modelling below), temperature increases for 2050 are up to 2.5°C for minimum temperature and up to 2.3°C for maximum temperature with increased winter rainfall (up to 21 per cent) but little change in summer precipitation.

The two UKHI scenarios of climate change have the same patterns of changes, being derived from the same equilibrium experiment. In 2025 annual temperature increases range from 0.5°C on the Atlantic coast to 1°C in continental Europe. In 2050, mean annual temperatures rise by up to 2°C over western and northern Europe. Increases of greater than 2°C occur in central and eastern Europe with a maximum increase of 3°C in eastern parts of the former Soviet Union. In 2050 annual precipitation changes suggest that Europe south of 48°N will get drier, the maximum reduction being 20 per cent in the southern Iberian Peninsula. Areas north of 48°N are predicted to get wetter, the maximum increase of 20 per cent occurring in the Baltic states. The same pattern of precipitation change is seen for 2025, the range of annual change being ±5 per cent. Summer PET increases in 2050 from up to 2 per cent in the western UK to over 12 per cent in Mediterranean areas. The increases by 2025 are a more modest 4 per cent in the Mediterranean.

An additional scenario was developed, based on the Geophysical Fluid Dynamics Laboratory (GFDL) transient experiment (Manabe *et al.*, 1991, 1992). The model decade 55–64 has been estimated to occur in the year 2064, assuming a mean climate sensitivity and the IS92a emissions scenario in MAGICC. MAGICC was also used to calculate CO_2 concentrations for the IS92a emissions scenario (617 ppmv in the 2050s). At Woodingdean, changes in monthly temperatures range from 1.2 to 3.5°C, while precipitation increases in

88

November to December (about 15 per cent), but decreases in July and August (30 per cent). Higher precipitation in the spring (12–24 per cent) probably benefits the growing crops.

Over Europe the GFDL pattern of climate change is somewhat distinct, with low correlations with the UKHI scenario (Barrow et al., 1995). The greatest winter warming is in northeast Europe (3°C). There is little precipitation change. The result is a generally drier scenario with less of a north–south gradient than for the UKHI. (Atmospheric CO_2 concentrations were assumed to be 350 ppmv in 1990, 423 ppmv in 2025 and 488 ppmv in 2050.)

These scenarios provide only one set of possible outcomes. The prospects for future drought hazard are uncertain, and the following results will vary for different climate scenarios. In particular, note that the scenarios do not include potential changes in climatic variability (although there is now some suggestion that variability will change; see, for example, Waggoner, 1990). The number of rain days, rainfall intensity, wind speed and relative humidity also were not changed.

4.3 Field-scale drought assessment

4.3.1 Methodology

Woodingdean, located in the South Downs in southeast England about 6 km southwest of Lewes, was selected to illustrate the sensitivity of crop production and farm-level gross margins to drought. The soils in the South Downs are shallow silty loams, about 20 cm above chalk, with 38 per cent stone content and 14 per cent slopes. Maximum temperatures are 20°C and average annual rainfall is around 830 mm. Rain-fed winter cereals are the dominant crops (Favis-Mortlock et al., 1991).

Physically based crop models attempt to simulate observed responses to physical laws within the crop–climate–soil system. Under new conditions, physically based crop models may be expected to be more reliable than more empirically based models, which make use only of relationships between the system's inputs and outputs (Wilmott and Gaile, 1992). However, such physi-cally based crop models are complex and require considerable data, thus they are often confined to small, well-documented, spatially homogeneous areas (Brignall et al., 1994). Examples of applications of field-scale crop models for studies of climate change impacts upon agriculture include Semenov et al. (1993), Semenov and Porter (1995), Armstrong and Castle (1992, 1995), Rosenzweig and Iglesias (1994), and Harrison et al. (1995).

The Erosion–Productivity Impact Calculator (EPIC) model was developed in the early 1980s by the USDA/ARS (Williams et al., 1990). Its original purpose was to estimate the long-term impact of soil erosion on US crop productivity for the 1985 US Resources Conservation Act appraisal. It has subsequently been applied to sites in the USA (Easterling et al., 1992c, d; Phillips et al., 1993),

France (e.g. Quinones and Cabelguenne, 1990; Cabelguenne *et al.*, 1990, 1993), the UK (e.g. Boardman *et al.*, 1990; Moxey, 1991; Favis-Mortlock *et al.*, 1991; Boardman and Favis-Mortlock, 1993; Favis-Mortlock and Boardman, 1995) and elsewhere.

EPIC is designed to estimate crop yield and soil erosion on a small area upon which soil properties and management are assumed to be homogeneous (Williams *et al.*, 1990). The model works on a daily time step, and is composed of physically based components to simulate plant growth, erosion and related processes. Sub-models additionally deal with hydrology, weather, nutrients, soil temperature, tillage, economics and plant environment control. A stochastic weather generator is built into EPIC (Nicks *et al.*, 1990) which can be used to produce synthetic sequences of daily weather data from monthly statistics. A subsequent enhancement to the model aims to simulate the direct effects of increased atmospheric CO_2 upon plant growth (Easterling *et al.*, 1992b; Stockle *et al.*, 1992a, b). However, since this modification does not consider interactions between (for example) increased temperature and enhanced CO_2, it must be regarded as preliminary (Kimball *et al.*, 1993).

Easterling and co-workers (Easterling *et al.*, 1992a, b, c, d; McKenney *et al.*, 1992; Rosenberg *et al.*, 1992) carried out an assessment of the potential effects of climate change upon agricultural production in the Missouri–Iowa–Nebraska–Kansas region of the USA which considered both indirect and direct effects of increased CO_2. The study noted substantial yield reductions, but suggested that modification of farming practices would be able to recoup such losses, at least partially.

Subsequently, Favis-Mortlock *et al.* (1991) used EPIC to estimate winter wheat yield under several equilibrium $2 \times CO_2$ climate scenarios at a site on the South Downs. This study also considered direct effects. Increased yields were indicated for most scenarios, those for the warmer-winter scenarios being due to sustained crop growth throughout the winter months. Realisation of these in practice, however, was considered doubtful.

The EPIC model also includes a simple accounting package to calculate the costs of inputs and compute returns (Williams *et al.*, 1990). Costs (including income) are divided into two groups: those that vary with yield and those which do not. All cost accumulators are cleared at harvest; thus all operations after harvest are charged to the succeeding crop. Farm overhead, land rent and other fixed costs are not considered by EPIC, nor are crop subsidies. This sub-model was used in this study to calculate gross margins.

EPIC's weather generator was used to produce sequences of daily data for both current and changed climates. The weather generator is based on a first-order Markov chain (Nicks *et al.*, 1990). While convenient, this approach assumes that generated sequences are statistically identical to sequences of measured daily data drawn from the same population. The model's ability to reproduce the mean and variance of observed distributions of daily rainfall is confirmed by Nicks *et al.* (1990). Other low-order Markov chain-based climate generators are known to underestimate the length of dry spells, leading to an underestimate of the effects of drought (Cole *et al.*, 1991; Racsko *et al.*, 1991).

A 14-year sequence of daily rainfall for Woodingdean was created using EPIC's weather generator, and the distribution of lengths of dry periods analysed (a dry day was taken as a day with zero rainfall). A similar analysis was carried out on a 14-year sequence of measured daily rainfall. The weather generator under-represents long dry periods. In the generated data, no dry spell was longer than 21 days, and 20 per cent of the dry periods were longer than 6 days. In contrast, the longest dry period in the measured sequence was 36 days, and 20 per cent of the dry periods were longer than 9 days. The implications of this are discussed later.

Basic data for a generic modern European winter wheat cultivar were supplied by John Porter (personal communication, 1992). Tillage operations for the UK site are representative of current agricultural practices, based on data of Favis-Mortlock (1994) and Nix (1994) (Table 4.4). Fertiliser applications were automatically scheduled by EPIC.

Three scenarios were devised for fertiliser usage:

- high: no fertiliser stress, unlimited applications;
- medium: fertiliser up to 200 kg ha^{-1}, applied when crop stress reaches 0.8;
- low: fertiliser up to 100 kg ha^{-1}, applied when crop stress reaches 0.8.

Table 4.4 Heat units, tillage and harvest dates for winter wheat

Heat units to maturity, 1970

Tillage	Date	Operation	Fraction of maturity
	30 Aug	Chisel plough	
	15 Sep	Disc	
	28 Sep	Drill wheat	
	1 Oct	Roll	
	—	Spray	0.25
	—	Spray	0.30
	—	Spray	0.35
	—	Spray	0.40
	—	Spray	0.45
	—	Harvest	1.10

Harvest dates

Year	Earliest	Median	Latest	Median shift
1990	20 Aug	29 Aug	14 Sep	—
2025	20 Aug	20 Aug	03 Sep	−9 days
2050	20 Aug	20 Aug	23 Aug	−9 days

Since harvest dates were specified by heat unit accumulation, they fell on different dates depending on prior temperature (Table 4.4). This forward shift of harvest would offer a farmer the opportunity to carry out subsequent tillage operations at an earlier date. This would in turn allow drilling of a following autumn-planted crop to be brought forward, allowing a longer period for crop establishment before growth ceases owing to winter cold. (Note that earlier drilling is not always a good thing since problems of competition by weeds, pests and diseases can be exacerbated.) An additional set of scenarios was devised whereby drill date was brought forward for the climate change scenarios. In these scenarios, post-harvest operations (including drilling of the next crop) were moved forward by the median shift in harvest date: 9 days for both 2025 and 2050.

Table 4.5 gives details of the costs of tillage operations and yield prices, which were used for the simulations. All are standardised to US dollars (assuming UK £1.00 = $1.66). All prices were selected to be broadly representative – agricultural machinery prices vary widely according to size and performance. Wherever possible, several prices were compared as a check for consistency. Note that depreciation is not considered in any machinery cost. All prices exclude any subsidies not directly included in farm prices. Average returns in the UK were US$1,507 per hectare in 1991 (Maché, 1993). The variable costs of seed, fertiliser and spray were US$318 (21 per cent), resulting in a gross margin of US$1,189.

4.3.2 Results

EPIC was run for 100 years for each scenario, without the erosion component. As a validation of the control runs, mean simulated yields were compared with

Table 4.5 Economic data for winter wheat, Woodingdean

Operation	Cost	Source	Assumptions
Spike harrow (US$ ha^{-1})	8.98	ABC (1995)	Light soil, medium-sized farm
Chisel plough (US$ ha^{-1})	19.69	ABC (1995)	Light soil, medium-sized farm
Disk (US$ ha^{-1})	18.68	Nix (1994)	Light soil, medium-sized farm
Roll (US$ ha^{-1})	17.10	ABC (1995)	Light soil, medium-sized farm
Drill (US$ ha^{-1})	24.65	ABC (1995)	Light soil, medium-sized farm
Spray (US$ ha^{-1})	8.37	ABC (1995)	Light soil, medium-sized farm
Harvest (US$ ha^{-1})	96.97	ABC (1995)	95% efficient
Winter wheat seed (kg^{-1})	0.72	Murphy (1994)	100 kg ha^{-1}
N fertiliser (US$ kg^{-1})	0.38	Murphy (1994)	Calculated
P fertiliser (US$ kg^{-1})	0.42	Murphy (1994)	Calculated
Spraying (US$ ha^{-1})	13.10	Murphy (1994)	Calculated; 2 kg ha^{-1} active ingredient per application
Yield price (US$ ha^{-1})	204.49	Murphy (1994)	1992/93 price

data for reported yields. For the high-fertiliser scenario the yields appear realistic, 5.8 t ha^{-1} simulated yields compared to 5–6 t ha^{-1} typical of the South Downs, 6.6 t ha^{-1} average for southeast England in 1983–93 and 5.8 t ha^{-1} average for the UK in 1973–93 (MAFF data). Note that yields on the thin, stony soils of the South Downs are near the low end of the UK average.

Observed (rather than generated) weather for 1975–88 was also used. Annual yields were compared with measured values for East Sussex, south-east England, eastern England and the whole of the UK (Figure 4.2). Years of high and low yield compare adequately; as might be expected, there is much less year-to-year variability of the area-averaged measured yields than for the single-field simulation. Simulated yields are generally lower than all measured yields, particularly in the less productive years. Again this is unsurprising: on the premium farmland of the eastern UK (for example), higher and more reliable yields would be expected than on the thin soils of the South Downs.

Simulated costs, returns and profits were compared with expected values (Table 4.6). As with yields, there is some difficulty in comparing costs, returns and margins for a farm on the South Downs with values for the prime land of eastern England (data for the southeast were not available). It appears that the simulations overestimate costs; nonetheless, returns and gross margins compare reasonably well to reported averages.

A time series of simulated gross margins was produced using measured weather data. This was compared to reported data for gross margins in the eastern

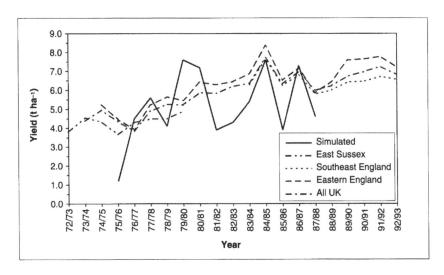

Figure 4.2 Simulated and measured winter wheat yields, 1975–88. Simulated values are for Woodingdean; measured yields are for East Sussex and southeast England (MAFF, personal communication, 1995), eastern England (Murphy, 1994) and the whole UK (MAFF data)

Table 4.6 Woodingdean: current simulated and measured average costs, returns and gross
margins (US$ ha^{-1})

Fertiliser usage	Cost	Return	Gross margin
Simulated			
High	475.38	1,207.79	732.41
Medium	315.06	862.13	547.07
Low	303.17	757.94	454.77
Measured			
—	240.09	960.62	720.53

Notes: Measured figures are 1974–93 means for eastern England (Murphy, 1994)

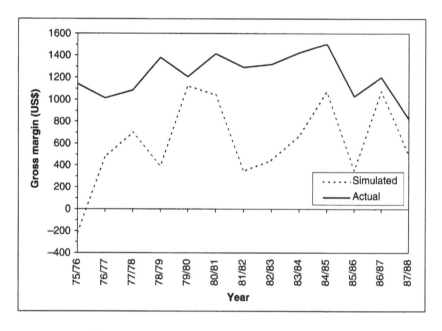

Figure 4.3 Simulated and measured gross margin for winter wheat, 1975–88. Simulated
values are for Woodingdean; measured values are for the eastern UK, calculated from Murphy (1994)

counties of England (Figure 4.3). A non-linear relationship between yields and
gross margin means that the lower yields estimated for Woodingdean (when
compared with yield for eastern England) result in an even more marked
discrepancy between simulated and measured values in individual years.

Probabilities for current winter wheat yield at Woodingdean, calculated using
EPIC with 100 years of generated weather, are shown in Figure 4.4 for the three

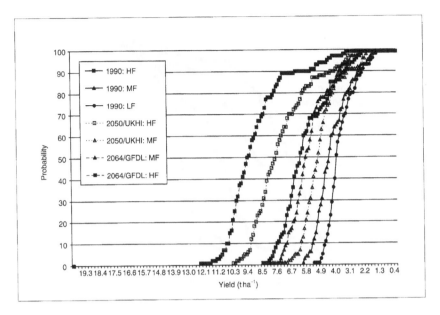

Figure 4.4 Simulated yield probabilities for three fertiliser scenarios on winter wheat at Woodingdean for 1990 and climate change scenarios (UKHI in 2050 and GFDL in 2064). N = 99

fertiliser usage scenarios. There is a greater increase in yield between the high- and medium-fertiliser scenarios than between the medium and low scenarios, but with a greater range of yields with high fertiliser usage. The distribution of gross margins follows the pattern for yields (Figure 4.5). Margins are highest with high fertiliser usage, though the superiority of this management approach is less obvious once the cost of the extra fertiliser is considered. As with yield, there is a greater range in margins associated with the high-fertility scenario, and a greater range of margins below the median (0.5 probability) than above.

Estimated median yields for the current climate and three climate scenarios are shown in Table 4.7. For all three climate scenarios (2025 and 2050 scaled to the UKHI experiment and 2064 from the GFDL transient experiment), simulated yields noticeably increase under each management regime. Further increases are predicted for rotations with earlier drilling. Figure 4.4 shows that the increased wheat yields for the 2050 and 2064 scenarios are accompanied by increased variability in the fertilisation scenarios. This is more marked in the high-fertiliser/ early drill date situation, where the range is over 8 t ha^{-1}. A similar but less obvious tendency also occurs in the 2025 scenarios.

Gross margins under current and changed climates are shown in Table 4.8. Trends for median gross margins differ from those for mean yields (Table 4.7). For example, mean gross margin increases strongly in 2025 for wheat in the

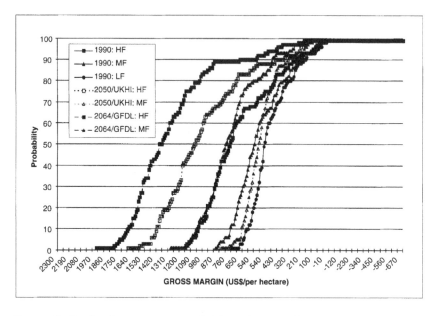

Figure 4.5 Simulated gross margin probabilities for three fertiliser scenarios on winter wheat at Woodingdean for 1990 and climate change scenarios (UKHI in 2050 and GFDL in 2064). *N* = 99

Table 4.7 Median yield (t ha^{-1}) for current climate and climate scenarios

Management scenario		*Present*	*UKHI*		*GFDL*
Drilling	*Fertiliser usage*	*1990*	*2025*	*2050*	*2064*
Present	High	6.2	6.9	7.7	9.2
	Medium	4.4	4.8	5.1	5.6
	Low	4.0	4.3	4.6	4.6
Earlier	High		7.5	8.6	10.3
	Medium		5.3	5.3	5.3
	Low		4.3	4.5	5.0

high-fertiliser scenario, but decreases for the medium- and low-fertiliser options without earlier drilling. When earlier drilling is adopted, however, margins improve markedly. All margins increase between the 2025, 2050 and 2064 scenarios, and earlier drilling continues to be beneficial.

Variability of wheat gross margins (like yield variability) increases markedly for the high-fertiliser scenario by 2050 (Figure 4.5), so that the range increases to over US$1,600 ha^{-1}. With earlier drilling, this is accentuated further: the range for the high-fertiliser option then rises to over US$1,800 ha^{-1}. Variability also

Table 4.8 Median gross margin (US$ ha^{-1}) for current climate and climate scenarios

Management scenario		Present	UKHI		GFDL
Drilling	Fertiliser usage	1990	2025	2050	2064
Present	High	810	970	1,110	1,400
	Medium	590	520	560	840
	Low	510	460	510	650
Earlier	High		1,080	1,270	1,620
	Medium		770	800	800
	Low		590	630	740

Table 4.9 Thresholds for drought impacts

Drought	Fertiliser usage	Yield	Gross margin
Moderate, 80%	High	4.3	420
	Medium	5.2	330
	Low	2.8	280
Severe, 90%	High	3.3	210
	Medium	2.5	190
	Low	2.3	150

Note: Drought is defined on the the basis of probability of yields and gross margins being exceeded in 80% (for moderate drought) and 90% (severe drought) of the simulated years for the present climate.

increases for the other fertiliser options in this case. Increased variability with climate change results largely from increases in high yields.

To return to the definition of drought, Table 4.9 presents thresholds for moderate and severe drought. Yields and gross margins that are exceeded 80 per cent of the time, on the basis of simulations of the present climate, are defined as moderate drought. The corresponding threshold for severe drought is 90 per cent. The thresholds are assumed to be valid for future economic conditions, at least as a measure of the risk of climate change. Obviously, farmers will adjust to the new climates, which could raise the thresholds of acceptable risk to a greater or lesser extent.

Probabilities for moderate and severe drought with climate change are given in Table 4.10 for yields and Table 4.11 for gross margins. The probability of exceeding 4.3 t ha^{-1} (the present threshold for moderate drought with high fertiliser use) goes up from 80 per cent to 85–94 per cent with climate change. Similar increases occur for other fertiliser levels, and even more so with earlier drilling. The probability of exceeding the severe-drought threshold also increases. These increases are matched by changes in gross margins.

The UKHI scaled to the 2050s and the GFDL transient run for 2064 are both beneficial for winter wheat. For the GFDL, the combination of warmer

Table 4.10 Probability of yields exceeding drought threshold

(a) Moderate drought

Scenario			Present	UKHI		GFDL
Drilling	Fertiliser	Yield threshold	1990	2025	2050	2064
Present	High	4.3	80	85	89	94
	Medium	3.2	80	84	89	93
	Low	2.8	80	86	91	93
Earlier	High		80	89	91	97
	Medium		80	89	90	95
	Low		80	87	90	96

(b) Severe drought

Scenario			Present	UKHI		GFDL
Drilling	Fertiliser	Yield threshold	1990	2025	2050	2064
Present	High	3.3	90	92	94	97
	Medium	2.5	90	90	92	97
	Low	2.3	90	90	91	97
Earlier	High		90	95	98	99
	Medium		90	93	94	98
	Low		90	93	94	98

temperatures, increased moisture and CO_2 fertilisation provides an environment that is twice as productive as the present one in the South Downs. Other scenarios might be less beneficial, but winter wheat is expected to do well in northern Europe (see Harrison *et al.*, 1995). Although these scenarios do not include changes in the variability of rainfall or rainfall intensity, they suggest that drought risks for winter wheat would be lessened.

4.4 Regional drought assessment

4.4.1 Methodology

Extreme drought is one of the most important causes of present agricultural losses in Europe. In the countries of the European Union significant yield losses of wheat due to drought occurred in five of the years between 1980 and 1992 (Russell and Wilson, 1994). Adverse effects from drought are often overcome by altering crop management, such as by the increased use of irrigation. However, in many countries in southern Europe there is already concern over the loss of

Table 4.11 Probability of gross margins exceeding drought threshold

(a) Moderate drought

Scenario			Present	UKHI		GFDL
Drilling	Fertiliser	Yield threshold	1990	2025	2050	2064
Present	High	420	80	84	88	94
	Medium	330	80	83	88	93
	Low	280	80	83	90	93
Earlier	High			86	91	97
	Medium			89	90	95
	Low			88	90	96

(b) Severe drought

Scenario			Present	UKHI		GFDL
Drilling	Fertiliser	Yield threshold	1990	2025	2050	2064
Present	High	210	90	93	93	98
	Medium	190	90	90	92	97
	Low	150	90	94	95	97
Earlier	High			94	99	99
	Medium			95	94	98
	Low			94	95	98

groundwater to irrigation (Kenny and Harrison, 1992). This is particularly evident in Spain, where national policies which aim to economise the volume of water currently consumed in agriculture have been developed (Ministerio de Obras Públicas Transportes y Medio Ambiente, 1993).

In contrast to the field-scale assessment above, the broad-scale evaluation seeks to portray the drought risk for an assemblage of current crops. While winter wheat is likely to benefit from climate change in Europe, the effects on other crops may be less beneficial (Harrison *et al.*, 1995). A process-based agricultural water deficit index was developed. Scenarios of climate change (see Section 4.2.3) were applied to the water deficit index.

The study area encompasses a large European region extending from Scandinavia in the north to north Africa in the south and from Ireland in the west to the Black Sea in the east. At this continental scale it is possible to analyse broad-scale aspects of drought susceptibility and risk, such as changes in the spatial distribution of severe drought, identification of sensitive areas and changes in comparative advantage between regions.

The current climatic database consists of mean monthly temperature and total monthly precipitation for the period 1961–94. Individual-year data and a 30-year average climatology (1961–90) have been interpolated to a 0.5 degree latitude by 0.5 degree longitude grid across Europe (Hulme *et al.*, 1994).

Temperature data were used to calculate potential evapotranspiration (PET) using the Thornthwaite method (Thornthwaite and Mather, 1955). Thornthwaite-derived PET was used as insufficient climatic variables were available to calculate PET using a more sophisticated formula, such as that of Penman (1948), which requires temperature, radiation, humidity and wind speed data. A previous study (Harrison, 1994) compared five methods of computing PET using the average 1961–90 European climatology: Thornthwaite, Samani–Hargreaves, Priestley–Taylor, Penman and Penman–Monteith. All methods gave similar results in northern and western Europe, but became increasingly different towards eastern Europe, owing to continentality effects, and southern Europe, owing to radiation effects. The Thornthwaite method was found to produce the lowest values of PET across the whole southern Europe region when compared with the other four methods. More specifically, Thornthwaite estimates of annual PET were approximately 40 per cent lower than the Penman estimates in southern Spain, southern Italy and Greece. This underestimation in PET in southern Europe must be borne in mind when interpreting the subsequent climate change analyses. It must also be emphasised that because the Thornthwaite method is empirical, and calibrated for present conditions, it is likely that it will have to be recalibrated if large mean changes in climate occur in the future (Kenny and Harrison, 1992).

Soils data were also required in the agricultural water deficit index. The available water-holding capacity of the soil (AWC) has been estimated using the methodology and database of Groenendijk (1989).

The aim of the broad-scale assessment was to develop an index of agricultural drought which is applicable to a range of crop types cultivated in Europe; that is, not specific to individual crops but representing water requirements of a combination of C_3 and C_4 crops. The development of an index that can be applied over a large geographic region is limited by the availability of data. As described in the previous section, only mean monthly temperature, total monthly precipitation and AWC data were available throughout the entire European region for 30 years. Thus, a process-based soil water balance model with a monthly time step has been developed which requires only mean monthly temperature and precipitation.

The soil water balance model is a layered 'bucket'-type model, in which the balance between total monthly precipitation (P), total monthly potential evapotranspiration (PET) and the available water-holding capacity of the soil is computed. The AWC is divided into three layers: a top, middle and lower layer representing 40 per cent, 40 per cent and 20 per cent of the total AWC respectively. Water is abstracted from the top layer at the full rate of PET, from the second layer at half the PET rate and from the third layer at one-quarter of the PET rate. This follows the methodology devised by MAFF (1967) and which forms part of the MORECS system (Gardner and Field, 1983). The model was initialised assuming that the soil profile was at field capacity. It was then run for five iterations using the average 1961–90 climatology to determine realistic start values for soil water content in the different regions of Europe. The model was then run continuously throughout the 34-year time series. Output produced by the model includes mean monthly actual evapotranspiration (AET), maximum annual potential cumulative soil water deficit (MAXCSWD), the duration of field capacity (FC) in months, excess winter moisture (RUNOFF) and monthly soil water deficit (SWD).

Drought is defined as occurring when water supply is limiting to the extent that crop growth is significantly affected. A comparison of actual and potential evapotranspiration from the water balance model indicates the degree to which water is available in the system to satisfy potential atmospheric demand. Hence if AET is significantly less than PET, insufficient moisture will be available for optimal crop production. Following this reasoning, the agricultural water deficit (AWD) index is defined as the ratio AET/PET throughout the period that encompasses the moisture-sensitive development phases of common European crops. The beginning and end of moisture-sensitive phases of crop development in Europe have been determined from expert knowledge and a survey of the literature.

The most moisture-demanding phases of crop development for the main grain crops in Europe (e.g. maize, wheat, oats, barley) are generally the period of flowering and subsequent grain filling. The timing of these phases varies throughout Europe depending on sowing time, crop variety and climatic factors. However, some generalisations are possible. For grain maize, flowering is particularly sensitive to moisture stress (Doorenbos and Kassam, 1979), and, despite wide variations in the timing and length of the growing season in Europe, Bignon (1990) states that flowering of maize occurs, almost universally, in July. For wheat and oats, flowering generally occurs in May in southern Europe (Narciso et al., 1992) and June in northern Europe (Hough, 1990), while grain filling takes place during June, July and August. Barley is more tolerant of drought than wheat and maize, but can still suffer from moisture stress during grain formation and filling. In general, this occurs between April and July, depending on whether the crop was winter or spring sown (Russell, 1990). Hence a universal definition of the beginning and end months for which grain crops in Europe are generally sensitive to moisture stress would be the months of April and August, respectively.

Possible negative effects of increased evapotranspiration on the water available to a crop must be balanced against improved water use efficiency from elevated concentrations of atmospheric carbon dioxide (CO_2). There are two principal effects of increased concentrations of CO_2 on crops: the rate of photosynthesis increases and the rate of transpiration decreases. Responses vary greatly between crop types; C_3 crops, such as wheat, generally experience a significant enhancement of photosynthesis and moderate reductions in transpiration. On the other hand, C_4 crops, such as maize, generally experience only marginal, or no, improvements in photosynthesis, but much larger decreases in transpiration rate compared to C_3 crops. Decreases in crop transpiration will affect the agricultural water deficit index and, hence, have been incorporated into the model. Goudriaan and Unsworth (1990) calculated a reduction in daily transpiration of 0.897 and 0.74 for C_3 and C_4 crops respectively under doubled CO_2 from 350 to 700 ppmv. Wolf (1993) also assumed transpiration would reduce by 10 per cent and 26 per cent for wheat and maize respectively for a doubling of CO_2 from 353 to 706 ppmv in an application of the WOFOST crop model. The agricultural water deficit index is non-crop-specific and, as such, an average reduction factor for transpiration of 0.82 was used for a doubling of CO_2. For CO_2 concentrations below this level a linear relationship was assumed.

The AWD index is first used to measure the susceptibility of a region to drought based on long-term soil moisture deficits for the average 1961–90 climatology. The soil water balance model produces AET values based on the availability of water in the soil. AET therefore is the amount of PET which is satisfied from soil water and precipitation. For the ratio AET/PET, a value of 1 represents water conditions that are non-limiting, and water availability diminishes as the index approaches zero. Based on this ratio, a classification of drought susceptibility was derived (Table 4.12).

4.4.2 Baseline results and validation

The agricultural water deficit (AWD) index was mapped for the 1961–90 average climatology as a measure of long-term drought susceptibility (Figure 4.6). Very severe AWD index values are found on the east coast of Spain and in parts of Sardinia and Sicily. Severe AWD values occur in southern and central Spain, southern Portugal, eastern and southern Italy, most of Greece, northern Finland and Sweden, and parts of Ukraine. Moderate AWD values occur over much of the remaining European continent, including southern and western France, northern Spain and Portugal, northern Germany, Denmark, southern and eastern England, most of Sweden, southern Finland, Russia and parts of Poland, Hungary, Bulgaria, Romania and the former Yugoslavia. Low AWD values are found on the margins of mountainous regions and include eastern and central

Table 4.12 Classification of long-term drought susceptibility

Drought susceptibility	AWD index
None	0.98–1.00
Low	0.96–0.98
Moderate	0.70–0.96
Severe	0.50–0.70
Desert	<0.50

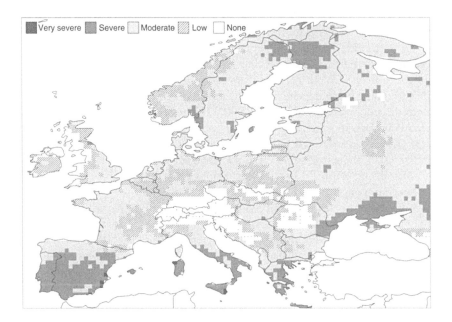

Figure 4.6 Average index of agricultural water deficit for the baseline (1961–90) climate

France, Belgium, western Germany, western Norway and parts of eastern Europe. Areas classified as experiencing no drought are confined to the high-altitude zones and the wet western fringe of Europe. The pattern of AWD is related to three factors: AWC, precipitation and temperature. Areas with low AWC and low rainfall are likely to have high AWD values. Northern Finland has low AWC, leading to severe AWD. Northern Scotland and western Norway also record low AWC, but in comparison have high precipitation values. Southern Spain, Italy and Greece have severe AWD as a result of high temperatures and summer rainfall totals consistently below 200 mm, and frequently below 100 mm. A further complicating factor may be the presence of snow over winter; northern Finland records air temperatures below 0°C up to 6 months a year on average, which will render the soil water model unreliable in this area.

Validation of the index is restricted by data availability across the large European region. Further, available agricultural statistics are highly influenced by management practices, and the averaging of statistics across administrative districts means that the actual impacts of extreme events are often diminished in the records. Long-term drought susceptibility can be validated against only more general information or period-averaged statistics. Information on published precipitation limits to crop growth and on the extent of irrigation in the European Union has been compared with the index. In virtually all areas of Spain and Greece 100 per cent of maize production, and significant proportions of other crops, are irrigated (Bignon, 1990). This corresponds with those areas indicated as severe in the AWD index in Figure 4.6. In France there is generally more irrigation in the west and south (up to 70 per cent of the acreage in the south) and less in the east (as little as 5 per cent of the acreage). This corresponds with those areas identified as experiencing moderate and low AWD respectively in Figure 4.6. In the UK, irrigation occurs in the drier eastern regions, the magnitude and timing being dependent on the moisture sensitivity of the crop and the expected economic return (Hough, 1990). This region coincides with the moderate AWD class. In western Great Britain, western Ireland and western Norway, cereal production is limited by excessive rainfall as the probability of achieving effective establishment of grain crops and their subsequent harvest is extremely low (Russell and Wilson, 1994). These regions coincide with those identified as experiencing no agricultural drought in Figure 4.6.

According to the time series of AWD values, most of Europe, with the exception of the Alps, Carpathians, Transylvanian Alps and west Scotland, have experienced moderate or severe droughts at least once during the 1961–90 period. Very severe drought is limited to southern Spain and Portugal, southeast Italy, Greece, Sicily and Sardinia. The very severe areas always record a moderate or worse drought. The AWD for individual years can be validated against known agricultural losses from drought; however, such data are not widely available.

Table 4.13 provides observations on drought events that have affected cereal yields in member countries of the EU. Although general (there is no allowance for regional variation within countries, and the reporting of statistics by the member countries may not be complete or consistent), this dataset provides information on the timing of critical droughts. Using the AWD index, it is possible to highlight some of the droughts referred to in Figure 4.1 and Table 4.13. Figure 4.1 shows average mean AWDs for selected regions and years with

Table 4.13 Years when significant yield loss occurred in the countries of the EU

Year	Belgium/Luxembourg	France	Germany	Ireland	UK	Netherlands	Spain	Italy	Portugal	Denmark	Greece
1983								D			
1984											
1985											
1986							D				
1987											
1988								D			
1989											
1990							D		D		D
1991											
1992										D	

Source: Russell and Wilson (1994)

notable droughts. In further comparison with country-level indices, some discrepancies may be noted. The index successfully identifies droughts of 1986 (Spain), 1990 (Greece, Spain and Portugal) and 1982 (Denmark). If the index is working successfully, there should have been yield losses in Portugal (1981, 1992) and Denmark (1983), but records were not available to prove this point. The Italian predictions are poor; this is due in part to the wide variation in AWD expected across the country within any one year because of large altitudinal variations within the 0.5 degree latitude/longitude grid cells. Using the minimum AWD for the country it is possible to detect the droughts of 1983 and 1988. However, using this index, yield reductions should also be recorded for 1981, 1985, 1987 and 1989.

Based on the validation exercise the following interpretation of the agricultural water deficit index applies. In regions with a severe AWD index, a crop would require approximately 100 per cent irrigation during moisture-sensitive developmental stages to produce a good yield. Hence if the necessary water resources are not available during critical periods of moisture stress, large crop losses may occur. In areas with a moderate AWD index, drought susceptibility is dependent on other factors, such as the soil properties, the moisture sensitivity of the crop and whether irrigation is an economically viable option. Areas classified as low or none in the AWD index experience only limited or no long-term susceptibility to agricultural drought respectively, as moisture deficits are generally within the water-holding capacity of most agricultural soils.

The index of agricultural water deficit illustrated in Figure 4.6 is based on the average climate for a 30-year time series. It shows the spatial pattern of long-term susceptibility to drought in Europe and identifies regions which persistently suffer from severe moisture deficits. However, drought is a temporal phenomenon which varies according to seasonal weather conditions. The year-to-year probability of drought, within the long-term susceptibility, is very important to agriculture as this will determine specific yield losses in any one year. Individual years from the period 1961–90 have been analysed to show the severity of specific drought episodes and their probability of occurrence.

Specific historic drought episodes are apparent in the 1961–90 time series. In particular, the years 1976, 1989 and 1990 exhibit moderate and severe agricultural drought over large regions of Europe. In 1976, total rainfall during the period April to August was much lower than expected. Over the study area the maximum amount received was 780 mm compared with 950 mm on average. This was by far the driest summer in the period 1961–90. The lack of rainfall particularly affected northwest Europe, while in the Mediterranean more rain fell than usual. For this region, 1965 or 1986 recorded very low rainfall totals (<100 mm over all Spain, Portugal, Italy and Greece). For the whole of Europe 1989 and 1990 were not significantly dry. The drought in these years particularly affected northwest France and the UK. The distribution is different from the average shown in Figure 4.6 and demonstrates the importance of regionalised rainfall patterns. The year-to-year variation in drought is high. Factors such as over-winter rainfall are taken into account in this index. The effects of the dry winter of 1975/6 in northern Europe are clear. The droughts of 1989 and 1990 were important in northern Europe as two moderate to severe droughts occurred in successive years. In terms of agriculture, UK cereal yields were below average, while yields of potatoes and sugar beet responded well to irrigation (Cannell and Pitcairn, 1993).

The probability of occurrence of very warm years increases greatly as mean temperatures increase, assuming that temperature variability will not change in the future (DoE, 1991). For example, the DoE (1991) computed that in central England the probability of a summer at least as hot as 1976 occurring in any given year increases from about 0.1 to 10 per cent by 2030, and to 33 per cent by 2050 with a 'business-as-usual' greenhouse gas emissions scenario. Similar changes in the probability of occurrence of the year 1990 are likely with increases in mean temperatures. This would imply that agricultural drought is likely to become a more frequent phenomenon with future climate change.

The probability of occurrence of a severe agricultural water deficit has been determined by calculating the number of years within the 30-year time series (1961–90) belonging to the severe AWD class. The classification of severe agricultural drought risk is shown in Table 4.14 and the spatial distribution of this risk in Figure 4.7. The southern Mediterranean countries record greater than 50 per cent chance of severe drought. Severe drought risks of over 20 per cent are found in northeast Scotland, southern and northern Sweden and northern Finland because of low values of AWC.

4.4.3 Broad-scale effects of climate change

The impacts of the climate change scenarios on agricultural drought depend on the relative changes predicted for temperature and precipitation and the balance between improved water-use efficiency of crops due to elevated CO_2 against increased atmospheric evaporative demand.

Changes in mean drought susceptibility for the 2025 UKHI scenario are fairly modest. Analysis of the AET/PET ratio values, on which AWD is based, high-lights two principal regions where generally lower ratio values are recorded: southern and continental Europe. This translates into changes in the AWD classification as follows. Parts of eastern and northern France experience changes

Table 4.14 Classification of severe agricultural drought risk

Number of years (out of 30)	Probability of a severe agricultural water deficit %
24–30	>80
15–24	50–80
6–15	20–50
0–6	<20

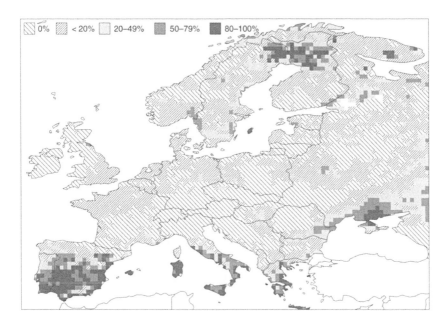

Figure 4.7 Probability of severe agricultural drought (AWD < 0.70), 1961–90

from low to moderate AWD. Changes in the severe AWD class occur in Spain, Greece, Romania, Russia, Norway and Hungary. Small areas of the UK (except south and southwest England), Scandinavia, the Baltic States, northern Germany, Netherlands, Denmark and Belarus experience the highest AET/PET ratios, indicating a lessening of drought risk. The moderate droughts currently experienced in southern Spain are predicted to generally worsen, either covering a greater area or worsening in severity. Severe drought can also be expected over larger areas of Spain than at present. Improvements in drought risk for north and northwest Europe can be explained by the effects of increased rainfall more than compensating for increased PET.

By 2050 more significant changes in the AET/PET ratio are apparent, following the same pattern of change identified in 2025 (Figure 4.8). Drought risk generally reduces in northern and northwestern Europe and increases in southern and continental Europe. The changes in the AET/PET ratio translate into only slight changes in the AWD classification. The largest areas of change in AWD are in Spain and the countries bordering the north of the Black Sea, where a larger area is classified as severe. In central Europe, from central France in the west to Romania in the east and the Balkans in the south, the area classified as no drought considerably reduces, being replaced by classification as low and moderate drought. Spain experiences an increase in the area where at least moderate or severe drought occurs all the time, and there is a

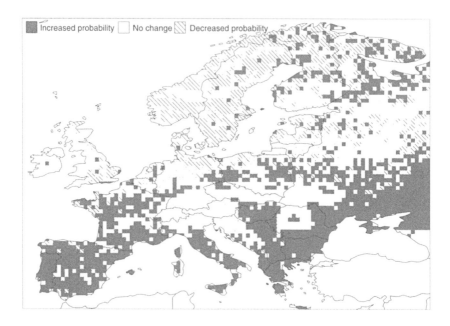

Figure 4.8 Change in the probability of severe agricultural drought (AWD < 0.70) for
UKHI climate scenario in 2050

dramatic increase in all parts of Europe in the area likely to have a severe
drought.

The probability of occurrence of a severe agricultural water deficit was determined for the two scenarios by calculating the number of years within the
30-year time series belonging to the severe AWD class. The spatial distribution of
severe drought risk is shown in Figures 4.8 and 4.9. Generally there is a slight
increase in risk, though in some areas a decrease is observed. Scandinavia, northern Germany and parts of southern England experience a reduced probability of
severe drought in both the UKHI (2050) and GFDL (2064) scenarios, as the
effects of increased rainfall offset the increased temperatures. In southern
Europe, particularly Spain, Portugal and Italy, the risk increases, with the
UKHI scenario being more severe than the GFDL scenario. In the UKHI
scenario, central and southeastern Spain have an increase in the area where there
is a 100 per cent chance of severe drought. Central Europe sees a decline in the
area where there is no risk of drought.

4.5 Responding to drought

Drought risk is the conjuncture of the probability of agricultural drought and the
vulnerability of European agriculture. Future changes in both the hazard and
vulnerability warrant the consideration of adaptive responses that will secure

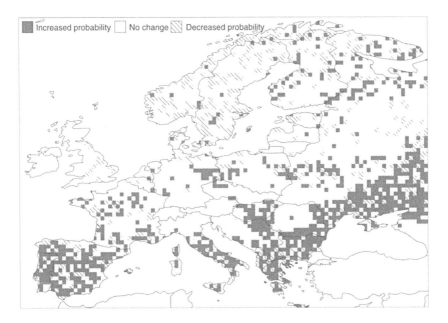

Figure 4.9 Change in the probability of severe agricultural drought (AWD < 0.70) for the GFDL climate scenario in 2064

agricultural incomes and ameliorate regional impacts of drought. Many such responses are warranted in the present; they may become even more urgent in the future owing to climate change and European agricultural, economic and environmental policies. This section presents an overview of such adaptive responses. Table 4.15 provides an overview of the kinds of agricultural adjustments that may be effective.

4.5.1 Farm-level adjustments

A longer potential growing period in general provides more options (Armstrong and Castle, 1995). These could include the introduction of new crops and varieties (Marsh, 1989), altered crop rotations, different timings of operations, and new technologies (Woolhouse, 1989). Additional applications of fertiliser, to take advantage of the carbon dioxide enrichment, may be effective.

While such options may increase high yields, there are also options to mitigate the effects of drought. If greater soil moisture deficits occur in summer, irrigation will be more attractive (Brown, 1990). However, irrigation may not be economically viable in, say, the South Downs, and is unlikely to be required for winter wheat. Also, the supply of water for irrigation may be uncertain in the future.

Box 4.1 The UK drought of 1995

A severe drought gripped the UK in 1995, raising fresh concern about climate change. The Department of the Environment commissioned a review of the cost of the drought, in part to draw lessons for assessing the impacts of climate change (Palutikof *et al.*, 1997).

The year was the warmest year on record globally. For the Central England record, beginning in 1659, new records were set for the summer (June to August was 3°C warmer than average), and the year (November 1994 to October 1995 was 1.6°C warmer than the 1961–90 normal. The preceding winter was very wet, and was followed by a dry spring and very dry summer. Rainfall for July and August in England and Wales was the lowest on record. The high temperatures and dry weather caused exceedingly high demands on water resources.

What does this extreme event have to tell us about climate change, drought and agriculture? In fact, the wet winter ensured that soil moisture levels were high and grain crops did well in the dry, summer weather. Major gains were recorded for wheat, barley and oilseed rape. Low yields and poor quality affected potatoes, other vegetables and minor crops. In contrast, livestock suffered, with costs estimated at around £200 million.

This study highlights the difference between national and farm-level impacts. Comparing national output during 1995 to a trend line from recent years suggests that the total benefit from wheat, barley, oilseed rape, sugar beet and potatoes was £59 million, including a loss of £40 million for potatoes. However, gross margins for a typical farm reflect higher prices (and higher inputs, although these were not large in total). If we use the gross margin, the drought benefited farmers to the tune of £624 million, with potato farmers experiencing the largest gains (£390 million). Of course, consumers paid for the higher prices. This example suggests that drought impacts need to be considered at a variety of scales and for different stakeholders.

Farmers actively responded to the drought. Some cited 'unsociable' hours as a major impact. They worked at night and early in the morning to avoid the heat and to get freshly harvested produce to markets. The use of herbicides, pesticides and fungicides was both increased and reduced. More irrigation was required. A major lasting impact has been incentives for farmers to develop small-scale, on-farm water storage to catch winter run-off, which required adjustments to government regulations.

If such droughts became more common, the impacts on agriculture could be very different. Persistent dryness, and drought over the winter and spring, would dramatically reduce yields. This was the case in 1976, and appears to be happening again in 1997. Each drought is different; linking a specific drought to climate change is tenuous. Future conflicts over water resources are likely in any case.

Source: Palutikof *et al.* (1997)

4.5.2 Regional and national adjustments

Changes in patterns of nutrient leaching may be expected under a changed climate, owing to different patterns of soil moisture deficit (Armstrong et al., 1994). This has implications for water quality. Any increased use of irrigation would have implications for management of water resources. Additionally, increased use of irrigation would tend to increase soil erosion (Boardman and Favis-Mortlock, 1993), with implications for water quality and long-term productivity.

According to present management practices all areas classified as severe or very severe would require 100 per cent irrigation to avoid crop moisture stress and produce a good yield. However, the use of irrigation is likely to become less of an option in many areas, given the present demand for water in relation to surface- and groundwater supplies (Kenny et al., 1993). If the necessary water resources are not available during critical periods of moisture stress, large losses in potential yield would occur. This will be crucial in countries such as Spain, where there is already concern over loss of groundwater to irrigation (Kenny and Harrison, 1992). Many areas are investing in improved water resource and irrigation systems to make better use of available water. This is likely to be a further requirement in the future.

A key issue for private utilities and government is planning for recurrent drought. As seasonal (and longer) climate prediction improves, there is no reason not to be prepared for severe drought. Public management of demand, allocation of water between users, and long-term investment in efficient water systems are required. The Australian experience provides one prototype for a mix of public and private responsibility for drought management (Drought Policy Review Task Force, 1990). In the USA, the insurance sector provides coverage for crop losses due to extreme weather events, including drought (Changnon and Changnon, 1989, 1990). At present, farm insurance in the UK is limited to buildings. If drought risk becomes more severe and farmers are expected to absorb inter-annual fluctuations in income, insurance, drought bonds and other income-smoothing instruments might become attractive.

4.6 Summary and discussion of impacts

4.6.1 Summary of site and regional impacts

The EPIC simulation results suggest that water shortage is a major constraining factor for wheat-growing on the thin Downland soils. There is a linear relation-ship ($R^2 = 0.73$) between simulated gross margins and the number of days when the crop experiences water stress at Woodingdean. By contrast, there is no relationship between temperature stress and gross margins.

The results for the changed-climate scenarios indicate that, as expected, wheat (a C_3 crop) responds strongly to increased atmospheric CO_2. On the water-

Table 4.15 Agricultural adjustments to climate change

Strategy and Adjustment	Mechanism	Costs	Timing for implementation	Constraints and issues
Land management				
Moisture management: conservation, irrigation, soil drainage, mulching, fallowing; Soil management: mulching, tillage, crop rotation, land drainage	Regulate soil water balance through incremental irrigation, drainage, control of evaporation and run-off; Enhance organic matter, use of fertiliser, control soil erosion. Drainage may be required in northern latitudes (e.g. St Petersburg)	Higher costs for additional irrigation works, water, operation and maintenance; Some additional labour and inputs	Gradual implementation with increased temperatures, often in response to drought	Water availability (surface and ground), water quality, terrain, alternative demands for water, investment capital and incentives
Crop variety and land use				
Cultivars; Rotations; Crop substitution; Cropped area; Crop location; Conversion to/from crops or pasture; Changes in specialisation; Livestock types and levels	Switch varieties, crops or rotations: longer-maturing varieties, heat and drought tolerant, requiring less vernalisation; More flexible cropping system with seasonal forecasts, spread risk; Switch location (regional or within farm) to new climates or soils; Change specialisation, e.g. arable/pasture production; Change resource intensity (e.g. stocking rate)	Costs include: development of cultivars; livestock breeding; restructuring for different farming systems; Marginal costs may be minimal if encompassed in normal agricultural investment	Costs are staged or incremental, but related to rate of climate change and possible effects of severe episodes; New cultivars require 10–15 years to develop	Lead time to develop new cultivars; Soil suitability and terrain in conversion to agricultural uses; Delayed response to new conditions; need for additional information and training; Availability of genetic material

Crop husbandry: planting and harvesting				
Timing of planting and harvest; Plant mixed varieties; Planting depth; Plant density	Earlier/later scheduling with changed growing season or to shift timing of heat stress; Flexible cropping system; Plant deeper in drier conditions; Thin crop in dry years to lower plant density and reduce competition for moisture	Few additional costs; Shifts in labour requirements during season	Gradual adjustment with little lead time; Possibly greater flexibility in response to seasonal or monthly weather forecasts	Availability of cultivars, changes in winter season, frost, soils, field accessibility due to wet conditions may limit applicability
Crop husbandry: Fertility and pest management				
Herbicides; Pesticides; Fertiliser application; Nitrogen-fixing crops	Control weeds to reduce competition for moisture nutrients, and light; Control pests and diseases that limit plant growth, yield or yield quality; Nature, quantity and timing of fertiliser affect plant uptake	Input costs increase in general; Considerable savings possible for some fertiliser/crop regimes, but increased costs in other regions	Gradual adjustment with short lead time and rapid responses, except for new crops and invading pests, diseases and weeds	Toxic and ecological concerns with fertiliser leaching; Information to respond to new pests, diseases and weeds, new crops
Economic adjustments: farm level				
Investment in agriculture: equipment and machinery; Farm inputs; Savings and storage; Labour and employment; Off-farm purchases; Food consumption	Increased investment in agriculture to increase yields; Increased food storage reduces variability in supply; Increased savings and purchases supplement storage; Off-farm employment to support increased investment and food purchases; Altered food consumption to cope with seasonal shortages, shifts to new varieties, economic crises, labour demands	Infrastructure for storage, marketing; Operation and maintenance costs for storage; Opportunity cost of off-farm employment; Costs of new technology; Additional costs in dry years for purchases, replanting, etc.	Gradual but variable related to yields; Storage facilities minor on-farm; Gradual shifts in employment, but sudden with extreme episodes	Limited by finance, technology, type of agricultural production, surplus, access to regional and international economy and trade

constrained Downs, the increased water-use efficiency enables the crop to make more effective use of water when it is available, and hence increase yields. Although increased evapotranspiration may well make less water available in the summer months, these simulations suggest that the increased opportunism of the C_3 crop will more than compensate for this on average. Nonetheless, wide variations about such an average are possible, depending on the timing of water shortages. Droughts of the severity of the present-day analogues, however, seem likely to occur less often.

Increased fertiliser appears to be needed to make fullest use of the CO_2 effect. If fertiliser is not unlimited, though, further increases in temperatures would appear – by means of higher evapotranspiration and hence increased water short- age – to erode the advantages gained by CO_2 fertilisation. This would explain the simulated slowing of wheat yield increase after 2025 for fertiliser-limited sce- narios at Woodingdean.

Only the simplest of management changes have been considered – bringing forward the drilling date for winter wheat – but it is clear that even this action has a large effect on yields and gross margins. Interactions between yield and the variable costs needed to produce this render the economic response more com- plicated, as (for example) increased fertiliser use presumably temporarily out- weighs increased yield in the 2025 medium- and low-fertiliser Woodingdean results.

With the scenarios of climate change, dry conditions reduce in northern and northwestern Europe and expand in southern and continental Europe. Spain, in particular, has a larger area classified as susceptible to severe drought, while a shift from no drought to slight to moderate drought occurs in central Europe. The probability of severe agricultural water deficits, drought episodes, follows a similar pattern. Scandinavia, northern Germany and parts of southern England experience a reduced probability of severe drought by 2050. In southern Europe, particularly Spain, Portugal and Italy, the risk increases. In central Europe there is a decline in the area where there is no risk of drought.

Compared with previous work, these scenarios are reasonably optimistic. In the case of winter wheat, slower rates of warming are compensated by carbon dioxide enrichment. At the spatial scale, a previous assessment used equilibrium scenarios of climate change without incorporating the effects of carbon dioxide (Harrison, 1994).

4.6.2 Research issues in multi-scale modelling

This exercise in field and spatial modelling of drought hazard illustrates a number of methodological issues concerning the selection and integration of different scales of analysis (see also Orr and Brignall, 1995).

Model processes

A multi-level approach recognises that scientific and policy issues are addressed on different scales. The choice of scale needs to match these issues. For example, at the field scale, effective farm management strategies need to be identified that provide secure incomes. At the regional level, planners need to know how extensive a drought may be and the interactions between regions in agricultural trade and development. Thus site models tend to focus on crop dynamics and farm management, while spatial models are oriented toward landscape dynamics.

Process-based models

An assumption is often made that process-based models are better than semi-empirical assessments. A process-based model such as EPIC should work adequately for conditions within the range for which it was developed. There is, however, some doubt regarding the model's responses for water-stressed crops. Williams *et al.* (1989) noted poor correlations between simulated and measured yield on some irrigated sites. Kiniry *et al.* (1992) describe a modified version of EPIC which better replicated observed crop behaviour under conditions of water stress. Cabelguenne *et al.* (1995) have developed a version of EPIC (EPIC-PHASE) which aims to forecast more accurately the water requirements of irrigated crops in order to optimise use of water resources. The standard version of EPIC (as used in this study) tends to underestimate the effects of water stress (Maurice Cabelguenne, personal communication, 1995). Thus the results for the thin and droughty soils of the South Downs may be optimistic. On the other hand, chalky soils are able to provide more water for plants than is suggested by the depth of the soil (Burnham and Mutter, 1993).

The relationships between increased CO_2 and crop growth in the EPIC model do not consider other environmental variables such as temperature and should be considered tentative (Kimball *et al.*, 1993). Additionally, they are derived from a relatively limited number of studies (Stockle *et al.*, 1992a). Although more efficient use of water under elevated levels of CO_2 is predicted by the model, it is unclear how far this is likely to be reflected in practice.

Interactions

The effects of increased temperature and CO_2 upon competing pests are not modelled (cf. Treharne, 1989; Porter *et al.*, 1991). This is a potentially serious omission; aphid attacks on wheat in 1995 (for example) were exacerbated by the warmer weather conditions. On the other hand, losses of yield and quality would diminish with better weather at harvest. Temperature will also influence nitrogen mineralisation; this will interact with water availability (Graham Russell, personal communication, 1995). The interaction of water availability and irrigation requirements was noted in the broad-scale analysis.

Spatial variability

The spatial variability of climate and soil needs to be taken into account. The frequency of drought at individual sites cannot be directly scaled up to the frequency of a widespread drought of similar magnitude. Soil type is a major factor influencing crop yields and response to drought. For example, Weir *et al.* (1984) present mean yields of winter wheat in 1979 and 1980 for several hundred sites on nineteen UK soil series. Yields varied by up to about 1 t ha^{-1} between the two years on individual soils. However, the range of yields among all soils was about 3.5 t ha^{-1} for each year. Even within a single field, soil properties are likely to vary and produce corresponding variations in crop yields (e.g., Evans and Catt, 1987). Robust approaches to handling this variability are still being developed.

Climatic data and scenarios

Long time series (preferably 100 years) are required in order to analyse drought frequencies, since extreme events are rare by definition. As noted previously, generated weather tends to mute the impact of climatic extremes, including dry spells (see Semenov and Porter, 1992, for an analysis of an improved weather generator).

The Thornthwaite method for calculating PET tends to underestimate PET in southern Europe. The climate scenarios used in this study are limited in the range of precipitation changes. For example, a more convective climatic regime would probably result in increased rainfall intensities during summer storms. This would mean that a greater proportion of such rainfall would run off (DoE, 1991, 1996), thus further reducing the amount of water available to crops in summer. A policy study should evaluate the full range of potential climate scenarios and outcomes.

Adaptation

The future is evolving, and part of the change is continual adaptation to climatic variations. In some respects, climate change impact assessment seeks to identify the new environments as targets for improved cropping systems. Methods for evaluating the full range of potential cultivars and cropping systems, including optimal strategies, have yet to be devised (Easterling *et al.,* 1992c; Mendelsohn *et al.,* 1994). Only a few such strategies (automatic scheduling of fertiliser and the changed drill date for some runs) were tested. The key unknown is forecasts of future technologies, especially in crop varieties and their requirements for fertiliser, pesticides, tillage, etc.

Integrating natural and social science

The field-level economic analysis is based on the assumptions of the crop simulation, in addition to uncertainties in farm economics. Yield prices, for example, are held static, whereas in reality prices fluctuate according to supply and demand; the model is thus uncoupled from the wider economy. Future costs are also assumed not to change. Note too that since fixed costs are ignored, any change in their future importance relative to gross margins is also ignored. At the wider scale, estimating changes across a variety of farm types in different regions, in order to understand the macro-economic effects of drought, would require a complete sample of agricultural systems. While such efforts have been compiled in the USA (in the MINK study for example; Easterling, 1992a, b, c, d), data on Europe are more difficult to compile, farming systems more diverse, and policy effects dominate between countries.

4.6.3 Conclusion

The risk of agricultural drought in Europe is likely to change with climate change. To judge from the limited, broad-scale analysis undertaken here, the drought hazard may alleviate somewhat in northern Europe and worsen in southern Europe. For specific crops, especially winter wheat, the combined effects of a CO_2-enhanced climate and temperature change, with little changes in precipitation, would lead to increased mean yields. A decreased risk of drought is likely for the South Downs in England, although with a rather greater variability of year-to-year yield than at present, particularly if earlier drilling dates are adopted. For winter wheat in southern France, climate change would also appear to produce positive effects (see Harrison et al., 1995).

Further research is warranted on a broad range of issues in multi-level modelling: from crop responses to elevated CO_2 levels in field situations, to site and spatial crop modelling, and integration with social and economic data and models.

Acknowledgements

We are most grateful to John Porter (Royal Agricultural and Veterinary University, Denmark) for providing data for the UK winter wheat cultivar; Andrew Moxey and David Oglethorpe (Newcastle-upon-Tyne), Graham Russell (Edinburgh), Eric Audsley (Silsoe) and Verel Benson (USDA Temple, Texas) for agro-economic information; Robert Coles for UK farming information; the National Rivers Authority (now the Environment Agency) for providing meteorological data; MAFF for providing yield data; and Jimmy Williams and the rest of the EPIC team for assistance with their model. Two reviewers provided helpful comments. The Climatic Research Unit provided spatial data and climate scenarios. This chapter is a substantially revised version of Brignall et al. (1996).

Note

1 These values are the average for western Europe, as calculated using the Penman equation and including concurrent effects on humidity, radiation and wind for one climate scenario (called the UKHI).

References

Agro Business Consultants Ltd (ABC) (1995) *Farm Machinery Costs*, Melton Mowbray: ABC.

Armstrong, A.C., and Castle, D.A. (1992) Potential impacts of climate change on patterns of production and the role of drainage in grassland. *Grass and Forage Science* 47: 50–61.

Armstrong, A.C., and Castle, D.A. (1995) 'Potential effects of climate change on agricultural production and the hydrology of drained grassland in the UK', in D.F.M. McGregor and D.A. Thompson, eds, *Geomorphology and Land Management in a Changing Environment*. Chichester: John Wiley, 139–151.

Armstrong, A.C., Matthews, A.M., Portwood, A.M., Addiscott, T.M., and Leeds-Harrison, P.B. (1994) 'Modelling the effects of climate change on the hydrology and water quality of structured soils', in M.D.A. Rounsevell and P.J. Loveland, eds, *Soil Responses to Climate Change*. Berlin: Springer-Verlag.

Barrow, E., Hulme, M., and Semenov, M.A. (1995) *Scenarios of Climate Change*. Oxford: Environmental Change Unit, University of Oxford.

Bignon, J. (1990) *Agrométéorologie et physiologie du maïs grain dans la Communauté Européenne*. EUR 13041 FR. Luxembourg: Office for Official Publications of the European Communities.

Boardman, J., and Favis-Mortlock, D.T. (1993) Climate change and soil erosion in Britain. *Geographical Journal* 159, 2: 179–183.

Boardman, J., Evans, R., Favis-Mortlock, D.T., and Harris, T.M. (1990) Climate change and soil erosion on agricultural land in England and Wales. *Land Degradation and Rehabilitation* 2, 2: 95–106.

Brignall, A.P., Hossell, J.E., Favis-Mortlock, D.T., and Rounsevell, M.D.A. (1994) Climate change and crop potential in England and Wales. *Journal of the Royal Agricultural Society of England* 155: 140–161.

Brignall, A.P., Downing, T.E., Favis-Mortlock, D., Harrison, P.A., and Orr, J.L. (1996) 'Climate change and agricultural drought in Europe: Site, regional and national effects', in T.E. Downing, A.A. Olsthoorn and R.S.J. Tol, eds, *Climate Change and Extreme Events: Altered Risk, Socio-economic Impacts and Policy Responses*, Amsterdam: Vrije Universiteit, 51–89.

Brown, P. (1990) Ban on watering crops 'will ruin farmers'. *Guardian*, 8 August.

Burnham, C.P., and Mutter, G.M. (1993) The depth and productivity of chalky soils. *Soil Use and Management* 9, 1: 1–8.

Cabelguenne, M., Jones, C.A., Marty, J.R., Dyke, P.T., and Williams, J.R. (1990) Calibration and validation of EPIC for crop rotations in southern France. *Agricultural Systems* 33: 153–171.

Cabelguenne, M., Jones, C.A., and Williams, J.R. (1993) Utilisation du modèle EPIC pour la recherche de strategies optimales d'irrigation dans l'hypothèse de resources en eau limitées: Application au maïs dans le sud-ouest de la France. *Comptes Rendus de l'Académie d'Agriculture de France* 79, 4: 73–84.

Cabelguenne, M., Jones, C.A., and Williams, J.R. (1995) Strategies for limited irrigations of maize in southwestern France. A modelling approach. *Transactions of the American Society of Agricultural Engineers* 38, 2: 507–511.

Cannell, M.G.R., and Pitcairn, C.E.R. (1993) *Impacts of the Mild Winters and Hot Summers in the United Kingdom in 1988-1990*, London: HMSO.

Changnon, S.A., and Changnon, J.M. (1989) Developing rainfall insurance rates for the contiguous United States. *Journal of Applied Meteorology* 28: 1185–1196.

Changnon, S.A., and Changnon, J.M. (1990) Use of climatological data in weather insurance. *Journal of Climate* 3: 568–576.

Cole, J.A., Slade, S., Jones, P.H., and Gregory, J.M. (1991) Reliable yield of reservoirs and possible effects of climatic change. *Journal of Hydrological Sciences* 36, 6: 579–597.

Department of the Environment (DoE) (1991) *The Potential Effects of Climate Change in the United Kingdom*, London: HMSO.

Department of the Environment (DoE) (1996) *Review of the Potential Effects of Climate Change in the United Kingdom: Conclusions and Summary*, London: HMSO.

Doorenbos, J., and Kassam, A.H. (1979) *Yield Response to Water*, FAO Irrigation and Drainage Paper 33, Rome: Food and Agriculture Organization of the United Nations.

Downing, T.E., ed. (1996) *Climate Change and World Food Security*, Heidelberg: Springer-Verlag.

Downing, T.E., Favis-Mortlock, D.T. and Gawith, M. (1994) *Climate Change and Extreme Events: Scenarios of Altered Hazards for Further Research*, Oxford: Environmental Change Unit, University of Oxford.

Drought Policy Review Task Force (1990) *Managing for Drought*, Canberra: Australian Government Publishing Service.

Easterling, W.E., Rosenberg, N.J., McKenney, M.S., and Jones, C.A. (1992a) An introduction to the methodology, the region of study, and a historical analog of climate change. *Agricultural and Forest Meteorology* 59: 3–15.

Easterling, W.E., Rosenberg, N.J., McKenney, M.S., Jones, C.A., Dyke, P.T., and Williams, J.R. (1992b) Preparing the Erosion–Productivity Impact Calculator (EPIC) model to simulate crop response to climate change and the direct effects of CO_2. *Agricultural and Forest Meteorology* 59: 17–34.

Easterling, W.E., McKenney, M.S., Rosenberg, N.J., and Lemon, K.M. (1992c) Simulations of crop response to climate change: Effects of present technology and no adjustments (the 'dumb farmer' scenario). *Agricultural and Forest Meteorology* 59: 53–73.

Easterling, W.E., Rosenberg, N.J., Lemon, K.M., and McKenney, M.S. (1992d) Simulations of crop responses to climate change: Effects with present technology and currently available adjustments (the 'smart farmer' scenario). *Agricultural and Forest Meteorology* 59: 75–102.

Evans, R., and Catt, J.A. (1987) Causes of crop patterns in Eastern Europe. *Journal of Soil Sciences* 38: 309–324.

Favis-Mortlock, D.T. (1994) Use and abuse of soil erosion models in southern England. Unpublished PhD thesis, University of Brighton.

Favis-Mortlock, D.T., and Boardman, J. (1995) Nonlinear responses of soil erosion to climate change: A modelling study on the UK South Downs. *Catena* 25, 1–4: 365–387.

Favis-Mortlock, D.T., Evans, R., Boardman, J., and Harris, T.M. (1991) Climate change, winter wheat yield and soil erosion on the English South Downs. *Agricultural Systems* 37, 4: 415–433.

Gardner, C.M.K., and Field, M. (1983) An evaluation of the success of MORECS, a meteorological model, in estimating soil moisture deficits. *Agricultural Meteorology* 29: 269–284.

Goudriaan, J., and Unsworth, M.H. (1990) *Implications of Increasing Carbon Dioxide and Climate Change for Agricultural Productivity and Water Resources*, ASA Special Publication no. 53, Madison, WI: American Society of Agronomy, Crop Science Society of America and Soil Science Society of America.

Groenendijk, H. (1989) *Estimation of the Waterholding Capacity of Soils in Europe: The Compilation of a Soil Dataset*, Simulation Report CABO-TT no. 19, Wageningen: CABO, Department of Theoretical Production Ecology, Agricultural University.

Gwynne, R.N., and Meneses, C. eds (1994) *Climate Change and Sustainable Development in the Norte Chico, Chile: Land, Water and the Commercialisation of Agriculture*, Oxford: Environmental Change Unit, and Birmingham University.

Harrison, P.A. (1994) 'The effects of climate change on agricultural drought in Europe', in T.E. Downing, D.T. Favis-Mortlock and M.J. Gawith, eds, *Climate Change and Extreme Events: Scenarios of Altered Hazards for Further Research*, Oxford: Environmental Change Unit, University of Oxford, 27–38.

Harrison, P.A., Butterfield, R.E., and Downing, T.E., eds (1995) *Climate Change and Agriculture in Europe: Assessment of Impacts and Adaptation*, Report no. 9, Oxford: Environmental Change Unit, University of Oxford.

Hossain, M. (1992) *Structure and Distribution of Household Income and Income Dimension of Poverty: Re-thinking Rural Poverty*, Dhaka: Bangladesh Institute of Development Studies.

Hough, M.N. (1990) *Agrometeorological Aspects of Crops in the United Kingdom and Ireland: A Review for Sugar Beet, Oilseed Rape, Peas, Wheat, Barley, Oats, Potatoes, Apples and Pears*, EUR 13039 EN, Luxembourg: Office for Official Publications of the European Communities.

Hulme, M., Conway, D., Brown, O., and Barrow, E.M. (1994) *A 1961–90 Baseline Climatology and Future Climate Change Scenarios for Great Britain and Europe: Part III, Climate Change Scenarios for Great Britain and Europe*, Norwich: Climatic Research Unit.

Huq, S., Ahmed, A.U., and Koudstaal, R. (1996) 'Vulnerability of Bangladesh to climate change and sea level rise', in T. E. Downing, ed., *Climate Change and World Food Security*, Heidelberg: Springer-Verlag, 347–380.

International Federation of Red Cross and Red Crescent Societies (IFRCRCS) (1994) *World Disasters Report 1993*, Geneva: IFRCRCS.

Karim, Z. (1996) 'Agricultural vulnerability and poverty alleviation in Bangladesh', in T.E. Downing, ed., *Climate Change and World Food Security*, Heidelberg: Springer-Verlag, 307–346.

Katz, R.W., and Brown, B.G. (1992) Extreme events in a changing climate: Variability is more important than averages. *Climatic Change* 21: 289–302.

Kenny, G.J., and Harrison, P.A. (1992) Thermal and moisture limits of grain maize in Europe: model testing and sensitivity to climate change. *Climate Research* 2: 113–129.

Kenny, G.J., Harrison, P.A., and Parry, M.L., eds (1993) *The Effect of Climate Change on Agricultural and Horticultural Potential in Europe*, Report no. 2, Oxford: Environmental Change Unit, University of Oxford.

Kimball, B.A., and Idso, S.B. (1983) Increasing atmospheric CO_2: Effects on crop yield, water use and climate. *Agricultural Water Management* 7: 55–72.

Kimball, B.A., Mauney, J.R., Nakayama, F.S., and Idso, S.B. (1993) Effects of elevated CO_2 and climate variables on plants. *Journal of Soil and Water Conservation* 48, 1: 9–14.

Kiniry, J.R., Blanchet, R., Williams, J.R., Texier, V., Jones, C.A., and Cabelguenne, M. (1992) Sunflower simulation using the EPIC and ALMANAC models. *Field Crops Research* 30: 403–423.

Kirschbaum, M.U.F., Bullock, P., Evans, J.R., Goulding, K., Jarvis, P.G., Noble, I.R., Rounsevell, M., and Sharkey, T.D. (1996) 'Ecophysiological, ecological and soil processes in terrestrial ecosystems: A primer on general concepts and relationships', in R.T. Watson, M.C. Zinyowera and R.H. Moss, eds, *Climate Change 1995. Impacts, Adaptations and Mitigation of Climate Change: Scientific-technical Analyses*, Cambridge: Cambridge University Press, 57–74.

Maché, R. (1993) Profitability of wheat production: Four farms considered. *The Furrow* 98, 1: 8–10.

McKenney, M.S., Easterling, W.E., and Rosenberg, N.J. (1992) Simulation of crop productivity and responses to climate change in the year 2030: The role of future technologies, adjustments and adaptations. *Agricultural and Forest Meteorology* 59: 103–127.

Manabe, S., Stouffer, R.J., Spelman, M.J. and Bryan, K. (1991) Transient responses of a coupled ocean–atmosphere model to gradual changes of atmospheric CO_2, Part I: Annual mean response. *Journal of Climate* 4: 785–818.

Manabe, S., Spelman, M.J. and Stouffer, R.J. (1992) Transient responses of a coupled ocean–atmosphere model to gradual changes of atmospheric CO_2, Part II: Seasonal response. *Journal of Climate* 5: 105–126.

Marsh, J. (1989) 'Economic dimensions of the "greenhouse effect" for UK agriculture', in R.M. Bennett, ed., *The 'Greenhouse Effect' and UK Agriculture*, Reading: Centre for Agricultural Strategy, 79–90.

Marsh, T.J., and Lees, M. (1985) *The 1984 Drought*, Wallingford, Oxfordshire: Institute of Hydrology.

Mearns, L.O., Rosenzweig, C., and Goldberg, R. (1992) Effects of changes in interannual climatic variability on CERES-wheat yields: Sensitivity and $2 \times CO_2$ general circulation model scenarios. *Agricultural and Forest Meteorology* 62: 159–189.

Mendelsohn, R., Norhaus, W.D., and Shaw, D. (1994) The impact of global warming on agriculture: A Ricardian analysis. *American Economic Review* 84, 4: 753–771.

Ministerio de Obras Públicas Transportes y Medio Ambiente (1993) *Plan Hidrológico Nacional, Memoria*, Madrid: Ministerio de Obras Públicas y Transportes, Secretaria de Estado para las Políticas del Agua y el Medio Ambiente.

Ministry of Agriculture, Fisheries and Food (MAFF) (1967) *Potential Transpiration*, Technical Bulletin no. 16, London: HMSO.

Mitchell, J.F.B., Manabe, S., Meleshko, V., and Tokioko, T. (1990) 'Equilibrium climate change and its implications for the future', in J.T. Houghton, G.J. Jenkins and J.J. Ephraums, eds, *Climate Change: The IPCC Scientific Assessment*, Cambridge: Cambridge University Press, 131–172.

Moxey, A. (1991) Estimation of bid prices for biotechnological frost protection of maincrop potatoes. *Agricultural Systems* 37, 4: 399–414.

Murphy, M.C. (1994) *Report on Farming in the Eastern Counties of England 1992/93*, Cambridge: Agricultural Economics Unit, University of Cambridge.

Narciso, G., Ragni, P., and Venturi, A. (1992) *Agrometeorological Aspects of Crops in Italy, Spain and Greece: A Summary Review of Durum Wheat, Barley, Maize, Rice, Sugarbeet, Sunflower, Soybean, Rape, Potato, Tobacco, Cotton, Olive and Grape*, EUR 14124 EN, Luxembourg: Office for Official Publications of the European Communities.

Nicks, A.D., Richardson, C.W., and Williams, J.R. (1990) 'Evaluation of the EPIC model weather generator', in A.N. Sharpley and J.R. Williams, eds, *EPIC – Erosion/Productivity Impact Calculator. 1. Model Documentation*, Temple, TX: US Department of Agriculture, 105–124.

Nix, J., ed. (1994) *Farm Management Pocketbook*, 25th edition, Wye: Department of Agricultural Economics Farm Business Unit, Wye College, University of London.

Orr, J.L., and Brignall, A.P. (1995) 'Integration of crop model results: Study recommendations for policy and further research. Methods for site and regional scale integration', in P.A. Harrison, R.E. Butterfield and T.E. Downing, eds, *Climate Change and Agriculture in Europe: Assessment of Impacts and Adaptation*, Oxford: Environmental Change Unit, University of Oxford, 391–400.

Palmer, W.C. (1965) *Meteorological Drought*, Washington, DC: US Weather Bureau.

Palutikof, J.P., Subak, S., and Agnew, M.D., eds (1997) *Economic Impacts of the Hot Summer and Unusually Warm Year of 1995*, Norwich: University of East Anglia.

Parry, M.L., Carter, T.R., and Konijn, N.T., eds (1988) *The Impact of Climatic Variations on Agriculture*, vol. 1: *Assessment in Cool Temperate and Cold Regions*, Dordrecht: Kluwer.

Penman, H.L. (1948) Natural evaporation from open water, bare soil and grass. *Proceedings of the Royal Society, London* 193: 120–146.

Phillips, D.L., Hardin, P.D., Benson, V.W., and Baglio, J.V. (1993) Non-point source pollution impacts of alternative agricultural management practices in Illinois: A simulation study. *Journal of Soil and Water Conservation* 48, 5: 449–457.

Porter, J.H., Parry, M.L., and Carter, T.R. (1991) The potential effects of climatic change on agricultural insect pests. *Agricultural and Forest Meteorology* 57: 221–240.

Quinones, H., and Cabelguenne, M. (1990) Use of EPIC to study cropping systems, II: Improved simulation of water use, growth and harvest index in corn. *Agricoltura Mediterranea* 120: 241–248.

Racsko, P., Szeidl, L., and Semenov, M. (1991) A serial approach to local stochastic weather models. *Ecological Modelling* 57, 1–2: 27–41.

Rahman, H.Z. (1992) *Crisis and Insecurity: The 'Other' Face of Poverty: Rethinking Rural Poverty*, Dhaka: Bangladesh Institute for Development Studies.

Rind, D., Goldberg, R., Hansen, J., Rosenzweig, C., and Ruedy, R. (1990) Potential evapotranspiration and the likelihood of future drought. *Journal of Geophysical Research* 95, D7: 9983–10004.

Rosenberg, N.J., McKenney, M.S., Easterling, W.S., and Lemon, K.M. (1992) Validation of EPIC model simulations of crop responses to current climate and CO_2 conditions: Comparisons with census, expert judgement and experimental plot data. *Agricultural and Forest Meteorology* 59: 35–51.

Rosenzweig, C., and Iglesias, A., eds (1994) *Implications of Climate Change for International Agriculture: Crop Modeling Study*, Washington, DC: US Environmental Protection Agency.

Russell, G. (1990) *Barley Knowledge Base*, EUR 13040 EN, Luxembourg: Office for Official Publications of the European Communities.

Russell, G., and Wilson, G.W. (1994) *An Agro-pedo-climatological Knowledge-Base of*

Wheat in Europe, EUR 15789 EN, Luxembourg: Office for Official Publications of the European Communities.

Salinger, M.J., Mullan, A.B., Porteous, A.S., Reid, S.J., Thompson, C.S., Coutts, L.A., and Fouhy, E. (1990) *New Zealand Climate Extremes: Scenarios for 2050 AD*, Wellington: New Zealand Meteorological Service for the Ministry for the Environment.

Semenov, M.A., and Porter, J.R. (1992) Stochastic weather generators and crop models: climate change impact assessment. Paper in *Effects of Global Change on Wheat Ecosystems*, at GCTE Focus 3 Meeting, Saskatoon, Canada.

Semenov, M., and Porter, J.R. (1995) Non-linearity in climate change impact assessments. *Journal of Biogeography* 22, 4–5: 597–600.

Semenov, M.A., Porter, J.R., and Delecolle, R. (1993) 'Simulation of the effects of climate change and development of wheat in the UK and France', in G.J. Kenny, P.A. Harrison and M.L. Parry, eds, *The Effects of Climate Change on Agricultural and Horticultural Potential in Europe*, Oxford: Environmental Change Unit, University of Oxford, 121–136.

Shaw, R.H. (1977) 'Climatic requirement', in G.F. Sprague, ed., *Corn and Corn Improvement*, Madison, WI: American Society of Agronomy.

Stockle, C.O., Williams, J.R., Rosenberg, N.J., and Jones, C.A. (1992a) A method for estimating the direct and climatic effects of rising atmospheric carbon dioxide on growth and yield of crops, Part I: Modification of the EPIC model for climate change analysis. *Agricultural Systems* 38: 225–238.

Stockle, C.O., Dyke, P.T., Williams, J.R., Jones, C.A., and Rosenberg, N.J. (1992b) A method for estimating the direct and climatic effects of rising atmospheric carbon dioxide on growth and yield of crops., Part II: Sensitivity analysis at three sites in the Midwestern USA. *Agricultural Systems* 38: 239–256.

Thornthwaite, C.W., and Mather, J.R. (1955) The water balance. *Climatology* 8: 1–104.

Treharne, K. (1989) 'The implications of the "greenhouse effect" for fertilizers and agrochemicals', in R.M. Bennett, ed. *The 'Greenhouse Effect' and UK Agriculture*, Reading: Centre for Agricultural Strategy, 67–78.

Viner, D., and Hulme, M. (1992a) *Climate Change Scenarios for Impact Studies in the UK*, Norwich: UK Department of the Environment and Climatic Research Unit, University of East Anglia.

Viner, D., and Hulme, M. (1992b) *Construction of Climate Change Scenarios by Linking GCM and STUGE Output*, Norwich: UK Department of the Environment and Climatic Research Unit, University of East Anglia.

Waggoner, P.E. (1990) Anticipating the frequency distribution of precipitation if climate change alters its mean. *Agricultural and Forest Meteorology* 47: 321–337.

Warrick, R.A. and Ahmad, Q.K., eds (1996) *The Implications of Climate and Sea Level Change for Bangladesh*, Dordrecht: Kluwer.

Weir, A.H., Bragg, P.L., Porter, J.R. and Rayner, J.H. (1984) A winter wheat crop simulation model without water or nutrient limitations. *Journal of Agricultural Science* 102: 371–382.

Williams, J.R., Jones, C.A., Kiniry, J.R. and Spanel, D.A. (1989) The EPIC crop growth model. *Transactions of the American Society of Agricultural Engineers* 32, 2: 497–511.

Williams, J.R., Jones, C.A. and Dyke, P.T. (1990) 'The EPIC model', in A.N. Sharpley and J.R. Williams, eds, *EPIC – Erosion/Productivity Impact Calculator, 1: Model Documentation*, Temple, TX: US Department of Agriculture.

Wilmott, C.J., and Gaile, G.L. (1992) 'Modelling', in R.F. Abler, M.G. Marcus and J.M. Olson, eds, *Geography's Inner Worlds*, New Brunswick, NJ: Rutgers University Press.

Wolf, J. (1993) *Effects of Climate Change on the Wheat Production Potential in the E.C.*, Wageningen: Department of Theoretical Production Ecology, Wageningen Agricultural University.

Wolf, J. (1994) 'Pot experiments in a greenhouse on spring wheat, faba bean and sugar beet', in P.A. Harrison, R.E. Butterfield and T.E. Downing, eds, *Climate Change and Agriculture in Europe: Assessment of Impacts and Adaptation*, Oxford: Environmental Change Unit, University of Oxford, 137–152.

Woolhouse, H. (1989) 'The "greenhouse effect" and crop production in the UK', in R.M. Bennett, ed., *The 'Greenhouse Effect' and UK Agriculture*, Reading: Centre for Agricultural Strategy.

5

FLOODING IN A WARMER WORLD: THE VIEW FROM EUROPE

John Handmer, Edmund Penning-Rowsell and Sue Tapsell

5.1 Introduction

A warmer world is likely to be one with more flooding than at present. There is general agreement among the various climate models that rainfall intensity will probably increase with increasing greenhouse gas concentrations (Watson *et al.*, 1996: 337). And intensity is the driving force behind floods regardless of changes in average annual rainfall (Fowler and Hennessy, 1995). Rainfall, and rainfall intensities, are expected to increase by between 10 and 20 per cent in mid-latitude areas during winter, to be greater throughout the year in high latitudes, and to increase in monsoonal regions (Weijers and Vellinga, 1995). In contrast, and probably of more significance, rainfall in arid and semi-arid areas is expected to decline or at best remain the same. Rivers depending on snowmelt are expected to experience a change in their seasonally high flow from spring/early summer to winter; warmer winters are likely to result in the snowpack melting periodically during winter rather than accumulating for a single spring thaw. A second reason for the change in seasonality is the increased likelihood of winter rain causing flooding. In addition, fewer snowmelt floods are expected in plain regions because of decreased snowfall – however, the predictions for alpine areas are considered to be highly uncertain (Watson *et al.*, 1996: 336–337).

Nevertheless, in terms of defining credible scenarios for flooding, this statement is described by the Intergovernmental Panel on Climate Change (IPCC) as a low-confidence result. The amount of increase in flooding is very uncertain and is likely to vary greatly between catchments. The IPCC (Watson *et al.*, 1996: 338) lists four main reasons for the uncertainty:

1   It is difficult to define credible scenarios at the catchment scale for changes in precipitation that produce floods;
2   It is difficult to model the processes that transform rainfall (or snowmelt) to floods;

3   Available climatic and hydrological records have limited information about flooding.
4   In many cases, it is difficult to differentiate the effects of climatic change from anthropogenic changes to land use.

Complicating factors include uncertainty over changes in flood seasonality, and the effect of changes to the El Niño–Southern Oscillation (ENSO) as global climate models (GCMs) do not simulate ENSO effects at present (Watson *et al.*, 1996: 338). ENSO is an important factor in the interannual variability of rainfall in the Pacific, Australia, Indonesia, South America, and to a lesser extent, adjacent regions.

We should be careful to define what we mean by 'flooding': the word 'flood' covers a continuum of events from the barely noticeable through to catastrophes of diluvian proportions. Our prime interest here is with floods that cause damage, but are nevertheless well short of the probable maximum event. We focus on damage, but recognise that floods are not simply negative. They are crucial in many regions for sustaining life in both human and non-human systems. Flood flows cause damage, but fill reservoirs, irrigate fields, flush out or dilute toxic substances (both natural such as blue-green algae, and synthetic) in streams; and flooding may underlie reproductive processes in arid areas. In arid and semi-arid areas in particular, an increase in flood frequency may have considerable benefits for both human activities and the endemic biota. A decrease in flood frequency would be likely to have detrimental effects – although the dynamics of such systems are very complex (Watson *et al.*, 1996: 349).

Strzepek and Smith (1995) have collated research results on climate change and developing countries. They examined the Indus, Mekong, Uruguay, Zambezi and Nile river basins, and conclude that increased flooding is unlikely to be a major issue. Instead, the critical factor is decreased annual average flow. However, results also illustrate the great uncertainty surrounding climate change and run-off modelling. Under the scenarios developed by some GCMs, flooding may become more severe in some river basins. Under the GFDL model, flooding in the Uruguay basin would reach 'catastrophic levels' (but would decrease under two other models). Flooding along the Indus is likely to increase 'because of melting of the Himalayan glaciers', but it would not bring economic benefits. Floods in the Mekong delta 'could start earlier and be larger, adversely affecting fisheries and agriculture' (Smith *et al.*, 1995: 14). For the Nile, increased evaporation due to increased temperatures outweighs other factors and is likely to lead to substantial reduction in flows (one GCM predicts increased flows) (Riebsame *et al.*, 1995: 86).

In higher-latitude areas, scenarios suggest that flooding is likely to increase; to what extent depends on the studies consulted. The IPCC (Watson *et al.*, 1996: 337) reports on work (not based on GCMs) in Belgium and Switzerland which suggests that the average annual flood will increase by about 10 per cent. Using GCM scenarios and hypothetical cases, large increases in 'flood frequencies and

126

the risk of inundation' were found for the Rhine (Watson *et al.*, 1996: 338; see also Kwadijk and Rotmans, 1995; Kwadijk and Middelcoop, 1994). From a study based on palaeoflood records for the upper Mississippi River, Knox (1993) concluded that small changes in temperature (1–2°C) and changes in average annual rainfall of around 10–20 per cent can result in large changes to flood frequency and magnitude.

Relatively small changes in climate can cause or exacerbate water resource problems, especially in arid regions (Smith *et al.*, 1995: 12). And the 1996 IPCC report shows that dry areas are more sensitive to 'climate variation' than wet areas. Overall, increases in flood frequency are unlikely to have effects as severe as an increase in droughts. More droughts would prove very costly in most regions, especially where water is already fully exploited.

Even though the consensus is that any increase in flooding induced by climate change will be small, there is concern over its potential impacts – in particular, on human health in developing countries, and especially among the residents of informal settlements.

This chapter, first, reviews the results from a soil water balance model to investigate the impact that climate change may have on precipitation and run-off in Europe – given the predictions of various GCMs. Second, the general pattern in terms of the nature of the impacts in relation to the climate shifts is considered. Third, specific impacts are examined for three case studies of locations vulnerable to flooding: the lower Thames catchment, UK; the Île de France; and the Meuse valley, Limburg, the Netherlands. We also consider the policy responses to increased flood severity induced by climate change in terms of conventional policy responses and more radical alternatives. These are considered in terms of both the individual and the agencies of government. The history and current status of flood hazard management in Britain is reviewed, and the role of the Environment Agency examined. This Agency, created in April 1996, has overall responsibility for flood management (known in Britain as flood defence), and needs to adopt the appropriate policy responses for dealing with climate change; its policy options are reviewed later in the chapter. The EA absorbed the National Rivers Authority, and like it must work with local authorities, as the Agency has no general land-use control powers.

5.2 Floods across Europe

To generalise, there are two main types of floods that affect Europe (Table 5.1). First, there is the rapid-rise thunderstorm-type event, affecting major urban areas and also, locally, large areas of southern Europe. Second, there are the slow-rise events characterised by floods on the large rivers of northern Europe, notably the Rhine, Vistula, Danube, Thames, Seine, Loire and Rhône. These large catchments are slow to respond to individual rainstorm or snowfall events, but respond to prolonged periods of rain or snowmelt, such as a period of sustained rainfall during frontal weather over a period of 1 to 2 weeks. For mountainous areas and

Table 5.1 A simplified typology of floods

Winter rainfall floods: Depressions tracking over Europe from the Atlantic with well developed warm fronts bring winter precipitation which, when heavy, continuous and prolonged, can lead to soil becoming saturated, in turn leading to high volumes of run-off. When this occurs, rivers may flow out of bank, causing flooding. The seriousness of this depends on the duration of the rainfall, on whether the catchments were previously saturated, and on the activities at risk.

Snowmelt floods: Rapid snowmelt in changeable weather conditions is sometimes an important contributory cause in this type of flooding. This is most likely to occur in spring, when warm southern airstreams track northwards into alpine areas, creating sudden snowmelt accompanied sometimes by heavy rainfall. This phenomenon can be very localised, and when this is the case, and urban areas occupy valley bottoms, the effects can be serious, especially since floodwater velocities can be high.

Summer convectional storm-induced floods: Intensive storms and floods are sometimes caused by convectional thunderstorms. These usually occur in summer months and appear to have become more frequent over heavily urbanised areas where the urban 'heat island' effect is pronounced. They also can be common in southern European areas, where prolonged hot periods of the summer months can end with sudden storms. The seriousness of this type of flooding is compounded by the fact that coastal and other flood-plain areas can be in seasonal use for camp sites in the areas affected.

Sewer flooding: Many older cities have inadequate sewerage systems, which means that normal intensive rainfall events can create abnormal flooding. Some new developments in Europe also have inadequate storm sewer systems because the developments are unplanned or even illegal.

Lowland waterlogging: High levels of water retention in lowland clay soils, together with low evapotranspiration rates in the cool, moist climate, result in waterlogging and flooding in many parts of lowland Europe. Areas can be flooded from rainfall in areas surrounded by embankments failing to find an outflow. Many lowland areas can be flooded for weeks or months, but in these areas agricultural practices have often responded by growing water-resistant and low-value crops. In some areas urban basement flooding has become common, owing to a rising water table associated with industrial and urban decline; which in turn has led to a decline in groundwater abstractions. The rising groundwater levels manifest themselves in basements as permanent waterlogging. This tends to be a small, localised problem rather than a serious hazard.

Sea surge and tidal flooding: Sea and tidal flooding is a major problem, and is closely bound to the problem of coastal erosion, which may subsequently lead to flooding. Many European coastal areas are low-lying and the main threat is from a combination of high tides, low atmospheric pressure and strong onshore winds producing tidal surges. Important parts of coastal Europe are also prone to flooding and erosion from the sea. Coastal flooding can, also arise from other mechanisms, including seepage of sea water at high tide through natural shingle embankments, through artificial banks created to protect land from the sea, through breaches of these embankments, and occasionally through ocean swell phenomena perhaps related to earthquake activity in the mid-Atlantic ridge area causing waves to overtop these embankments. Secular sea-level rise, isostatic land-level fall and local land subsidence are all factors which add to tidal and sea flooding problems.

Dam-break flood risk: Finally, there are low probability–high consequence flood risks associated with large dams. Climate change may lead to larger extreme floods, in which case spillways may become underdesigned in terms of current standards – and are more likely to fail.

Source: Penning-Rowsell and Fordham (1994)

---

parts of the north of the continent a third type of flooding, that from snowmelt, may be important.

There has been considerable analysis of flooding and flood hazard, in Britain and elsewhere in Europe (e.g., Handmer, 1987; Saul, 1992), the USA (Rosen and Reuss, 1988; Moore and Moore, 1989), New Zealand (Ericksen, 1986), Australia (Smith, 1996), and elsewhere. This literature approaches floods from a number of different perspectives, including hydrology, hydraulics, economics, geography and environmental management. The EUROflood project has recently investigated the social and economic aspects of floods, and evaluated responses in terms of flood mitigation investment, flood warning schemes, evacuation, and the modelling of different types of floods at different scales of analysis across Europe (Penning-Rowsell and Fordham, 1994).

Damaging floods are not new to Europe, and as more and higher-value human activity is located on flood plains, losses resulting from flooding will rise – unless floods become less frequent through the use of engineering works or because of changes in the physical phenomena. Therefore, any impacts of climate change will be superimposed on what is already a steadily rising European flood loss potential.

Some terminology is useful here. We are concerned in this chapter with the damage and destruction resulting from floods, in particular from extreme events. These damages are generally differentiated as to whether they can be expressed in money terms or not; the former are known as 'tangible' losses and the latter as 'intangible' losses or impacts. Secondly, these impacts can be differentiated as to whether they are direct (that is, direct damage caused from physical contact between flood waters and susceptible assets) or indirect (for example, the consequential or secondary effects of flood damage on industrial production caused by factory premises being closed during flood events).

This typology is shown in Table 5.2. It is generally recognised that the more extreme flood events have greater indirect effects than lesser events, and that those experiencing the more extreme events are more likely to suffer more intense intangible effects (death; stress-induced ill-health; and so on). However, this generalisation is by no means a universal rule, and certain minor events can have major indirect effects (particularly when they cut important communication links within the economy), as well as local but major intangible effects.

The impacts of floods are not restricted to those individuals and groups that suffer damage. Institutions of government are also involved, and these vary by country. Generally, flood problems are tackled by basin agencies in a number of

Table 5.2 A typology of flood damages

	Direct damages	*Indirect damages*
Tangible	Damage to infrastructure, buildings and contents from contact with water	Loss of production and retail trade consequent upon flooding and flood damage
Intangible	Damage to items of cultural significance and personal memorabilia	Damage to health consequent upon flooding, inconvenience and disruption

Source: Penning-Rowsell and Chatterton (1977)

countries (e.g., in France and the UK), and in theory these should deliver integrated strategies. Practice falls well short of this, however. One of the impacts of climate change-induced flooding could be the need for institutional change to cope with the greater hazard.

5.3 A soil water balance model

5.3.1 Methods and data

A soil water balance provides a process-based approach to the assessment of soil moisture status throughout the year (see also Brignall *et al.*, Chapter 4, this volume). One of the outputs from such models can be the amount of 'excess rainfall' occurring over winter. This varies as a function of available water storage capacity in the soil, rainfall and temperature. The soil moisture conditions from the previous season are important, and in some years deficits may be carried over, resulting in little 'excess rainfall'. The model used here is not hydrological; the run-off component is simply the precipitation which is not required to restore field capacity and is not lost by evaporation.

The soil water balance model is defined as a layered 'bucket'-type model, in which the balance between total monthly precipitation (P), total monthly potential evapotranspiration (PET) and the available water-holding capacity of the soil (AWC) is computed. The AWC is divided into three layers: a top, middle and lower layer, representing 40, 40 and 20 per cent of the total AWC respectively. In each month a soil water deficit accrues if PET > P. Water is abstracted from the top layer at the full rate of PET, from the second layer at half the PET rate and from the third layer at one-quarter of the PET rate. When the profile is empty no more extraction occurs. Once P > PET, the profile fills until the profile is full, or at field capacity. Any additional precipitation is then run-off. The model was initialised assuming that the soil profile was at field capacity. It was then run for five iterations using the average 1961–90 climatology to determine more realistic start values for soil water content in the different regions of Europe. The model was then run continuously throughout the 34-year time series.

5.3.2 European scenarios

The soil water balance model derives values for the anticipated mean change in precipitation per grid cell for the European region. The difference between the baseline period (1961–90) and 2050 is characterised by a clear latitudinal stratification with areas north of about latitude 47° receiving more precipitation, and vice versa. There is a substantial reduction in rainfall in southern Europe (large areas receiving between 10 and 20 per cent less rainfall on an annual basis), and large areas of northern Europe receiving rainfall totals greater by 15 per cent as compared with the baseline period.

It may seem surprising that the pattern is uninterrupted by typography or the relationship between land and sea. As far as topography is concerned, the reason may be that it is simply not well represented in the baseline model, and in the GCMs it is largely ignored. Topography is far more important for flood analyses than for general climate modelling. However, the land and sea interface is incorporated into atmosphere–ocean coupled models.

While there may be a reduction in annual precipitation in southern Europe over this period, if a greater proportion of that rainfall comes as summer thunderstorms, then flood situations could continue to be as bad as they are now, or become worse. Similarly, if the increased rainfall in northern Europe were to be largely summer rainfall, then this would not affect the flooding situation there since this flooding is generally characterised by events resulting from winter rainfall creating floods down the large northern European rivers (e.g. the Thames, the Rhine/Meuse and the Vistula). Higher temperatures would mean less snowfall and therefore a reduced contribution of snowmelt to river flows. This could lead to winter rainfall generating river flow earlier in the year, leading to a relative increase in winter run-off and a reduction in run-off in early spring (Arnell *et al.*, 1994).

In fact, the seasonality information that we have suggests that the northern European rainfall increases are from increased winter rainfall. Greater uncertainty remains as to what may be the flooding situation in southern Europe, since even with the decreased rainfall totals there could be greater storminess and hence more severe flooding. More detail on changes in the seasonality of climate shifts will be important to refining any projections concerning flood severity there. Other research (DoE/MO, 1989) also suggests that soil moisture (controlled by the balance between rainfall and evaporation) will be lower in summer across the mid-latitudes of northern continents, which will have implications for fluvial flooding by increasing infiltration and reducing run-off.

In terms of run-off, and therefore the flood-inducing mechanisms, Figure 5.1 portrays average run-off for the baseline period, while Figure 5.2 shows the shift between the baseline period and a scenario of climate change for 2050. The differences between the two climate change scenarios (UKHI minus GFDL) are shown in Figure 5.3. Although the GFDL scenario is somewhat more extreme, both scenarios show a similar pattern. Serious reductions in run-off are predicted

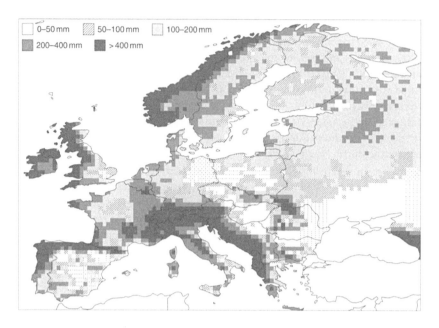

Figure 5.1 Baseline 1961–90 average run-off (mm)

for southern Europe, such that most rivers appear to be dry (i.e., run-off reduced by 100 per cent). A 'continental' phenomenon is apparent, whereby precipitation reductions are greatest in central Iberia, the Danube basin and southern Russia, and the reductions stretch northwards towards the Baltic, despite the fact that precipitation increases are quite marked throughout most of Europe north of a line between Nantes and the Crimea. In terms of Britain, the difference between the baseline period and the year 2050 shows increases in run-off in south and eastern England and, surprisingly, a small reduction in run-off at the extreme southeastern part of Scotland.

Since increased run-off will be largely a winter phenomenon, rivers such as the Rhine, the Thames and the main rivers of Scandinavia may see substantially increased severity of flooding. The rivers of southern Europe, by contrast, should have their flood frequencies and magnitudes reduced, perhaps quite markedly. In southern Europe the scenarios suggest that we appear to be moving towards a drier climate, and flood problems there may decrease. However, as pointed out above, this may not occur if there is an increase in storminess, where less rain falls but it falls with greater intensity – as is typical of many arid and semi-arid areas.

The situation in 2025 entails fewer changes than for 2050, but the pattern for both precipitation and run-off is similar. There are minor reductions in run-off throughout most of southern Europe, of the order of between 0 and 10 per cent,

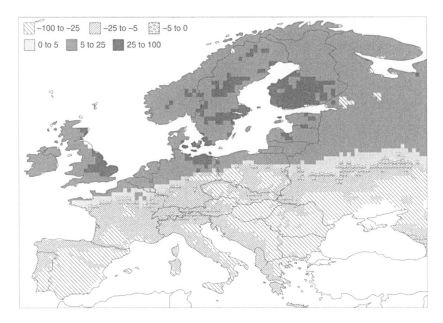

Figure 5.2 Difference between average run-off for the baseline period and 2050 (per cent). Based on the UKHI scenario (Mitchell *et al.*, 1990) scaled to global-mean temperature change expected in a non-interventionist scenario (IS92a)

with substantially greater reductions in run-off in the large land masses of the Iberian Peninsula and, broadly, the Danube basin. There is a small reduction in run-off in the extreme southeast of Scotland. Most of northern Europe shows increased precipitation and increased run-off, although while precipitation increases significantly over much of northern Germany and the north Polish plain, run-off here declines. The same is true concerning parts of northern France, and southern Russia (north of the Black Sea). Most of inland Scandinavia shows marked increases in run-off (between 10 and 15 per cent), but the coastal areas do not follow the same pattern. Why this is the case is not clear.

Clearly, as Riebsame (1989) states, the implications of climate change for flood control and flood hazard mitigation deserve much greater attention.

5.3.3 UK scenarios

As far as Britain is concerned, precipitation and run-off are shown to increase in all areas except southeast Scotland. This could indicate increased flooding problems with catchments such as the Trent and Thames, particularly since the precipitation increase is a winter phenomenon.

In central/eastern and southeastern England the 2050 scenario indicates increased run-off of between 25 and 100 per cent for the majority of the Trent

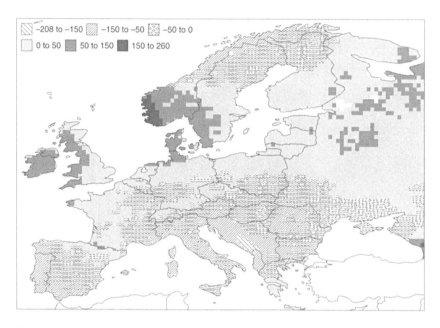

-208 to -150 -150 to -50 -50 to 0
0 to 50 50 to 150 150 to 260

Figure 5.3 Difference between run-off in two scenarios of climate change (mm). Based on the UKHI (Mitchell *et al.*, 1990) scenario for 2050 minus the GFDL transient scenario (Manabe *et al.*, 1991, 1992) for the 2060s

catchment, and between 15 and 25 per cent for at least two-thirds of the Thames catchment (the remainder seeing increases of between 25 and 100 per cent). Both rivers either have significant flood control schemes in place (e.g., the Trent at Nottingham) which may become seriously underdesigned if the river flood regime changes, or already have significant flood problems (e.g., the Thames at Maidenhead and between Datchet and Teddington).

So far only two studies of the potential implications of climate change for groundwater recharge and storage have been carried out in the UK (Hewett *et al.*, 1992; Wilkinson and Cooper, 1993), and there appear to have been few studies elsewhere. Climate change predictions suggest that it is likely that river flow regimes in the UK will be more seasonal than at present. In groundwater catchments, summer flows are controlled by groundwater recharge during winter, and it is possible that flows in lowland chalk streams could become higher in summer than at present, even if summers become warmer and drier (Arnell, 1991). On the other hand, lower spring rainfall would have major implications for the availability of groundwater resources during the summer. The effect of climate change on groundwater recharge depends on the extent to which the recharge season may be shortened and how this is compensated by an increase in rainfall during winter. As groundwater recharge largely takes place in winter, and prolonged steady rain is more effective at recharging groundwater than short-period intense rainfall, recharge would be reduced in the future if winter rainfall were to be concentrated into shorter periods (Arnell *et al.*, 1994).

In summary, the soil-water analysis shows increased precipitation and run-off in northern Europe during winter, thus reinforcing and increasing the severity of the flood regimes there. In southern Europe there is a significant reduction in rainfall and the flooding situation there should be reduced in severity unless there is a shift away from winter rainfall, towards substantially increased summer storminess. Here problems seem more likely to emerge from increased droughts.

5.4 Shifts in the magnitude and severity of floods: Annual average damages and event damages

Changes in flood frequency are likely to alter the levels of flood damage and may thereby place demands on flood management systems – including mechanisms such as insurance which deal with the redistribution of losses. Changes in climate resulting in increased storminess or increased rainfall totals may have profound effects on the fundamental relationship which describes floods: the relationship describing the frequency of floods of different magnitudes (Saul, 1992). Magnitude may be expressed by flood-water discharge or flood height at a specified point (but note that the relationship between discharge and height will depend on local topography). The less frequent (the lower the probability of occurrence) the flood, the larger its magnitude. This relationship summarises the physical phenomenon of flooding by indicating the number of major events experienced at any one site; the full relationship also gives the statistical model of these major events and the more minor events at the same location. Floods are typically referred to in terms of their probability or frequency of occurrence: as the 100-year flood – that is, that flood with 1 chance in 100 of occurring in any year, also known as the 1 per cent or 1 : 100 flood, or in Australia the flood with a 1 per cent annual exceedence probability. This describes a flood that would be expected to occur on average once in every 100 years over a very long period – hence another term for the concept, the 'return period'; for a 1 per cent flood the return period would be 100 years.

There is clearly an underlying assumption that the flood-producing aspects of climate and topography are constant over very long periods of time; to assess the magnitude of relatively rare events like the 1 per cent flood takes a long record. This is an important underlying assumption of both flood record analysis and flood damage assessment (see below). The assumption here tends to be that climate changes resulting from global warming are introducing substantial changes to a stationary situation. In fact, there is good evidence that the flood-producing aspects of climate are highly variable over periods of decades (for eastern Australia see Smith and Greenaway, 1983). In addition, in some areas short-term variability – also referred to as reliability – may be enormous. Arid areas exhibit this characteristic dramatically, where long periods of no flow may be followed by extensive flooding. McMahon *et al.* (1992) show that stream flow and therefore flooding in Australia and southern Africa varies greatly over the short term – in addition to the longer-term variability mentioned above.

The effects of land-use change adds further uncertainty to frequency/magnitude calculations. For all their apparent precision, and the attention paid to the flood record, calculations of rare floods are rather abstract and sit within very wide confidence bands.

It is also quite possible that climate change will affect the shape of the frequency–magnitude distribution. There appear to be at least two possibilities here. One is that the rarer or more extreme events, such as the 1 : 100 flood or even rarer events, will be affected disproportionately. Areas where this occurs would have a very substantial difference between frequent floods (such as 2- or 5-year events) and rare events (such as the 1 : 100 flood) (Figure 5.4, curve a). At other locations, the whole magnitude–frequency relationship may shift, so that floods of all degrees of severity will change (generally increase) in frequency (Figure 5.4, curve b). This second type of change has occasionally been documented from historical flood data; for example, Smith and Greenaway (1983) have shown that in southeastern Australia the flood frequency–magnitude regime has shifted, with a dramatic increase in flood damages.

Under the first scenario above, it is possible that the probable maximum flood

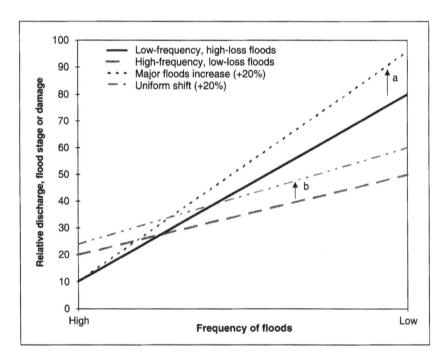

Figure 5.4 Contrasting types of shift in flood frequency and losses with climate change

(PMF) will increase, perhaps substantially. This could have potentially expensive implications for structures designed to withstand the PMF, such as major dams – unless a lower standard is accepted and the structures are left as they are, while the increased risk is handled through improved emergency planning.

The second type of change – a uniform shift of the whole magnitude–frequency relationship – could be a function of all rainfall events (or other flood-inducing causes) being larger across the frequency spectrum, whereas a flexing of the magnitude–frequency curve could mean that only the less frequent storms are bigger (or smaller).

The impact of any shift in the flood regime is likely to be greatest for places where there is little short-term variability in stream flow and where there is relatively little difference in discharge between frequent and rare floods.

These two different effects would have widely different economic impacts (Table 5.3).

Changes to the whole magnitude–frequency relationship will have a profound effect on the calculated annual average damage (Table 5.3, column A). This annual average damage calculation is heavily biased towards smaller and medium-sized events since their probability of occurrence is greater. For example, in the model used here a change in the magnitude of all floods by 20 per cent (whether judged by stage, damage or discharge) for a given frequency will have the effect of increasing annual average damages by 20 per cent (Table 5.3, column A). On

Table 5.3 The effect of different types of shift in the magnitude–frequency relationship on annual average damages

Return period (years)	Base case	A. Uniform shift		B. Bigger events get bigger still		C. Smaller events only get bigger	
	Damage (£)	Overall increase factor	Damage (£)	Overall increase factor	Damage (£)	Overall increase factor	Damage (£)
5	1,500	1.20	1,800	1.00	1,500	1.20	1,800
10	3,300	1.20	3,960	1.04	3,432	1.16	3,828
25	15,000	1.20	18,000	1.08	16,200	1.12	16,800
60	36,000	1.20	43,200	1.12	40,320	1.08	38,880
100	58,000	1.20	69,600	1.16	67,280	1.04	60,320
200	100,000	1.20	120,000	1.20	120,000	1.00	100,000
Annual average damages (£)	2,692		3,231		2,922		3,001
Change over base case (%)			20.0		8.5		11.4

the other hand, if just the major events are increased in this way by 20 per cent, then the annual average damages will increase by only 8.5 per cent (Table 5.3, column B), other things being equal.

The converse is somewhat different. If we leave the magnitude of the 100-year event unchanged, but change the magnitude of the 5-year event by +20 per cent, and change the intermediate events proportionately, then the annual average damage increases by 11.4 per cent (Table 5.3, column C). Quite clearly, changes to the more minor events have a greater effect on the annual average damages than do the more major events, other things being equal.

But why worry about annual average damages? Because this statistic is the summation of the tangible damages from all flood frequencies and magnitudes that go towards calculating the economics of flood mitigation schemes: scheme benefits are calculated by summing the proportionate contribution from the total range of events, and discounting this sum over the lifetime of the scheme or policy according to standard cost–benefit procedures (Penning-Rowsell and Chatterton, 1977; Dixon and Hufschmidt, 1986). Thus shifts to the whole magnitude–frequency relationship will have a major effect upon the economic viability or otherwise of flood mitigation investment.

However, changes to the major events alone have an impact on a different audience, in policy terms. Those concerned with insurance – and, in particular, reinsurance – are concerned not only about the average event in any one year, but more importantly with the maximum total single event loss that could be experienced within the lifetime of an insurance policy. Therefore, if climate change were radically to increase the severity of large events such as the 50- or 100-year flood event within a catchment, this could take insurable losses beyond the threshold usually covered by retail insurance companies, into the domain of reinsurance. For example, if the changes for the river Thames increased the flood depths for the 100-year event by just 100 mm, the increase in damages caused by that event (on the Windsor to Teddington reach) would be from £49.3 million to £64.3 million, or an increase of over 30 per cent (Tables 5.4 and 5.5).

This analysis shows that the impact of climate change in terms of flood impacts is different depending upon the nature of that climate change. If increased storminess affects only the major events (as opposed to the frequent minor and medium-sized events), then the long-term average economic impact is small, but that impact could be of great significance to the reinsurance industry. If, on the other hand, it is the minor but damaging events which increase in frequency, leaving the major events much the same, then the economic case for significant increases in flood mitigation investment may be advanced – as these investment decisions are driven by the minor and medium-sized floods and their local economic impact. Note that 'economic' losses are those incurred by a specific economy, for example the UK national economy, and may be quite different from 'financial' losses, which are those incurred by insurance companies, other commercial enterprises, households and so on (Penning-Rowsell et al., 1992).

Table 5.4 The relationship between flood return period and impacts for the Datchet to Teddington river Thames flood plain

Return period of flood event (year)	Number of properties affected	Event damage (£000s)
5	442	1,215
9	989	2,810
25	5,250	13,036
56	8,622	30,973
101	10,447	49,384
204	12,435	85,404

Table 5.5 The impact of increasing flood depths by 100 mm for the Datchet to Teddington river Thames flood plain

	101-year flood			204-year flood			PMF
	Properties affected	Event damage (£m)	Capital sum (£m)	Properties affected	Event damage (£m)	Capital sum (£m)	Capital sum (£m)
A. +10 cm	11,576	64.3	34.8	13,578	103.8	41.4	51.1
B. At 1995	10,447	49.3	23.1	12,435	85.4	30.2	38.1
C. A/B (%)	10.8	30.4	50.6	9.2	21.6	37.1	34.1

Notes: PMF is probable maximum flood. Capital sum is the future annual average sequence of flood damages expressed as current value, using the standard approach of economics

5.5 Flood damages and climate change: Three case studies

It is evident from the soil-water analysis that areas of northern Europe could experience increased precipitation and run-off between the baseline period and the years 2025 and 2050. In some cases, in certain areas, these increases are quite significant. In addition, the seasonality data indicate that these are increases in winter rainfall events, which, by and large, are the major causes of the more extensive and extreme floods in northern Europe.

Given these increases in precipitation and run-off it is instructive to look at the sensitivity of current flood plains – and their river catchments – to increased flood volumes and flood water levels. This is particularly instructive where flood problems have already been identified, and policies and solutions are being debated, as in the three areas examined below.

5.5.1 The lower Thames catchment, UK

The Thames Region of the Environment Agency has overall responsibility for flood hazard management – known as 'flood defence' (before April 1996 the

National Rivers Authority, Thames Region, had responsibility, and before 1989 the Thames Water Authority) (see National Rivers Authority, Thames Region, 1995). The Agency is a construction body. It has no general land-use power, and can only make recommendations to local planning authorities. The Thames basin in southeast England occupies about 5,000 square miles (12,000 sq km), and is home for some 12 million people. It includes most of London.

The last major flood in the lower Thames occurred in 1947. Since then the flood hazard has increased for two main reasons. First, there has been a significant increase in catchment urbanisation, which has probably exacerbated run-off volumes and speeds, although not to a large extent (given the size of the catchment). In addition, much development has also occurred on the flood plain, particularly to the west of London in the area between Windsor and Teddington (Figure 5.5). Much of this development occurred immediately after the Second World War, when land-use planning systems were not as sophisticated and rigorous as they are today. The problem is exacerbated as many riverside properties were once small-scale and low-value recreation facilities (little more than shacks built adjacent to the river), but have since been modernised and substantially improved so that flood damage potential there is now very high.

The flood hazard potential for the river reach between Datchet and Teddington has been analysed through collecting data on the numbers of properties at

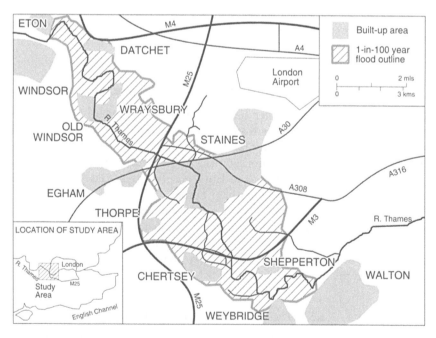

Figure 5.5 The flood plain between Windsor and London, UK and areas of maximum flood impact

risk in the area, the ground and floor levels, and the water level predicted by the mathematical hydraulic model developed by Halcrow Water (Penning-Rowsell and Tunstall, 1996). Table 5.4 indicates the relationship between flood probability (or return period) and the number of properties likely to be affected. In this river reach some 442 properties would be affected by the 1 : 5 flood (5-year flood). A 1 : 25 event would flood some 5,250 properties and a repeat of the event which occurred in March 1947 would flood some 8,622 properties. Here, therefore, we have an area which is susceptible to major flooding, but has not experienced such flooding for nearly 50 years – and where, not surprisingly, the perception of flood risk by local residents is low (Fordham and Tunstall, 1990). A more extreme event (here designated as the 204-year flood) would inundate some 12,435 properties, and last for approximately 1 week. This could constitute a national disaster, in that many thousands of people would need evacuation from the area and there are doubts about the effectiveness of existing emergency planning.

The river already has considerable flood alleviation capacity built into it, notably in the form of a regulation system of weirs and minor embankments. In addition, parts of the river are dredged to allow navigation, which contributes to the ease of drainage of the flood waters through the relevant river reach. These measures, however, are relevant only for frequent flood events.

With regard to climate change, Table 5.5 presents estimates of what could happen given increased flood volumes through this river reach such that flood levels were uniformly increased by 10 cm above current estimates for the specified return periods. Quite clearly this is a very generalised analysis, and it is unlikely that all flood return periods would be affected in the same way; nevertheless, it is instructive to investigate the sensitivity of the area to increased flood severity in this way.

The results in Table 5.5 indicate that adding 10 cm to flood levels for extreme events (here the 1 : 101 and 1 : 204 floods) has a significant effect upon both the number of properties affected and the event damages caused by those floods. For the 1 : 101 flood the number of properties affected rises by 10.8 per cent, but the event damage increases by 30.4 per cent. This difference is partly attributable to the fact that increased flood levels tend to flood relatively fewer extra properties in this area, and also because the land-use survey of the area stretched only to the extremities of likely known existing flood conditions. If we convert the future annual average sequence of flood damages to a current 'capital sum' using the standard approach of economics (and including assumptions about the discount rate and time period), we can see that the sum increases by 50.6 per cent for the situation with flood levels 10 cm higher. From a policy perspective, the significance of this is that it is worth spending 50.6 per cent more on flood mitigation. The increase also reflects the fact that lower return period floods have their flood depths increased, and by proportionately a greater extent.

The analysis for the 204-year event indicates that property numbers flooded increase again, but by a lesser amount. This is probably due to the fact that all the areas in the hazard survey are flooded. Some damages could be expected beyond the currently defined flood plain. Nevertheless, event damage increases from £85 million to £104 million, simply by the addition of the 10 centimetres of extra flood water. The capital sum again increases, by 37 per cent, and overall the situation is highly sensitive to quite small variations in flood level which could be induced by climate change. The policy ramifications of this are discussed below.

5.5.2 *The Île de France region, France*

The previous case study describes a detailed project appraisal undertaken for part of the lower Thames catchment in the United Kingdom. In many respects this gives a good insight into the effects climate change might have, if that change resulted in higher flood-water levels for a given event of a designated return period.

However, the Thames case study is concerned with a small area, and could be atypical of the effects of climate change on flood damage potential. Another case study, at a regional scale, is the Île de France area, covering a large part of the Paris basin drained by the river Seine and its tributaries. The soil water balance results for precipitation and run-off (Figures 5.1 and 5.2) show increased precipitation in northern France, although to a lesser extent than in southern England. A regional scale analysis of flood vulnerability has been undertaken in this area of France, in an attempt to gauge the vulnerability of Paris and its region to flooding (Debar *et al.*, undated). The study investigates the vulnerability of this area to floods of different depths, although the occurrence probabilities of the floods were not determined for the case-study locations subject to the analysis. A computer model used to assess damages in this way was developed initially for the river Thames by Green *et al.* (1987), and it was designed to operate with rudimentary data inputs, such as the land use for the area in question, in a generalised form, topography in the form of flood outlines for a restricted range of flood events, and aggregate potential flood damage data. Using Markov chain analysis, the model determines the likely pattern of development in the areas at risk from flooding, without field analysis and with data only from cartographic sources. The full methodology is described in Green *et al.* (1987), and the model is designed to be used in the planning pre-feasibility stage to determine whether there is a flood problem or not.

Here we are concerned only with the sensitivity of flood damages to extreme flood water levels, rather than with annual average damages. This is summarised in Table 5.6, which gives the stage–damage relationships for seven sub-areas within the study area as a whole. The results for each sub-area are standardised in relation to the depth of flood water at which damage begins (0.5 m). The sensitivity of these areas to extreme floods is indicated in Table 5.7, which gives the incremental change in damage with increases in flood depth (or stage). Thus, for example, the increase in damage from a flood depth of 1.5 to 2.0 m in the

Table 5.6 Stage/damage results for the Île de France case study sub-areas (£)

Flood depth (m)	Paris V	Ivry	Gournay	Villeneuve-la-Garenne	Geunevillers	Villeneuve-St. Georges	Vitry
0	0	0	0	0	0	0	0
0.5	83	5	1	27	14	32	16
1.0	670	47	7	240	210	240	140
1.5	1,200	160	23	570	1,100	390	440
2.0	1,400	320	51	910	2,000	450	1,500
2.5	1,500	520	80	1,100	2,700	520	5,200
3.0	1,600	918	120	1,300	3,200	850	8,300
3.5	1,600	1,400	140	1,400	4,100	1,100	9,300

Table 5.7 Incremental increase in flood damage with increased flood stage (as a percentage: see Table 5.6)

Flood depths (m)	Paris V	Ivry	Gournay	Villeneuve-la-Garenne	Geunevillers	Villeneuve-St. Georges	Vitry
1.5–2.0	17	100	122	60	82	15	241
2.0–2.5	7	63	57	21	35	16	246
2.5–3.0	7	76	50	18	19	63	60
3.0–3.5	0	53	17	8	28	29	12

sub-area of Gournay-sur-Marne is 122 per cent; the estimated damage increases from £23,000 to £51,000 with that increase in flood stage.

What can be seen from Table 5.6 is that the incremental change in flood damage at low flood depths (1.5 – 2.0 m of flood stage) is quite substantial, reflecting increased depth of flooding and, more importantly, increased numbers of properties in the flood risk areas suffering from flooding. However, as flood depth increases progressively up to a depth of 3.5 m the incremental increase in flood damages decreases. Thus, in the Gournay-sur-Marne example, the increase in flood damage from 3.0 to 3.5 m is only 17 per cent (from £120,000 to £140,000). (Note that the model makes no assumptions regarding building collapse as flood depth increases.) The overall pattern is nearly uniform across all the sub-areas, showing decreased incremental growth in flood damage with flood stage increases. The only exceptions are Geunevillers, where there is a different pattern in the middle part of this flood depth range, and Villeneuve-St. Georges, where the pattern is quite different. In the Villeneuve-St. Georges case, as flood water depth increases, flood damage continues to increase quite rapidly even at the maximum modelled flood stage; the increase in damages is greater per unit depth at high flood stages than at lower stages. Why this is the case is not clear, but what is evident from these seven examples is the consistent pattern described above: decreasing incremental growth in flood damage as floods get deeper.

The implications of these results are similar to those found in the lower Thames Valley. That is to say, major floods cause major damage, but incremental growth in the major floods (by 10 cm in the case of the lower Thames and by 50 cm in the case of the French case studies) does not result in disproportionately more damage. No key threshold is passed which suddenly leads to greatly increased damage; rather, total damage increases for each incremental increase in flood stage, but at a declining rate.

5.5.3 The Meuse Valley, Limburg, the Netherlands

After the Meuse floods of the early 1990s, a Dutch Parliamentary Commission was appointed, which commissioned research on hazard, damage and risk management. Part of the research was the impact that climate change would have on flood hazard in Limburg, in the south of the Netherlands (this is summarised by Schuurman, 1995; see also Langen and Tol, Chapter 6, this volume, for a history of flood management in the Netherlands).

In the Dutch study, a model of the hydrology of the Meuse catchment was linked to a model of the economic assets at risk. The hydrological model combines a two-dimensional process model of water discharge with a statistical model of flood height. Precipitation is taken from a stochastic weather generator. The damage model is a geographic information system, with a resolution of 10 cm elevation. Damage functions are fitted to the damage of the 1993 flood; no account is taken of future development of the area. Unfortunately, only one climate change experiment with the model has been reported. From this experiment it was concluded that a 10 per cent increase in winter precipitation, associated with a 2°C temperature rise, could lead to an approximate doubling of annual average damage from Dfl. 9.9 million to Dfl. 21.8 million in the Meuse Valley. A sum of Dfl. 12 million corresponds to about 0.002 per cent of Dutch GDP.

However, the current level of vulnerability appears to be quite serious by Dutch standards, and the floods of 1993 and 1995 invoked a strong response. (It is important to bear in mind that major river dikes in the Netherlands are designed to cope with a 1 : 1250 flood; the Meuse Valley in Limburg is an exception, as the soil is such that dike-building is impossible: the design level for flood protection is 1 : 250).[1] Substantial improvements to the river channel are planned. Alternatives being considered are: first, a deepening of the river bed; second, natural area redevelopment; and third, building quays along the river. In the first alternative, the summer bed of the Meuse would be deepened. A disadvantage of this strategy is that the groundwater table will fall, which would be detrimental for 'natural' values and agriculture. Three variants of the second alternative have been developed. This alternative comprises a widening of the summer bed, a deepening of the winter bed, and the digging of additional canals.

The main advantage is the chance for wildlife areas to redevelop. The main disadvantages are the costs, and the fact that the three measures have little influence on flood heights. Therefore, the summer bed would also need to be deepened and quays built where necessary. In the third alternative, only quays are developed. The main advantages of this option are the relatively low costs and the speed with which it can be implemented. The proposed responses would decrease annual average damage by a factor of between two and ten. The additional damage due to climatic change (that is, a 10 per cent increase in winter pre-cipitation for a 2°C temperature rise) would then amount to Dfl. 1–4 million, depending on the river-bed improvement strategy chosen. Table 5.8 displays the results; the effect of urbanisation is also shown.

The situation along the Meuse thus appears to resemble the Thames projec-tions rather than those for the Seine. A small increase in the hazard (10 per cent increase in winter precipitation) could result in a large increase in the annual damage (100 per cent). However, no indication of the overall shape of the damage function is reported. The case study of the Meuse also illustrates the importance of flood risk management: while climate change could double the risk, management has an influence of up to a factor of ten.

5.5.4 Summary from the case studies

The case studies described above cover major flooding problems in areas liable to increased run-off resulting from climate change (according to the results from various scenarios).

Table 5.8 Annual average expected damage from Meuse floods in the year 2050 (Dfl. million)

	Now	Increased urbanization (+5%)	Climate change	Urbanization and climate change
Current risk management	9.9	10.6	21.8	27.1
Deepening summer bed	0.7	0.7	1.5	1.9
Wildlife area development (1)	0.6	0.7	1.4	1.8
Wildlife area development (2)	0.9	0.9	1.9	2.4
Wildlife area development (3)	3.3	3.6	7.3	9.1
Building quays	3.5	3.8	7.4	9.6

Source: Schuurman (1995)

Notes: In all scenarios assets at risk are assumed to remain as they were in 1993. Now is no change in hazard. Climate change is a 10% increase in winter precipitation and a 2°C increase in temperature

Our results indicate that an increase in the severity of these events is likely to be serious but not catastrophic, in that the incremental change in flood damage with increased flood stage is not huge. However, an increase in the frequency of more minor events could be very significant in policy terms, but the analysis of such minor events is outside the scope of this research. Why these would be significant is that the whole flood regime (and the economic benefits of flood mitigation) is driven by the periodicity of the small and medium-sized floods, rather than by the magnitude of the rarer events. The situation would be quite different if for example, a major levee[2] or dike was breached or overtopped at a critical flood level.

5.6 Policy responses: Decisions in the face of increasing uncertainty

Overall, it appears that climate change will bring relatively minor increases in flooding. Nevertheless, a major increase in hazard is likely with concomitant policy demands, due to increased wealth in flood-prone areas and because of rising expectations for hazard protection by the people of Europe. Although great uncertainty is a key attribute of climate change predictions, in one important aspect the uncertainty with respect to flooding is bounded compared to many other hazards. Flooding is generally confined to reasonably well defined areas: flood plains in the case of riverine floods. Nevertheless, it is worth bearing in mind that should flooding decrease, adjustment to the loss of water in many areas of the world, and much of Europe, would be far more challenging than coping with marginal increases.

It is in this context of uncertainty that policy-makers have to devise viable strategies. Our efforts need to be focused on developing policy responses that are effective in the face of continuing and perhaps expanding uncertainty; for example, the predicted direction of change could switch, especially at the local level. In addition, responses should be in the form of processes capable of incorporating new knowledge as it emerges.

In approaching this we need to consider three distinct elements: existing urban development, future developments and residual risk (future hazard being composed largely of new tangible developments, increased knowledge and perceptions, and climate change). It is tempting to concern ourselves only with future development, as this is the easiest ground on which to make policy recommendations. However, the greatest reductions to flood damage will almost certainly come from improvements to the flood-related resilience of existing development, as new development is generally only a small fraction of the existing stock of buildings and other infrastructure.

In this respect we need to understand the vulnerability of human populations to flooding. If flooding – in the physical sense of more frequent and/or severe episodes of high water – increases, there is little humanity can do about it. But we can, and do, modify the impact of flooding. For any given physical event, in this case excess water, the damage will depend on the coping capacity or vulnerability of the affected community.

5.6.1 Adjustment to flood hazard

A key distinction can be drawn between 'tangible' and 'intangible' coping mechanisms. In this context intangibles are defined as items of a psychological or spiritual nature regardless of the existence of markets.

Most people probably adopt intangible approaches for dealing with risk – that is, any risk – most of the time. These would include the psychological defences against threats, through denial, rationalisation and personal invulnerability (Marks, 1990). Intangible approaches also include prayer, other rituals such as committees of inquiry and scientific investigations, and cultural change within organisations. Many writers on risk, for example, now regard cultural change within key organisations as a critical element in improving our capacity to deal with hazards. Broader intangible aspects of flood hazard management include an ability to learn from experience, to react appropriately to warnings – whether official or not – and the promotion and preservation of a certain degree of self-reliance.

The most obvious tangible actions are to build engineering structures, and to move settlements out of flood-prone areas. So, for our purposes, flood adjustments can be classified as follows. Note that they are conventionally seen as either structural or non-structural, as set out in the list below. Non-structural adjustment, are:

- individual action, both tangible and intangible;
- land-use planning;
- institutional or cultural change;
- emergency response and loss redistributive mechanisms, again both tangible and intangible; and
- doing nothing (although nothing may be done, inevitably some emergency response and loss-sharing mechanisms would come into play following a major flood).

Structural adjustments are:

- engineering works, such as dams, bypass channels and levees.

A critical need is to broaden the definition of flood-hazard management strategies away from the relatively narrow group discussed above. These traditional approaches – generally weakly applied apart from structures and emergency responses with compensation – see the hazard as something separate from society and as amenable to technocratic solutions. In the spirit of the founder of the flood-hazard field in geography, Gilbert White (1945), we need to consider 'all potential adjustments for coping with floods' – including the institutional structures which govern our society, and ultimately affect our coping capacity.

In adjusting to the flood hazard our main aim should be to reduce vulnerability to flooding – and given the immense uncertainty surrounding future flood patterns, we would seek approaches which focus on the development of integrated measures, with the emphasis on management rather than elimination of the flood hazard. The strategies should:

- have small opportunity costs;
- be 'reversible';
- be flexible and adaptable to changing circumstances; and
- be 'fail-safe' when their design level is exceeded.

Flood adjustments which cease to function at their design standard provide declining protection. Paradoxically, they may become more dangerous if designed to a high standard that encourages a false sense of security – the 'levee' effect. The continuing need to maintain a capacity to cope with the residual risk is easily overlooked.

Land-use measures may become increasingly attractive, in part because of their low capital requirements. However, such measures are generally tied to some 'regulatory' or 'standard' flood, such as the 1 per cent event. As floods become more frequent, this 1 per cent event becomes more frequent, perhaps becoming the 2 or 3 per cent event. The regulatory process needs to be flexible enough to cope with such inevitable changes. Even in the absence of climate change, the designated flood's expected frequency is almost certain to change as knowledge increases. The use of performance-driven rather than prescriptive regulations is likely to be more satisfactory, owing to the inherent flexibility and room for innovation in the performance approach.

However, there is a policy dilemma. The reality about land-use measures is that flood hazards are frequently only one of many considerations, and may rate well below other local issues and problems. Land-use planning for many hazards and environmental issues highlights a fundamental dilemma: on the one hand, there is a desire to promote growth, and on the other, a need for caution in the face of hazards. This is part of a broader tension between the interests of economic development and those trying to control development for environmental and community reasons. Both sides are supported by considerable bodies of legislation and policy (Fowler, 1991), and, often, powerful interest groups. Implementation of sound hazard and environmental policies is often subject to strong resistance from elements of the commercial and political worlds. The lesson here is that success will often depend on negotiation and bargaining rather than prescription.

5.6.2 *Managing the residual risk*

Management of the flood hazard, no matter how proficient, cannot eliminate the chance of loss. The better managed the hazard, the less visible is the residual hazard; but it still exists, although often ignored. Management involves warning systems and emergency planning and response. Critically, it also involves loss redistribution. If a decision is taken – either deliberately or by default – to do nothing, then loss redistribution would generally come into play following a damaging event.

The costs of hazards can be transferred to the future – to be borne by another government or generation – rather than spread among the present residents of, say, London or Paris. It is often argued that future technological developments will enable the degradation to be reversed at low cost; this is speculation and in the case of flooding is most unlikely. Furthermore, the approach violates a fundamental principle of sustainable development: that of inter-generational equity. Instead the emphasis should be on a precautionary or preventive approach (Dovers and Handmer, 1995).

Typically the burden of loss redistribution is shared between the private, public and voluntary sectors: this includes material, advice, counselling and referral services. As compensation is very much geared to clearly defined events – provided there is a recognisable event – those affected would probably be able to obtain relief from insurance (if they were insured), and should obtain some support from government welfare and private charitable groups. In the absence of overarching co-ordination mechanisms, there is no question that media attention helps people (and politicians) to become generous. In turn, media attention is greater for sudden spectacular events, especially those that make dramatic television pictures. In some countries this *ad hoc* approach is replaced with a more co-ordinated, and much fairer, system of loss redistribution. Even with minor changes, some groups within society are likely to be more vulnerable and affected relatively severely. Efforts should be made to identify such groups, and to examine ways of enhancing their resilience.

5.6.3 *More radical alternatives*

When disaster occurs, tremendous efforts are made to 'get back to normal', even where 'normal' is clearly unsustainable. Post-disaster aid may worsen the situation during the next flood. For example, a flood-prone area which is gradually being abandoned may receive a substantial injection of reconstruction funds, new infrastructure, etc. following a particularly severe flood. Very rarely are opportunities taken to implement change following a major event – although less obvious changes to building regulations or insurance premiums may occur and very occasionally wholesale relocation takes place. The attraction of many structural measures, for example, is that they facilitate continued or expanded flood-plain occupancy, avoid the need for any social change, and provide an important but dangerous intangible dimension: a false sense of security.

The approaches discussed in the section above do not entail major changes to how we think or act as a society, a feature that may be a major advantage in that the approaches are all well tried, with existing mechanisms for their implementation. Unfortunately, this strength may also be a weakness in that the development of new institutional arrangements with different underlying assumptions cannot be accommodated; and change other than at the margins is not considered. In addition, the approaches tend to treat the problem as a discrete issue separate from other aspects of society. Potentially worst of all, these approaches of incremental adjustment may give the reassuring impression that the problem is being addressed, when in fact the main aim is preservation of the status quo through treatment of symptoms (for Britain, see Penning-Rowsell et al., 1986).

A strategy for a sustainable future in the face of global environmental change is likely to require some fundamental changes to our society. The threat has been evolving and will continue to evolve both physically and through changing knowledge and perceptions. It would be reasonable, then, to expect that the way we deal with flood hazards has evolved too. And of course this is happening to some extent. We may have to accept a substantially lower standard of protection against floods. For example, in a wetter climatic regime or at least one with more intense rainfall, the 1 per cent flood may increase in frequency to the 2 per cent flood, and we may have to cope with this by expanding the role and scope of loss redistribution and warnings and response.

5.6.4 Need for institutional reform

Dovers (1995) argues that global complex environmental problems are not amenable to the conventional policy toolkit, since this has been developed largely for the relatively simple 'end of pipe' problems and is not good at dealing with uncertainty. The major current institutional change, that towards the corporatisation and privatisation of public administration, may be increasing vulnerability to hazards (Hood and Jackson, 1992). Many important changes may occur in people's minds, and collective attitudes, rather than through expression in government policy: for example, public attitudes to the natural function of flood plains and falling tolerance of major structural solutions. We might also find a rising expectation (or at least acceptance) that market mechanisms will be used in flood-plain management.

We need to move away from the current 'fine tuning' approach with its focus on symptoms, and towards an approach capable of tackling the underlying problems; away from the institutional rigidities which appear to lock us into a continuation of past development practices (Handmer and Dovers, 1996). That said, improved 'reactive' approaches are necessary in an era of great uncertainty. In terms of flooding, the inevitable outcome of these practices and rising affluence is rising losses in the future – even in the absence of climate change –

combined with a rising disinclination to offer insurance cover. Peaceful radical institutional reform is unusual but is occasionally demonstrated at the national level. Handmer and Dovers (1996) discuss examples in the context of sustainable development.

Some specific suggestions for modest reforms could include:

- Proactive approaches would be developed for both existing and future risk. An enhanced capacity to cope with surprise would be developed; and where economically feasible, a precautionary approach would be used. Strong application of the approach might see river channels, and at least parts of their flood plains and associated wetlands, being left to serve their natural functions as floodways and flood-water storage areas. This would have many environmental, aesthetic and recreational benefits as well. The onus of proof would shift so that those wanting to develop such areas would have to show that, given an increase in future flood severity, their proposals would not worsen flood impacts. Where existing development is concerned, it might be necessary to develop relocation plans ready to take advantage of the situation immediately following a damaging flood. Such suggestions are often made; but implementation would require major changes in attitude.
- Strengthening local capacity to manage flood hazards. This covers issues of process, legal power, financial resources, priorities, technical expertise in hydraulics, economics, negotiation, etc.
- Eliminating the systemic bias in favour of engineering structures created, largely, through government grants and subsidies in that direction and through the mission and professional ethos of the organisations responsible for flood management; and a parallel acceptance that some urban developments should be removed from high-risk flood-plain areas.

Among other things, implementation of such suggestions would require an active acknowledgement that natural systems have some priority over human and current economic imperatives.

5.7 History of British flood hazard management

Some of these policy issues, including the difficulty of achieving real change, the dominance of engineering measures and the role of insurance, are well illustrated by the situation in Britain. Flood authorities there speak of 'flood defence', 'land drainage' and 'flood alleviation'. Change is occurring, but they have rarely used the broader terms 'flood-plain management' or 'flood hazard management', which have become universal elsewhere in the English-speaking world. The emphasis has been on flood prevention through flood-water control.

In the post-war period, British flood hazard management has been characterised by substantial government investment in structural works to protect both urban development and agricultural areas. This has been undertaken on a case-by-case cost–benefit approach, rather than through the setting of some socially desired standard – an approach set out recently in the Environment Act 1995. There are attempts to control the location of development or to make it less susceptible to flooding. Management of the residual hazard is shared between the private and public sectors; compensation for flood damage has long been largely in the hands of the private sector through the inclusion of flood cover in house-hold (and small-business) insurance policies. Government compensation for flood damage has generally been *ad hoc* (Handmer, 1990). Hazard and disaster management has been low on the political agenda, with the central government responding weakly to strong pressure from local government for legislative change and guidelines (Handmer and Parker, 1992). Yet land drainage for agricultural 'improvement' was high on the political agenda until the mid-1980s at least, and still commands a powerful lobby. There are no signs that this general pattern is likely to change, and in some important respects it has the disadvantage of inertia emanating from a long institutional history.

The approach of engineering solutions with central government support was established in the fifteenth century (Wisdom, 1979). In terms of the modern period, the Land Drainage Act 1930 – part of the general expansion of govern-ment at the time – established the basis for flood control (and land drainage). Since then the arrangements have been through many changes and restructuring, culminating with the privatisation of much of the water sector in 1989. The details of these changes, reasons for them, and the philosophies involved are reviewed in numerous publications (e.g. Kinnersley, 1985, 1994; Penning-Rowsell *et al.*, 1986; Hassan *et al.*, 1996). Flood control and management were not privatised, and in many respects the framework established by the 1930 Act is still in place (aspects of the organisations, the grant structure, precept system and local committees). Today, the Environment Agency (cre-ated in April 1996) has lead responsibility for these functions, which it exercises through its eight operating regions.

Far less attention has been given to the non-structural approaches of land-use planning and warning systems. This is despite the fact that Britain has long had a comprehensive, and prescriptive, system of land-use planning and building regu-lation (known as 'development control'). Land-use approaches to flood hazard management are not considered systematically, although the Environment Agency comments on development applications and may urge local authorities – which have planning power – to limit development in hazardous areas. A series of circulars sets out procedures for controlling flood-plain development. The first, from the Ministry of Town and Country Planning, came in 1947; the most recent from the Department of the Environment and others, in 1992. These documents emphasise the need for liaison and consultation, and rely on exhorta-tion. 'The Government looks to local authorities to use their planning powers to guide development away from areas that may be affected by flooding' (DoE/MAFF/WO, 1992: para 4). Traditionally flooding has been in the hands of engineers, and planners had other priorities.

There is now a new emphasis on strategic planning and away from an *ad hoc*, piecemeal approach to flood-plain development. Parker (1995) argues that flood plains were often developed so that other environmental priorities could be met, such as preservation of the 'green belt' around London, and the protection of agricultural land. However, flood plains without concrete have many environmental benefits quite separate from the avoidance of flood damages.

Although the technical side of flood detection and prediction has received much attention, it is only recently that the crucial issues of warning message dissemination and response have been properly considered across the country – although some authorities have been working on this issue for years. There is now a national effort to improve and upgrade flood warning services, led by the Environment Agency. The Ministry of Agriculture, Fisheries and Food (MAFF) now also gives flood warnings priority.

Flood insurance in Britain is unusual on two counts. First, it is entirely within the purview of the private sector; and second, no attempt is made to link insurance with flood hazard mitigation (with the exception of some large industrial and commercial enterprises). In contrast, the flood insurance schemes operating in the USA and France involve both the government and private sectors and are linked strongly with attempts at mitigation (Handmer, 1990). (The success or otherwise of such attempts is another matter.) In Britain, insurance may encourage flood-plain occupancy as the costs of flooding are spread among all insurance policy-holders, with the result that flood-free locations subsidise those that are not; and mitigation opportunities are lost.

Since the First World War, flood cover has been included in standard household *contents* insurance cover. Following severe flooding in many parts of Britain in 1960, this was extended in 1961 to cover *buildings* on request as a result of negotiations between the British government and the insurance industry. Other parts of this agreement were that comprehensive policies (which include flood cover) would be marketed more widely and that flood cover would be made more readily available to small businesses. Co-operation was helped by the industry's fears about the start of nationalisation (Arnell, 1987). After further serious flooding in 1968 showed that coverage was still patchy, the industry strengthened flood cover and increased its marketing activity. By the mid-1980s flood cover was a more or less universal inclusion in standard household policies – even though by then nothing was less likely than nationalisation. (At present, some 85 per cent of buildings and 75 per cent of contents are covered by household insurance.) Lending institutions support this system by requiring flood insurance for home loans. Today, flood cover is automatically included in comprehensive household policies. This characteristic has made the system work by eliminating adverse selection, and the need to charge actuarially based premiums.

Early attempts to charge higher premiums for flood-prone areas following the 1961 agreement were abandoned. According to Arnell (1987) and Porter (1970), increased competition eliminated differential premiums, although an increased excess may be charged. In the past few years, excesses of between five and ten thousand pounds have been imposed on householders following flooding.

This approach implies that the potential flood loss for the insurance industry is minor – yet 'Estimates show that a ten billion pound market loss is not out of the question' (Wright, 1996: 44). Not surprisingly, the industry is now moving towards loading premiums and 'excess' requirements in flood-prone areas through exploiting the opportunities offered by geographic information systems (GISs) and the extensive databases compiled over the past few decades. Coupling this capacity with flood-loss models is helping to give new precision to reinsurance requirements and therefore cost. However, it appears that much insurance for commercial enterprises is not easy to locate spatially; and there is concern that increases in premiums may decrease total income (Wright, 1996: 45). At present, there appears to be no likelihood of the industry withdrawing from the business, although there is anecdotal evidence of individual residents being declined cover.

5.7.1 The Environment Agency perspective for Britain

Climate change will necessarily require policy responses by those government authorities or bodies charged with responsibility for the water environment, particularly flooding. For example, Huff and Changnon (1987) offer evidence that broad-scale climate change has caused increased riverine flooding in the midwestern USA and argue that hydrologists have been hesitant to recognise this trend and to incorporate it into planning and operations. Phillips and Jordan (1986) have also argued that increased rainfall has led to dramatically increased flooding in the Salt River basin in the southwest USA.

In England and Wales, the Environment Agency has the responsibility for protecting the water environment, through its core responsibilities for water resources, water quality, fisheries, conservation, recreation and navigation, flood 'defence', and a wide waste and pollution management role. The implications of climate change for the EA's fluvial flood mitigation function are outlined in Table 5.9. These include policy implications for flood-plain inundation from increased rainfall and run-off, the integrity of riverine flood structures, and the implications for development control in flood plains.

The effort required by the EA to fulfil at least some of its duties is likely to increase with climate change. For example, it may become more expensive to maintain existing levels of service. The problems are much broader than this, however, and the Agency is now considering the implications of climate change across its functions, in particular those related to water. This activity has been given impetus by the recent unusually hot and dry summers, and concern that this may be the new pattern.

The EA has adopted the precautionary approach in considering its response to climate change, owing to the many uncertainties involved. However, it is recognised that its response in one area may have major implications for other functions; for example, flood defence works have conservation implications. Each EA

Table 5.9 Implications of climate change for fluvial flood defence by the Environment Agency

---

Hydrological components affected by climate change: Each component of the hydrological system can be expected to change as the climate changes. There is a need to distinguish between the direct effect on river flows of changes in rainfall and evaporation inputs and the indirect effects of changes in land cover, plant water use and soil structure consequent upon climate change. Practically all catchment-scale impact studies have concentrated on the direct effects of climate change and have ignored the potential effects of changes in plant physiology and catchment vegetation mix on evaporation; nor have they considered possible changes in soil structure. Higher temperatures could lead to a loss of organic matter and a decrease in the soil's ability to hold moisture; higher temperatures could also encourage clay soils to crust, shrink and crack. Increased waterlogging, on the other hand, encourages the development of gleyed profiles and limits the effect of mineralisation on organic matter loss. Estimates of the change in potential evaporation also need scenarios for changes in humidity, wind speed and net radiation, as well as for temperature, which do not yet exist. Moreover, research is needed on weed growth, erosion and sedimentation patterns and their effects on river flow regimes.

Fluvial flood-plain inundation: The risk of fluvial flooding is likely to increase. An increase in winter rainfall and a possible increase in peak summer rainfall intensities and 'flash' floods would increase the risk of riverine flooding. The effect would vary considerably between catchments as possible changes are very uncertain at present. A flood that currently has an annual occurrence probability of 1 : 100 may in thirty years' time have an annual occurrence probability of 1 : 50.

Integrity of riverine flood defences: Peak discharge rates of rivers are important as they are responsible for flooding. Changes in discharge rates can be due to changes of a single parameter. However, the combined effect of several parameters may be considerably larger. Riverine flood works may be affected by changed erosion and sedimentation. Possible increases in weed growth will mean the discharge capabilities of channels will reduce and maintenance requirements increase. The level of risk in flood-plain areas depends on the management of surface water and groundwater in the rest of the catchment. The effect of a given climate change on flood frequencies will vary between catchments. The major implication of climate change for fluvial flood control is that protection levels fall below target standards of service (for urban areas currently at least the 1 : 50 flood, for agricultural land as low as the 1 : 5 flood). A change in climate might mean that a reach moves from a satisfactory flood protection status to an unsatisfactory one.

Development control: The EA generally recommends against development in flood-prone areas. An increase in extreme rainfall totals would imply a change in the performance of structures built to mitigate the downstream effects of development. Flood-plain land is allocated to one of five land-use classes according to density of development; each class has a specified level of service. Any new scheme must satisfy benefit–cost criteria laid down by the government. Climate change will probably not have much effect on EA flood-plain development control policies and the issue is currently of low priority.

Flood emergencies: Climate change is likely to have relatively little effect on the performance of flood forecasting systems. However, the systems might be used more frequently. Warning times could also be reduced, with implications for the ability to issue flood warnings and take emergency action. This is not a high-priority issue at present.

Urban storm drainage: Responsibility in this area is shared between water companies and local authorities. An increased frequency of high-intensity short-duration rainfall events would result in frequent surcharges of storm drainage systems and increase storm surcharge. This would have implications for 'off-river' flooding and for flood risks in receiving channels as well as for water quality. Urban storm drainage designs are needed that can be upgraded as more information about storm characteristics becomes available.

Source: Adapted from Arnell *et al.* (1994), Arnell and Reynard (1993), Hesselmans (1993), CCIRG (1991, 1996)

function therefore needs to consider and cost the implications for other functions when developing its response to climate change. With respect to the climate parameters, the greatest uncertainty for the EA obviously applies where the function is sensitive to changes in rainfall.

The uncertainties associated with climate change should be reviewed for their potential implications for flood estimation procedures (Arnell *et al.*, 1994). The general principles underpinning the EA's response to future climate change are likely to include multi-functional responses, possibly through the medium of catchment management plans; flexibility in water management (owing to an uncertain future); identification of critical thresholds for important Agency activities; and continual monitoring and review of further predictions of change (Arnell *et al.*, 1994). Both structural and non-structural approaches should be incorporated into the strategy, as mentioned in Section 5.5.1.

5.8 Assessment and conclusions

The future world is likely to have elevated greenhouse gas concentrations, be warmer, have increased rainfall intensity and more flooding. At least, this is the general prediction from the various climate models. Our analysis is in partial agreement. Overall, scenarios developed by the soil water balance analysis indicate that probable climate change over the next 50 years is likely to reinforce the existing pattern of rainfall and run-off. The predicted changes are also generally fairly small, and there is great uncertainty over certain details which may be critical in determining the specific rather than average changes in flood behaviour. However, in some regions increased run-off does seem to be highly significant, and adds to existing high levels of winter precipitation and run-off in these areas. The scenarios show increased precipitation and run-off in northern Europe, while southern Europe is likely to become drier.

There is no clear analysis yet of the relationship between this increased run-off and flood hazard problems, since the 'connecting' hydrologic and hydraulic models are not yet developed. The amount of increase in flooding is very uncertain and will almost certainly vary greatly between catchments, for reasons set out by the IPCC and reproduced in the introduction to this chapter. But the case studies analysed – in southern Britain, northern France and in the south of the Netherlands – indicate that flood damage potential is very sensitive to flood levels in the lower parts of major river catchments. Therefore, there is at least a case to be answered which suggests that flooding and flood damage situations are likely to get more serious in northern Europe with the process of climate change, but it is less certain that the same situation will occur in southern Europe.

Southern Europe seems likely to experience decreased run-off volumes. Only if increased storminess in summer thunderstorm conditions there far outweighs the reduction in overall precipitation could flooding situations in southern Europe be exacerbated by the climate change scenarios. This highlights that there is more to the long-term flood outlook than estimated aggregate rainfall and run-off. Fowler and Hennessy (1995) point out that 'warmer conditions tend to be associated with more intense precipitation events' and therefore with more flooding. At any given point the atmosphere is capable of delivering much heavier rain in summer than winter, but generally for shorter periods of time. A situation where the average annual precipitation declines significantly but the flood problem worsens is not inconceivable. However, we do not know that this will be the situation in southern Europe.

Given that the predicted changes are not major, it is difficult to justify substantial policy changes. In fact, it is most likely that land-use changes – independent of general environmental change – will have a more significant impact on flood regimes and therefore on the flood damage potential. Nevertheless, some general points should be made, and we make one major policy recommendation. If southern Europe becomes drier, as predicted, the implications may be much broader than simply flooding, and may extend to all aspects of river management. For example, it may not be possible to use rivers for disposal of effluent, as the necessary dilution factor may not be achievable. To generalise from this point, the responses of water management bodies throughout Europe to climate change need to consider the implications for all their functions, in addition to flood hazard management.

In the absence of evidence of significant predicted rise in flood damage potential, it is tempting to argue that the correct response is to pursue existing policy options. This presumes that the existing approach is adequate. However, in many circumstances the current policy responses are ineffective or difficult to implement for a variety of reasons including cost, public acceptability, environmental impacts and equity considerations. Current climate conditions and the apparently increasing flood hazard demand improved policy implementation of traditional approaches, and the exploration of more radical measures. Even with minor changes, some groups within society are likely to be more vulnerable and affected relatively severely. Efforts should be made to identify such groups, and to examine ways of enhancing their resilience.

Overall, we recommend that policies be framed to cope with uncertainties – and to avoid complacency. An enhanced capacity for adaptability and flexibility is likely to be an essential policy tool in the face of increasing uncertainty, not simply in the hydrological environment but also in the socio-economic contexts. Policies and associated institutional arrangements need to be able to take advantage of improvements in our knowledge base, while at the same time retaining the ability to cope with major surprises from the largely unknowable residual risk. This will require substantial strengthening of local capacity, and the removal of systematic incentives for engineering works.

Acknowledgements

We would like to thank Tom Downing, Paul Brignall and colleagues at the Oxford Environmental Change Unit for the opportunity to work on this project, and for supplying the data for the climate change scenarios and the European soil water balance. We thank Arnaut Schuurman of the Vrije Universiteit for his help in preparing the Meuse case study. Chris Haggett of the Environment Agency and Dennis Parker of Middlesex University provided useful comments on British flood hazard management. This chapter is an extended version of Penning-Rowsell et al. (1996). We also thank the editors for their work.

Notes

1 Dutch and English translations to river protection terms vary. 'Quay' is used here to translate *kade*, although in English a quay refers to a wharf for loading or unloading vessels. A levee is an embankment to prevent a river from flooding bordering land, but is generally a bank, mound or ridge, whereas a *kade* is always paved. Dikes, commonly used for river flood protection in the Netherlands, are of little help in south Limburg because the soil structure is such that water would seep underneath the dike.
2 A levee is an embankment to prevent a river from flooding bordering land. It is generally a bank, mound or ridge.

References

Arnell, N. (1987) 'Flood insurance and floodplain management', in J.W. Handmer, ed., *Flood Hazard Management: British and International Perspectives*, Norwich: Geobooks, 117–133.

Arnell, N.W. (1991) Climate change and river flow regimes in the United Kingdom. Paper read at Climate Change and the Hydrological Cycle, British Hydrological Society National Meeting, 2 December 1991, at the Institution of Civil Engineers, London.

Arnell, N.W., and Reynard, N.S. (1993) *Impacts of Climate Change on River Flows in the UK*, London: Institute of Hydrology Report to the Department of the Environment Water Directorate.

Arnell, N.W., Jenkins, A., and George, D.G. (1994) *The Implications of Climate Change for the National Rivers Authority*, R&D Report 12, London: Institute of Hydrology.

Climate Change Impacts Review Group (CCIRG) (1991) *The Potential Effects of Climate Change in the United Kingdom*, London: HMSO.

Climate Change Impacts Review Group (CCIRG) (1996) *Review of the Potential Effects of Climate Change in the United Kingdom*. London: HMSO.

Debar, P., Gazull, L., and Levinet, C.N.D. (undated) Premiers éléments statistiques sur les zones inondées par type MOS. Unpublished.

Department of the Environment in association with the Meteorological Office (DoE/MO) (1989) *Global Climate Change*, London: HMSO.

Department of the Environment, MAFF and the Welsh Office (DoE/MAFF/WO) (1992) *Development and Flood Risk*, Circular 30/92, London: DoE.

Dixon, J.A., and Hufschmidt, M.M. (1986) *Economic Valuation Techniques for the Environment: A Case Study Workbook*, Baltimore: Johns Hopkins University Press.

Dovers, S. (1995) A framework for scaling and framing policy problems in sustainability. *Ecological Economics* 2: 57–76.

Dovers, S., and Handmer, J.W. (1995) Ignorance, the precautionary principle and sustainability. *Ambio* 24: 92–97.

Ericksen, N.J. (1986) *Creating Flood Disasters?*, Wellington, NZ: National Soil and Water Directorate.

Fordham, M., and Tunstall, S.M. (1990) *Thames Perception and Attitude Survey: Datchet to Walton Bridge*, London: Flood Hazard Research Centre, Middlesex University.

Fowler, A.M., and Hennessy, K.J. (1995) Potential impacts of global warming on the frequency and magnitude of heavy precipitation. *Natural Hazards* 11: 283–303.

Fowler, R. (1991) 'Environmental law in Australia', in J.W. Handmer, A.H.J. Dorcey and D.I. Smith, eds, *Negotiating Water: Conflict Resolution in Australian Water Resource Management*, Canberra: CRES, Australian National University, 73–91.

Green, C.H., N'Jai, A., and Neal, J. (1987) *Thames Overview Pre-feasibility Study*, London: Flood Hazard Research Centre, Middlesex University.

Handmer, J.W. (1987) *Flood Hazard Management: British and International Perspectives*, Norwich: Geobooks.

Handmer, J.W. (1990) *Flood Insurance and Relief in the United States and Britain*, Working Paper no. 68, Boulder, CO: Natural Hazards Center, University of Colorado.

Handmer, J.W., and Dovers, S. (1996) A typology of resilience: Rethinking institutions for sustainability. *Industrial and Environmental Crisis Quarterly* 9, 4: 482–511.

Handmer, J.W. and Parker, D.J. (1992) Hazard management in Britain: Another disastrous decade? *Area* 24, 2: 113–122.

Hassan, J., Paul, N., Tompkins, J. and Fraser, I. (1996) *The European Water Environment in a Period of Transformation*, Manchester: Manchester University Press.

Hesselmans, G.H.F.M. (1993) *Climate Change: Recent Findings*, Delft: Delft Hydraulics.

Hewett, B.A.O., Harries, C.D., and Fenn, C.R. (1992) Water resources planning in the uncertainty of climate change: a water company perspective. Paper read at Symposium on Engineering in the Uncertainty of Climatic Change, London, 28 October 1992.

Hood, C., and Jackson, M. (1992) 'The new public management: a recipe for disaster?', in D. Parker and J. Handmer, eds, *Hazard Management and Emergency Planning*, London: James & James, 109–125.

Huff, F.A., and Changnon, S.A.J. (1987) Temporal changes in design rainfall frequencies in Illinois. *Climatic Change* 10: 195–200.

Kinnersley, D. (1985) *Troubled Water: Rivers, Politics and Pollution*, London: Hilary Shipman.

Kinnersley, D. (1994) *Coming Clean: The Politics of Water and the Environment*, Harmondsworth: Penguin.

Knox, J.C. (1993) Large increases in flood magnitude in response to modest changes in climate. *Nature* 361: 430–432.

Kwadijk, J., and Middelkoop, H. (1994) Estimation of impact of climate change on the peak discharge probability of the river Rhine. *Climatic Change* 27: 199–224.

Kwadijk, J., and Rotmans, J. (1995) The impact of climate change on the river Rhine: A scenario study. *Climatic Change* 30: 397–425.

McMahon, T.A., Finlayson, B.R., and Haines, A.T. (1992) *Global Runoff: Continental Comparisons of Annual Flows and Discharges*, Cremlingen-Destedt: Catena Verlag.

Manabe, S., Stouffer, R.J., Spelman, M.J., and Bryan, K. (1991) Transient responses of a coupled ocean–atmosphere model to gradual changes of atmospheric CO_2, Part I: Annual mean response. *Journal of Climate* 4: 785–818.

Manabe, S., Spelman, M.J., and Stouffer, R.J. (1992) Transient responses of a coupled ocean–atmosphere model to gradual changes of atmospheric CO_2, Part II: Seasonal response. *Journal of Climate* 5: 105–126.

Marks, D. (1990) 'Imagery, information and risk', in J.W. Handmer and E.C. Penning-Rowsell, eds, *Hazards and the Communication of Risk*, Aldershot: Gower, 19–29.

Mitchell, J.F.B., Manabe, S., Meleshko, V., and Tokioka, T. (1990) 'Equilibrium climate change and its implications for the future', in J.T. Houghton, G.J. Jenkins and J.J. Ephraums, eds, *Climate Change: The IPCC Scientific Assessment*, Cambridge: Cambridge University Press, 131–172.

Moore, J.W., and Moore, D.P. (1989) *The Army Corps of Engineers and the Evolution of Federal Flood Control Policy*, Boulder, CO: Institute for Behavioral Sciences.

National Rivers Authority, Thames Region (1995) *Thames 21: A Planning Perspective and a Sustainable Strategy for the Thames Region*, Reading: National Rivers Authority Thames Region.

Parker, D.J. (1995) Floodplain development policy in England and Wales. *Applied Geography* 15, 4: 341–363.

Penning-Rowsell, E.C., and Chatterton, J.B. (1977) *The Benefits of Flood Alleviation: A Manual of Assessment Techniques*, Aldershot: Gower.

Penning-Rowsell, E.C., and Fordham, M. (1994) *Flood across Europe: Hazard Assessment, Modelling and Management*, London: Middlesex University Press.

Penning-Rowsell, E.C., and Tunstall, S. (1996) 'Risk and resources: Defining and managing the floodplain', in M. Anderson, D. Walling and P.D. Bates, eds, *Floodplain Processes*, Chichester: John Wiley, 493–533.

Penning-Rowsell, E.C., Parker, D.J., and Harding, D.M. (1986) *Floods and Drainage: British Policies for Hazard Reduction, Agricultural Improvement and Wetland Conservation*, London: Allen and Unwin.

Penning-Rowsell, E.C., Green, C.H., Thompson, P.M., Coker, A.M., Tunstall, S., Richards, C., and Parker, D.J. (1992) *The Economics of Coastal Management: A Manual of Benefit Assessment Techniques*, London: Belhaven.

Penning-Rowsell, E.C., Handmer, J., and Tapsell, S. (1996) 'Extreme events and climate change: floods', in T.E. Downing, A.A. Olsthoorn and R.S.J. Tol, eds, *Climate Change and Extreme Events: Altered Risk, Socio-economic Impacts and Policy Responses*, Amsterdam: Institute for Environmental Studies, Vrije Universiteit, 97–127.

Phillips, D.H., and Jordan, D. (1986) 'The declining role of historical data in reservoir management and operations', in *Climate and Water Management*, proceedings of a meeting at American Meteorological Society, Boston, 83–88.

Porter, E.A. (1970) Assessment of flood risk for land use planning and property insurance, unpublished, University of Cambridge.

Riebsame, W.E. (1989) *Assessing the Social Implications of Climate Fluctuations: A Guide to Climate Impact Studies*, Boulder and Nairobi: Department of Geography and Natural Hazards Center, University of Colorado and UNEP World Climate Impacts Programme.

Riebsame, W.E., Strzepek, K.M., Wescoat, J.L., Perritt, R., Gaile, G.L., Jacobs, J., Leichenko, R., Magadza, C., Phien, H., Urbiztondo, B.J., Restrepo, P., Rose, W.R., Saleh, M., Ti, L.H., Tucci, C., and Yates, D. (1995) 'Complex river basins', in K.M. Strzepak and J. B. Smith, eds, *As Climate Changes: International Impacts and Implications*, Cambridge: Cambridge University Press, 57–91.

Rosen, H., and Reuss, M. (1988) *The Flood Control Challenge: Past, Present and Future*, Chicago: Public Works Historical Society.

Saul, A.J. (1992) *Floods and Flood Management*, London: Kluwer.

Schuurman, A. (1995) *An Insurance for the Meuse*, W95/19, Amsterdam: Vrije Universiteit.

Smith, D.I. (1996) 'Flooding in Australia. Progress to the present and possibilities for the future', in *Proceedings of NDR96 Conference on Natural Disaster Reduction*, Institution of Engineers, Barton, ACT, 11–22.

Smith, D.I., and Greenaway, M. (1983) Flood probabilities and urban flood damage in coastal northern NSW. *Search* 13, 11–12: 312–314.

Smith, J.B., Strzepek, K.M., Kalkstein, R.J., Nicholls, R.J., Smith, T.M., Riebsame, W.E., and Rosenzweig, C. (1995) 'Executive summary', in K.M. Strzepek and J.B. Smith, eds, *As Climate Changes: International Impacts and Implications*, Cambridge: Cambridge University Press, 1–18.

Strzepek, K.M., and Smith, J.B., eds (1995) *As Climate Changes: International Impacts and Implications*, Cambridge: Cambridge University Press.

Watson, R.T., Zinyowera, M.C., Moss, R.H., and Dokken, D.J. (1996) *Climate Change 1995: Impacts, Adaptations and Mitigation of Climate Change: Scientific-Technical Analyses*. Contribution of Working Group II to the Second Assessment Report of the Intergovernmental Panel on Climate Change, Cambridge: Cambridge University Press.

Weijers, E.P., and Vellinga, P. (1995) *Climate Change and River Flooding: Changes in Rainfall Processes and Flooding Regimes Due to an Enhanced Greenhouse Effect*, Amsterdam: IVM, Vrije Universiteit.

White, G.F. (1945) *Human Adjustment to Floods*, Research Paper 29, Chicago: Department of Geography, University of Chicago.

Wilkinson, W.B., and Cooper, D.M. (1993) The response of idealised aquifer/river systems to climatic change. *Hydrological Science Journal* 38: 379–390.

Wisdom, A.S. (1979) *The Law of Rivers and Watercourses*, 4th edition, London: Shaw & Sons.

Wright, A. (1996) An example of a liberal system: Flood and subsidence insurance in the UK. *SCOR Notes: Are Catastrophes Insurable?*, vol. 43–45, Paris: SCOR Reassurance.

6

A CONCISE HISTORY OF RIVERINE FLOODS AND FLOOD MANAGEMENT IN THE DUTCH RHINE DELTA

Andreas Langen and Richard S.J. Tol

6.1 Introduction

The Dutch have a long and well documented history of coping with one particular natural hazard: floods, coming both from the sea and the rivers. The risk of sea floods attracts more attention, but this chapter focuses on river floods, examining the interactions between social, political and economic developments and riverine floods over a period of a thousand years. Particular attention is paid to the river Rhine. The chapter shows how the river left its marks on the Dutch and the Dutch their marks on the river. Flood risk was often a limiting factor to human activity, and floods often triggered economic, social and political developments. The river Rhine is nowadays largely shaped by humans (see Figure 6.1 for the current situation), so much so that Stol (1993) claims that the battle of the Dutch against nature is in fact a battle against processes they themselves started earlier. This chapter is largely based on Langen (1995).

6.2 The origin of diking and its development until 1350

Until the twelfth century, riverine floods are reported only sporadically in the Netherlands (Gottschalk, 1971), although the sources are doubtful (Van de Ven, 1993). The number of flood reports gradually increased during the thirteenth century, and in the fourteenth century river floods become a recurrent phenomenon and an acknowledged problem. From that time on, reports on the establishment of flood control systems are found.

The cause of the first floods (in the twelfth century) dates back a few centuries. It is often thought that, in the Netherlands, dikes are a necessity for agriculture, but it is in fact the other way around: dikes are a consequence of agriculture, not a condition for it. Until the tenth century, the (then few) Dutch dwelt on higher

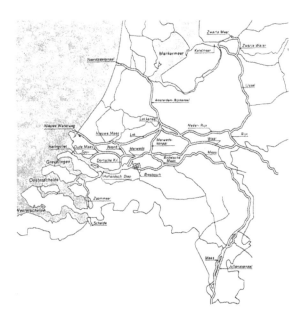

Figure 6.1 The delta of the river Rhine in present times. Most of the water of the Rhine (Rijn) is discharged through the IJssel, the Netherrhine (Neder-Rijn), Lek and the Waal. The Old Rhine mouths between the Nieuwe Waterweg and the Noordzeekanaal; the Old Rhine was a major mouth before the Lek (literally 'leak') emerged. The river Meuse (Maas) is also shown
Source: Van den Ven (1993)

ground, near to but safe from the rivers. The somewhat drier climate during the medieval optimum, newer ploughing techniques and new draining techniques allowed the population to expand and cultivate the many moorlands and fenlands, which it did rapidly, particularly in the countship of Holland and the diocese of Utrecht. The 'Great Cultivation' or 'Internal Colonisation' slowed down at the end of the twelfth century.

The twelfth century was characterised by a series of heavy sea surges which greatly altered the shape of the coast. The reason for these sea floods and the first river floods is to be found in the cultivation practices during the tenth and eleventh centuries. Cultivation required draining of the moors and fens. Draining led to dehydration and, in turn, oxidation of the high moorlands. Subsequently, settlement and subsidence resulted in a relative rise of sea and river levels. This implied increased flood risks and hence the need for flood protection such as dikes. Figure 6.2 depicts the process of cultivation–subsidence–dike-building.

The relative rise of the sea increased flood risk from both sea and rivers, and the higher water table hampered draining of tillage fields. This led to a conversion of

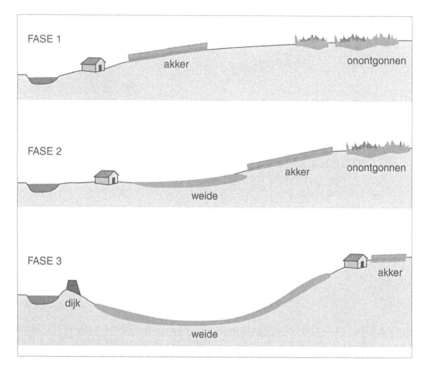

Figure 6.2 Cross-section of a typical Dutch river through history. Cultivation leads to
subsidence, so that first arable lands are converted to pasture land, and
second dikes need to buit. *Fase* (phase), *akker* (arable land), *onontgonnen*
(uncultivated); *weide* (pasture land), *dijk* (dike)
Source: Van den Ven (1993)

arable land to pasture land near the coast, and further cultivation upstream.
Draining was needed there as well, and subsidence of the land was the result.
River flood risk consequently increased, as both the flood hazard and the stock at
risk grew. Additional reasons for increased flood risks are large-scale deforesta-
tion in Germany and the fact that a dike downstream results in water accumulation upstream in case of high water.

In the thirteenth century, the first comprehensive dike systems were built in
Holland and Utrecht. The first local *waterschappen* (waterships) were then established. Waterships are the institutions concerned with water management,
including drainage and dike-building. The Dutch word *waterschap* refers to
both the water board and the area for which it is responsible. Examples of early
waterships are de Lekdijk Bovendams (south of Utrecht), the Grote Waard
(south of Gorinchem) and the later Rijnland (north of Leiden), which were
established prior to 1250.

Occupied as they were with water control, the so far relatively independent (internal) colonist villages sought legal security and stability with the count of Holland and the Bishop of Utrecht. This is the first sign of how flood risk helped build central authority. In turn, this laid the ground for regional waterships (from 1250 onwards in the west) and the codification of dike maintenance in *dijkbrieven* (dike letters); the oldest known dike letter, for Heusden (west of Den Bosch), is dated at 1273.

As the floods slowly crept upstream, so did flood control and the corresponding institutions. The establishment of waterships, dike rights, and the centralised power of the counts of Gelre and Kleve[1] followed about fifty years after the developments in the downstream regions of Holland and Utrecht. The counts of Gelre also introduced *weteringsbrieven* (watercourse letters), written concessions to build draining canals, thus regulating land cultivation. The first known letter is dated 1316 for the Tieler- and Bommelerwaarden (between Den Bosch, Gorinchem and Tiel).

The first half of the fourteenth century saw many disasters all over Europe. The Netherlands was no exception, witnessing severe floods in 1313 and 1315, and the famine of 1314–17, which killed 5 to 10 per cent of the population. The disastrous situation acted as a catalyst for the further formalisation of water control. By 1350, the Dutch river delta had closed dike systems along all major rivers.

6.3 Interactive processes: waterships, dike rights, land reclamation, climate variability and flood disasters until 1800

The story becomes more complicated after 1350. First some general history. The mutually independent countships were united, under Philip the Good of Burgundy, in 1433. The mutual independence was largely maintained, however, although the States General provided a platform for consultation. The House of Habsburg (Charles V) inherited the Burgundian territories around 1500. The Netherlands subsequently became part of Spain under Charles's son, Philip II. Between 1568 and 1648, an eighty-year war of independence was fought, resulting in the establishment of the Republic of the United Netherlands, a loose federal political structure. During the period of the Republic (1568–1795), water control was the terrain of the provinces and the local and regional waterships. The central government (the State Council and the Prince of Orange) interfered only for military purposes. Not until 1754 was a general inspector for the Rhine and other rivers in Holland appointed by central government in The Hague.

A series of major sea surges between 1400 and 1600 resulted in great land losses. During this period, the Dollard, the Lauwerssea, the western Shallows, the Zuyder Zee and the Biesbosch were formed. This in turn led to a major shift in the distribution of water between the branches of the Rhine; the IJssel slowly silted up whereas the Waal discharged more and more water. This process was accompanied by a number of river floods. In addition, climate all over western

Europe worsened from about 1480 on (the beginning of the Little Ice Age). In the second half of the fifteenth century, a new phenomenon emerged: ice-blocking. Sandbanks in and dikes along the river hampered discharge of ice, which occasionally resulted in huge dams of ice behind which meltwater accumulated. Ice-blocking was caused by a combination of gradual cooling, and rivers becoming narrower and more meandering. The river also became more shallow, as sediments were deposited only between the dikes. Consequently, discharge capacity was reduced; flood risks increased. The land behind the dike continued to settle and subside, and did so even faster with the introduction of novel draining techniques, such as the windmill and sluices, in the seventeenth century. Dikes grew higher and higher. Consequently, the pressure on them increased, and floods became more frequent and more severe. Secondary protection, notably seepage quays, had to be installed.

Another form of secondary protection, the cross-dike, was introduced by the end of the thirteenth century. An early example (1284) is the Diefdijk between the rivers Lek and Linge, between Culemborg and Leerdam (see Figure 6.3). Cross-dikes do not protect directly against river floods, but against dike bursts upstream. This reduced damage downstream, but enlarged it upstream, in the east. This was not particularly in the interest of the people in the eastern areas. On occasion, cross-dikes needed to be watched to prevent piercing. Obviously, the resulting atmosphere was not the most fertile conceivable for co-ordinated flood control. Only political reform, notably the establishment of larger political units, could reverse this process.

Dike maintenance was the duty of the landowners whose land was protected by the dike. The duty was divided according to the amount of land and the relative cost of dike maintenance (sharply bending dikes are more expensive to

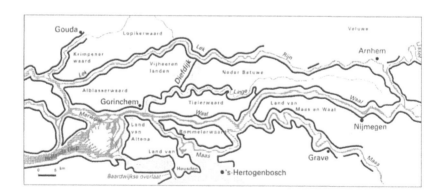

Figure 6.3 The situation along the rivers Rhine and Meuse around 1750. Thick lines depict dikes. The Diefdijk protected the Vijfheerenlanden and the Alblasserwaard against dike burst in the Neder Betuwe. At two places, the Waal and Meuse merge, occasionally leading to dangerously high water levels
Source: Van de Ven (1993)

maintain than straight ones). Commons are shared property, so the dike duty was shared accordingly. Cities were exempted from dike maintenance. Because of the link between land ownership (or rent) and dike maintenance (see also the next section), properties were thoroughly assessed in the 'dike roles'.

Dike control was in the hands of the 'dike chair' (the proper English word is water board), which consisted of an executive (the 'dike count'; the proper English word is dike reeve), a college of judges (the 'yard council'; the proper English is again dike reeve), and an administrative body (the 'dike writers'), responsible for the dike roles. The dike chair resided with the landlord. The dike count inspected the dikes three times a year, in spring, summer and autumn. During the spring inspection, the necessary repairs were established. The land-owner had to take care of these. In the summer inspection, it was checked whether the repairs had taken place. Severe fines were the price of negligence. In addition, labour was needed to repair the dike, while the dike chair resided in the nearby inn. Of course, all expenses were to be paid by the landowner, although the dike count was willing to lend him the money, at a 100 per cent or 200 per cent interest rate. Landowners often went bankrupt as a result. During the autumn inspection, additional dike reinforcements needed for the winter were checked.

The enforcement of water control was even harsher in times of emergency. Additional dike inspections were carried out. Although the dike count could raise extraordinary taxes, a large part of the costs of dike repair had to be borne by the landowner. If he was incapable of doing so, his properties would be sold and he would often be forced to leave the area (had he not already done so). Should his properties be insufficient, any part of his former property, regardless of who owned it, could be claimed by the dike count to cover the costs of dike restora-tion and maintenance. Note that, despite the dike roles, land transactions were not as well documented as nowadays; a committee of seven neighbours of the former landowner were to decide on these matters, which of course occasionally led to social tensions. Another problem with the procedure was the lack of control on the dike chair, which was enhanced by the fact that the dike count was often in the best position to buy the land. Indeed, many large landowners of later days had dike counts as ancestors.

Not only did land property imply dike-building and maintenance, dike-build-ing also implied land property. The morphology of the Rhine was then more dynamic than it is today. New land emerged all the time along the banks and in the river bed. New land along the river banks belonged to the owner of the adjacent old land. Hence, it was profitable to protect the new land as well; indeed, such activities could be defended as dike reinforcement. New land in the river bed was the property of the landlord. To alleviate shortages, entitlement to reclaim this new land was often sold. However, land reclamation in the river bed is disadvantageous for the opposite bank and for water discharge capacity. In addition, the river became more convoluted and narrower, which stimulated ice blocking and hampered shipping. These problems were recognised early on (the

Figure 6.4 Around 1690 a dangerous situation arose when the Waal discharged 90 per cent of the water of the Rhine. A lot of improvements were implemented, of which the Pannerden Canal was the most important. The situation was under control by around 1790

Source: Van de Ven (1993)

oldest date is 1602), but could not be solved without a strong central authority and a detailed map of the river, which took another 250 years.

Over the years, the river Waal discharged more and more of the Rhine water, at the expense of the rivers Netherrhine and IJssel, which became almost dry. The IJssel in particular was a major transportation route. The causes were the complex interactions between floods, land reclamation, and failed attempts to force the Rhine back to the Netherrhine.[2] The situation improved only after the mouth of the IJssel was improved (after 1705) and the Pannerden Canal was opened in 1707 (see Figure 6.4). The latter project was the first move towards Rijkswaterstaat, the central water board.

6.4 Catastrophic floods and major river-bed improvements in the nineteenth century

In the century from 1750 to 1850, the Netherlands was challenged by a series of extremely serious river floods. River floods now dominated sea floods in the political and public agenda. The fact that the problems were human-induced was by then already widely recognised. The saying that God created the world, but the Dutch created the Netherlands, was considered flattering but false. The main things the Dutch created were problems for the next generations.

Against the background of the Little Ice Age, land reclamation and silting up, the three main problems were the poor condition of the river beds and dikes, the unstable distribution of the water between the river branches, and the insufficient number of river mouths.

The dikes were the result of centuries of *ad hoc* protection, land reclamation and dike repairs. The dike system that had grown over the centuries was not optimal for water control. The dikes themselves were often weak and low. The river beds were too narrow because of land reclamation, and too shallow because of silting up.

The Pannerden Canal was an overwhelming success. Its aims were to alleviate the water shortage in the Netherrhine and IJssel, and to reduce the flood risk along the Waal. The canal shifted the major flood risks from the Waal to the IJssel and the Netherrhine. The water distribution was still unstable, however. Only by 1800 was this problem under control. After long negotiations and under Prussian pressure, the provinces of Gelderland, Utrecht and Overijssel jointly implemented large-scale structural improvements which stabilised water distribution. Two positive effects emerged from this. First, the confidence of the engineers was enhanced. Second, awareness grew that co-operation and co-ordination were in the general interest.

The rivers Waal and Meuse shared their mouth, so high water in the Rhine led to floods along the Meuse in spring and winter. The shared last part of the Waal and Meuse, the Merwede, passes along the Biesbosch, a coastal wetland. More and more water was discharged through the Biesbosch, so that the Merwede silted up. However, the Biesbosch froze in wintertime, blocking the water discharge.

Plans for flood control abounded. Further enforcement of the dikes was considered expensive and ineffective. A popular alternative was sideways diversions, canals connecting streams which should lead to a more even distribution of flood discharge. In addition, flood water was to be led to relatively lightly populated and unproductive land. This shifted rather than solved the problem. A more structural approach is due to Cornelis Velsen in 1749. The flood plains, he realised, should be widened and freed from the many obstacles (which were causing ice-blocking). The rivers should be straightened to speed up water discharge.[3] Two new mouths, one for the Waal and one for the Meuse, should be created. Technical, financial and organisational problems prevented these plans from being realised for about a century. It took a whole new political structure to establish these improvements.

The loose structure of the Republic of the United Netherlands ceased to exist in 1795. It was replaced by the centralist Batavian Republic under Napoleonic France, and later (1815) by the Kingdom of the Netherlands. Many things changed. A central agency for water control was established, which initially focused on early warning, relief and restoration. The waterships maintained control but were given directives from The Hague. The French law which shifted dike maintenance from a part of a dike (an individual landowner) to an entire dike system (the population protected by it) was one of the few Napoleonic laws that were not retained by the Kingdom, under pressure of the waterships.

Structural improvements were only studied and discussed. One committee after another was formed, but little happened. The major impediments were the high costs and the miserable state of government finances; the dense population, which was resistant to large infrastructure works; the fear of a redistribution of flood risks from Holland and Utrecht to the eastern and southern provinces, and the political instability resulting from resistance to the formation of a centralist state and from the independence war of Belgium.

Some progress was made, however. Private dike maintenance was replaced by public maintenance in 1838, despite resistance in the countryside, where floods were regarded as acts of God and taxes as an invention of, if not the devil, then untrustworthy townsfolk. Maps of the river beds were made, so that illegal land reclamation could be combated.

The big turnaround came in 1850. During the period 1820–50, artisan engineers were gradually complemented with scientifically (and militarily) trained ones; after the reorganisation of the corps of engineers in 1849, the latter took over control. The year 1848 brought about a new constitution. In 1849 a stable government was finally established under Prime Minister Thorbecke, and the state finances were reorganised. In addition, the transport sector, strongly supported from Germany, pressed hard for navigable waterways, as demand for transport increased with industrialisation.

Eventually, structural improvements were rapidly planned and implemented. The plans were based on the success story of the river Elbe in Germany. The river beds were straightened and deepened (with steam-powered dredging equipment). Dikes were reinforced. New mouths were dug. Sideways discharges were closed. Initially, the improvements met resistance, which rapidly waned after the floods of 1855 and 1861. Problems with draining were solved with the introduction of steam mills; the works lasted for ninety years. Around 1900, intensification took place, because of river transport. The last great flood occurred in 1926. The last sideways discharge was closed in 1943.

6.5 Conclusions

Figure 6.5 summarises 750 years of river floods in the Netherlands. Note that the data are very tentative, as the sampling procedure changed over the years. Figure 6.2 summarises the main process of land subsidence and dike-building. A couple of common threads can be found in the stories above. First, technological development played a dual role, both enhancing and reducing flood risk. Second, flood management has been reactive over the years; risk was reduced only after a severe flood. Management was also *ad hoc*, and directed at the short term and the direct surroundings. In hindsight, flood management was often counterproductive in the longer term. Comprehensive, long-term flood management was possibly only after the establishment of a strong central government, and in combination with improvements in the opportunities for river transport and military defence. Foreign interference and pressure also helped establish effective flood control. These conclusions argue that a comprehensive, dynamic look at hazards and vulnerability is the only road to fully understanding and managing hydrological risks.

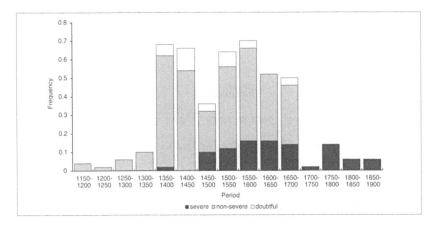

Figure 6.5 Summary of river floods in the Netherlands between 1150 and 1900 (50-year averages)

The role of climate in all this is interesting. The root cause of riverine flooding in the Netherlands lies in the expansion of human population and activity about a millennium ago, during the Medieval Optimum, when climate in Europe was favourable. Subsidence, land reclamation and short-term protection enhanced flood risks, but climate gradually cooled, moving towards the Little Ice Age. The structural improvements of the nineteenth century coincide with another period of climate change, this time warming.

At the end of the twentieth century, flood risk appears to be on the rise again, as witnessed in the Meuse floods of the past 15 years and the near miss with the Rhine in early 1995. The reasons are decades of relative neglect of the dikes and the river bed, changes in the watershed, increasing human activities near the river, and perhaps also climate change.

Acknowledgements

The authors hereby express their gratitude to Xander Olsthoorn, Angela Liberatore and Matthijs Hisschemöller for helpful comments and constructive discussion. Most of the research for this chapter was done while Andreas Langen was visiting IVM in the context of the ERASMUS exchange programme of the Commission of the European Union. This chapter is based on an earlier version (Langen and Tol, 1996).

Notes

1 The countship of Gelre was situated east of Utrecht, north of the rivers; the countship of Kleve was situated further east, extending into what now is Germany. The Netherlands was then part of the Holy Roman Empire.

171

2 Interestingly, one of the main reasons for re-establishing the discharge of the Nether-rhine was national defence. In 1672, the water line failed to protect against the French invaders because of lack of water in the Netherrhine.

3 Water discharge is so rapid nowadays that drought has emerged as a problem in the south of the Netherlands.

References

Gottschalk, M.K.E. (1971) *Sea Surges and River Floods*, 3 vols, Assen: Van Gorcum (in Dutch).

Langen, A. (1995) *History of River Floods in the Dutch Rhine Delta*, W95/16, Amsterdam: Institute for Environmental Studies, Vrije Universiteit (in German).

Langen, A., and Tol, R.S.J. (1996) 'A concise history of riverine floods and flood management in the Dutch Rhine delta', in T.E. Downing, A.A. Olsthoorn and R.S.J. Tol, eds, *Climate Change and Extreme Events*, Amsterdam: Institute for Environmental Studies, Vrije Universiteit, 129–138.

Stol, H. (1993) *Rising Water, Falling Land: History of the Netherlands and the Water*, Utrecht and Antwerp: Kosmos–Z&K (in Dutch).

Van de Ven, G.P. (1993) *Liveable Lowland: History of Water Control and Land Reclamation in the Netherlands*, Utrecht: Matrijis (in Dutch).

7

HURRICANE IMPACTS IN THE CONTEXT OF CLIMATE VARIABILITY, CLIMATIC CHANGE AND COASTAL MANAGEMENT POLICY ON THE EASTERN US SEABOARD

Roger S. Pulwarty

7.1 Introduction

The power of hurricanes has been amply demonstrated in recent years, as major storms have pounded Caribbean island nations and the US mainland. Hurricane Andrew, which struck south Florida on 24 August 1992, ranks as the most costly storm disaster in US history, with long-term estimates of losses exceeding $30 billion. Estimates of losses if Miami, just 30 km north, had been struck are in the range of $40–60 billion. A continuation of the same track would have put New Orleans under 5 to 6 m of water. The meteorological conditions that would create the difference in the two courses of movement (the actual course and a direct hit on Miami) are essentially undetectable with present observing systems (Sheets, 1994). In a careful analysis, Pielke and Landsea (1998) show that a repeat of the 1926 hurricane impacting S.E. florida and Alabama would result in losses exceeding $70 billion (1995 $).

The exposure of human communities to natural phenomena has been increasing as a result of population growth and movements into areas of risk (White, 1945; Burton and Kates, 1964; Pielke and Pielke, 1997; Pulwarty and Riebsame, 1997, among others). The result has been an increase in economic losses in recent decades. It remains crucial to identify the relative contributions of physical, social, economic and political factors to these losses. Glantz and Price (1994) point out that regional and national institutions that develop strategic plans for disaster preparedness often do not take into consideration the possible implications of climate change, or at least climate variability on greater than year-to-year time scales. This chapter seeks to (1) evaluate some of the changes in hurricane frequency that have occurred over the past century, (2) assess the state of knowledge of the effect of climate change on hurricane intensity, frequency and damage potential, and (3) place these changes in the context of important socio-economic trends and the evolution of coastal policy on the eastern seaboard of the USA. In particular, the chapter highlights not only the changing level of risks related to the tremendous growth in coastal population throughout this region, but the importance of understanding the nature of such changes. In addition to hurricanes and tropical storms, the east coast is subject to northeasters, which track down the coast from the north-east. While these storms can be devastating because of their slow movements and long lifetimes, our focus here is strictly toward hurricanes.

Three factors are usually identified for study in assessments of the hurricane hazard: (1) weather generators specifying the frequency, magnitude, and other characteristics such as storm surge and wind speed; (2) the population-at-risk (exposure), specifying the number and geographic distribution of people and buildings in areas prone to hurricane impact; and (3) vulnerability, commonly specified as the susceptibility of the population-at-risk to injury or damage when an event of a given severity occurs (e.g. Friedman, 1977). The number of people in a particular place indicates the degree of exposure, and the likelihood of a particular event in a particular location represents the probability of occurrence. Exposure, as a component of vulnerability, was identified as early as 1964 (Burton and Kates, 1964) and in numerous publications by Gilbert White and others (see Berke *et al.*, 1993; Jarrell *et al.*, 1992).

Adaptation, in varying degrees, is key to the human ecology of vulnerability to natural disasters. Four forms of adaptation have been described in the hazards literature (Alexander, 1991):

1   persistent occupation of the hazard zone despite the risk involved: (a) with comprehensive measures for risk mitigation and hazard abatement, (b) with only warning and evacuation measures, or (c) without any protection measures (the state of maximum vulnerability);
2   co-habitation with the damage caused by past disasters;
3   abandoning damaged or destroyed structures but relocating within the risk zone;
4   migration to safer zones (a) planned or (b) unplanned.

While these have been identified for a number of years (see Burton *et al.*, 1993), we are still bound by the question raised by White (1945) in the context of flood-plain management: 'Why are certain adjustments to the risk of floods preferred over others, and why, despite investment in those adjustments, are societal losses increasing?' (see also Mileti *et al.*, 1995).

Because of the great potential for hurricane-caused destruction of property and loss of life, the science of hurricane forecasting has always been intimately connected with preparedness for natural disasters, and hurricane forecasters have a long history of working closely with emergency management authorities, usually on an event-by-event basis (Pielke and Pielke, 1997). The trend towards greater urbanisation in hurricane-prone areas, growth of ageing populations, the necessity for environmental protection, and other changes in land-use practices suggest that the threats posed by conceivable variations in hurricane activity should be continually reappraised. It is noted, however, that methods for collecting data on social vulnerability and for merging that information with data on physical risks and environmental vulnerability are at an early and evolving stage of development (Mather and Sdasyuk, 1991; Pulwarty and Riebsame, 1997).

This review draws heavily on Diaz and Pulwarty (1997) and studies therein, Platt (1994); Beatley et al. (1995), Clark (1992), Popkin (1990) and the disaster mitigation literature in general. A review of this literature in the context of vulnerability assessment and sustainable development is provided by Pulwarty and Riebsame (1997) and Anderson (1995).

The major questions addressed in this study are:

- What features of climate control the frequency and intensity of extremes, i.e. in this case Atlantic hurricanes?
- How have these features varied in the past and how are they expected to change under climate change scenarios?
- Can we expect increased hurricane activity in the coming decades to levels that were more prevalent during earlier periods?
- What is the damage potential of past hurricanes under present-day conditions?
- Do the regional and national level changes support the assumptions concerning event frequency and population density that have become the basis for strategies of preparedness and preventative planning?

7.2 Climatic controls on hurricane frequency and intensity

Storm frequency is very sensitive to many features of the general circulation of the tropical atmosphere. While the Atlantic hurricane season runs from 1 June to 20 November (183 days), the greatest hurricane activity usually occurs from July to October. The necessary components for hurricane initiation and intensification are warm ocean temperatures (at least 26°C), low vertical shear in the troposphere, instability for cumulonimbus storm formation, and conditions for large-scale spin such as provided by monsoon troughs and easterly waves (Simpson and Riehl, 1981; Pielke, 1990). In addition, for intensification there needs to be a large-scale upper-tropospheric anticyclone over the surface low to maintain and enhance surface pressure reduction (Pielke, 1990). The time lag of the peak season, i.e. behind the summer heating maximum in June, is due to the thermal inertia of the oceans. Gray et al. (1997) and others show that variations of hurricane activity in the tropical Atlantic over the past hundred years appear to be linked (1) to mode-like variations of regional and global sea surface temperatures (SSTs), and (2) to related trends in global air temperature, pressure anomalies and atmospheric circulations.

Landsea *et al.* (1996) summarise the features of hurricane activity in the Atlantic Basin over 1944–96, the most reliable period of record, as follows:

- There has been no significant change in total frequency of tropical storms and hurricanes.
- There is a strong decrease in the numbers of intense hurricanes.
- There has been no change in the strongest hurricanes observed each year.
- There has been a moderate decrease in the maximum intensity reached by all storms over a season.
- The period 1991–4 is the quietest 4-year period on record, i.e. since 1944. The entire period produced just one intense hurricane (Andrew), though this hurricane set the record for the largest losses.
- The year 1995 produced 19 named storms with 11 hurricanes. 1996 produced 13 storms with 9 hurricanes. Note: the most active season on record is 1933, with 21 named storms.
- No hurricanes were observed over the Caribbean Sea during the period 1990–4. The year 1995, however, produced three hurricanes in that region alone.

It must be noted that the two very active years of 1995–6 should not be construed as a change in long-term trend. Hurricanes are classified by their damage potential according the Saffir–Simpson scale (Table 7.1) developed in

Table 7.1 The Saffir–Simpson scale of hurricane damage potential

Category	Winds (kph) (mph)	Central pressure (mb)	Surge (m)	Damage
1	119–153 74–95	>980	1–2	Primarily to unanchored mobile homes. Some coastal road flooding
2	155–177 96–110	965–979	2–3	Moderate damage to roofs. Escape routes flood 2–4 hours before arrival of centre.
3	179–209 111–130	945–964	3–4	Some structural damage. Mobile homes destroyed. Lowlands (<2 m) flooded up to 8 miles (13 km) inland
4	211–250 131–155	920–944	4–6	Some structural failure. Major beach erosion. Terrain lower than 3 m flooded, requiring evacuation of areas up to 6 miles (10 km) inland
5	>250 155	<920	>6	Complete failure on many building types. Major damage to lower floors on all structures below 3 m msl and within 500 yards (450 m) from shore. Massive evacuation necessary of residents 5 to 10 miles (8 to 16 km) from shore.

Sources: Simpson and Riehl (1981); Hebert *et al.* (1993)

the 1970s. A tropical storm is named when its sustained wind speed reaches 39 mph (62 kph). While only 20 per cent of US hurricanes have been major (Saffir–Simpson > 3), these account for 75 per cent of all damage (Pielke, 1990).

Only two category 5 hurricanes are known to have made landfall on the US coast during the twentieth century, Camille in 1969 and the unnamed Florida Keys Labor Day storm in 1935. Hurricane Andrew was category 4 at landfall. The period of hurricane activity since 1900 has shown remarkable interdecadal variations. The 'decadal' time scale denotes temporal variations lasting of the order of 10 to 50 years. Estimates indicate that if the activity in the 1970s had been similar to the two decades prior, then losses for that decade would have risen to about $3 billion a year in the USA (Sheets, 1994). Over the long-term, the average annual economic impact of damages in the continental United States has been recently estimated at $4.8 billion (1995 $) (Pielke and Landsea 1998). The time frame of the observed multi-decadal variability in hurricane characteristics such as frequency and intensity, and shifts in the tracks, is comparable to the time intervals associated with economic development plans and management strategies on the Atlantic seaboard (Diaz and Pulwarty, 1997). This is discussed further below.

7.2.1 Seasonal to decadal-scale variations

Apart from any modification of the climatology of Atlantic hurricanes from anthropogenic effects, there is abundant evidence that variations in hurricane activity have occurred on decadal (and longer) time scales. This variability is important in its own right and can either exacerbate or mask any potential anthropogenic effects in the future. Several investigators have documented the likelihood of the existence of low-frequency changes in the occurrence, intensity and principal tracks of Atlantic hurricanes during the past century (Shapiro, 1982a, b) and over a longer period (Walsh and Reading, 1991). Gray and collaborators provide a comprehensive review and analysis of these historically recorded changes in Atlantic hurricane activity in the post-Second World War period (Gray *et al.*, 1997). Figures 7.1, 7.2 and 7.3, illustrate the significant differences that have occurred in the frequency with which these storms have affected different segments of the Atlantic and Gulf of Mexico coasts of the USA. In the first two decades of the twentieth century, hurricane landfall occurred frequently along the Gulf of Mexico and the Caribbean Sea (Figure 7.1). During the next two decades (Figure 7.2), the pattern shifted to the Atlantic coast of the USA, with few storm systems affecting the Gulf region of Texas and Louisiana (Hebert and Taylor, 1979a, b). The 1950–80 period was one of relatively infrequent landfalls by intense hurricanes, with the 1970s experiencing the lowest incidence of hurricanes of any decade.

Landsea and Gray (1992) have documented several large-scale climatic relationships that affect the environment within which Atlantic tropical disturbances develop into tropical storms and hurricanes. Some of these are summarised from the extensive 'Gray' literature in Table 7.2. Gray, Landsea and colleagues (see

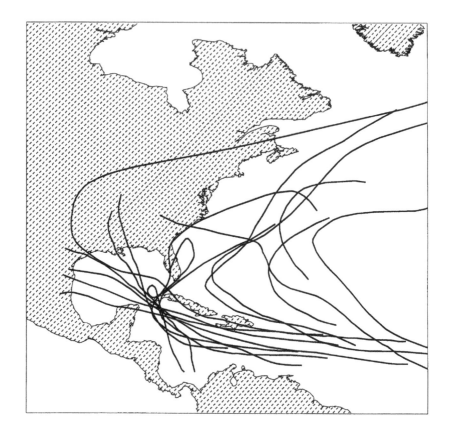

Figure 7.1 Tracks of moderate (category 3) and stronger hurricanes for the period 1900–
10. The figure illustrates the more frequent occurrence of hurricanes in the Gulf
of Mexico and the Caribbean during some decades of the twentieth century.
Reprinted with permission from Springer-Verlag Publishers, Heidelberg.

Landsea *et al.*, 1994) use these and other indices (thirteen in total) to provide
early seasonal forecasts of activity by December, June and August.

The El Niño–Southern Oscillation (ENSO) phenomenon is the single largest
modulator of interannual variations in tropical storms numbers. Wu and Lau
(1992) suggest that El Niño (the ENSO warm phase) acts to reduce the number
of Atlantic hurricanes while the opposite is true in the cold, La Niña, phase. The
number of hurricanes in the Atlantic Basin appears to decline by up to half during
ENSO as compared with that for neutral years (O'Brien, cited in Bengtsson *et
al.*, 1997). Another example of a process which may affect the number of Atlantic
storms is the level of soil moisture in the Sahel region, as has been suggested by
Gray (1984).

According to Gray and colleagues, the effect which seems to dominate as a
unifying process for these changes is decadal variations in the Atlantic Ocean

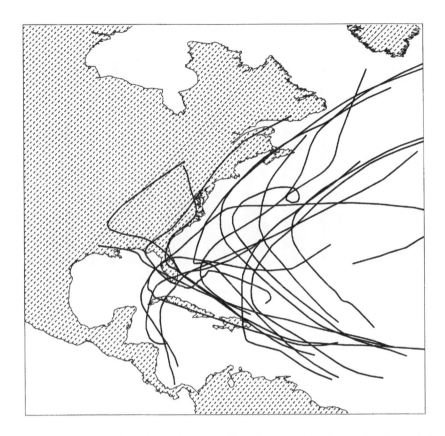

Figure 7.2 Tracks of moderate (category 3) and stronger hurricanes for the period 1920–30. The figure illustrates the shift of hurricane tracks toward the mid-Atlantic coast, compared to 1900–10. Reprinted with permission from Springer-Verlag Publishers, Heidelberg.

thermohaline circulation. During the late 1960s a notable decrease of upper-layer salinity appeared, spreading to large areas of the North Atlantic Ocean and persisting for at least 10 years. This salinity anomaly is presumed to have been caused by anomalous influx of sea ice from the Arctic. The salinity anomaly resulted in a reduction in the net rate of North Atlantic deep water formation with a collateral reduction of the compensating northward conveyor circulation of warm, salty water in the ocean surface layer. Weakening of the net northward Atlantic thermohaline transport during the past 25 years is probably the cause of the cooling of large areas of the North Atlantic Ocean and concurrent warming in much of the South Atlantic (e.g. Brewer *et al.*, 1983; Weaver *et al.*, 1994). The effects of these changes in the regional distribution of tropical SST anomalies are particularly evident as areas of cooling along the coast of northwest Africa (see Hastenrath, 1990). In the Sahel region, surface pressure has been influenced

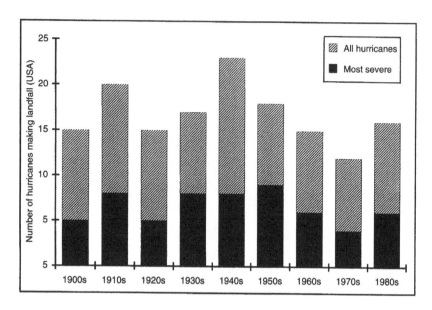

Figure 7.3 Number of hurricanes making landfall in the conterminous USA. Lower shade bars denote landfall frequency for the most intense (category 3–5) storms

Source: After Karl *et al.* (1995). Reprinted with permission from Springer-Verlag Publishers, Heidelberg

Table 7.2 Major indices used in seasonal hurricane forecasting by Gray and colleagues

Index	Association
El Niño-genesis Southern Oscillation (ENSO)	Tropospheric vertical shear increase inhibiting genesis and intensification of storms. Cold events (La Niña) enhance activity
West African Sahel rainfall, temperature and pressure	During Sahel drought, Atlantic hurricane activity intense. Note that this may be collinear with ENSO events (see Goldenberg and Shapiro, 1996)
Stratospheric quasi-biennial oscillation (QBO) direction	During the 12–15 months of the east-phase QBO, Atlantic activity is reduced. The causal mechanism has not been demonstrated
Caribbean sea-level pressure anomalies (SLPA)	When the SLPA is negative (i.e. low pressure in the Caribbean), activity is enhanced
Caribbean 200 mb zonal wind anomalies (ZWA)	While westerly ZWA correspond to ENSO or West African conditions, they also provide an independent measure of tropospheric vertical shear during years of neutral ENSO
Atlantic sea surface temperature (SST)	Warmer SSTs result in lower pressures and higher winds

Source: Gray *et al.* (1992, 1993, 1994)

by increasingly hot, dry regional land-surface conditions rather than by cooler ocean temperature changes. In Gray's view, the combined direct effect of these two complementary changes in the regional surface pressure gradient has been a decreased southwest monsoonal surface wind circulation, a weaker monsoon and less vigorous easterly waves propagating into the tropical Atlantic.

The ocean surface also cooled in much of the North Pacific, while strong SST warming occurred in much of the southern-hemisphere Atlantic, Indian and Pacific Ocean areas. This global distribution of altered ocean surface temperatures has been directly linked to altered patterns of Atlantic and West African surface pressure and monsoon circulations. Burpee (1972) demonstrated that easterly waves were generated by an instability of the African easterly jet. The jet arises because of the temperature contrast between the Sahara Desert and the cooler Gulf of Guinea coast. About 60 per cent of category 1 and 2 storms are generated by easterly waves, while nearly 85 per cent of major hurricanes have their origin as waves (Landsea et al., 1996). The nature and controls of inter-annual variations in easterly waves are as yet undetermined.

Since the West African monsoon is the source of Atlantic easterly wave activity, these effects are at least partly responsible for the recent 25 years of greatly reduced intense hurricane activity in the Atlantic. Trends to more El Niño-like conditions throughout the global tropics are also implicated in decreased hurricane activity on longer time scales. The prominent multi-decadal changes in West African rainfall and intense Atlantic hurricane activity appear to be a consequence of the variable Atlantic thermohaline circulation (Gray et al., 1997). The nature of the tie to increased El Niño activity needs more study, but historical and geological evidence indicate that similar trends, of both comparable and greater amplitude and duration, have occurred many times in the past. Hence these multi-decadal changes are not necessarily associated with recent anthropogenic climate alterations.

There is some evidence from 1994–5 data that the past 25-year mode of atmospheric response may even now be changing. There is, as well, a strong inverse relationship between the number of named storms and the seasonal character of the large-scale atmospheric pressure field (Diaz and Pulwarty, 1997). It is hoped that data being collected in new research, most notably the National Oceanic and Atmospheric Administration's (NOAA's) Atlantic Climate Change Program (ACCP), will identify additional factors which reflect impending changes in the ocean circulation and closely linked climatic conditions.

7.2.2 Anthropogenic climate change

The frequency with which tropical cyclones occur is a product of the prevalence of known necessary conditions for their formation and the frequency and strength of disturbances that have the potential of initiating tropical cyclones. The question of possible influences of global warming on hurricane frequency, distribution and intensity has received much recent attention (Houghton et al., 1996).

Emanuel (1987), through theoretical arguments, suggested that there exists a rigorous upper bound on the intensity of hurricanes, as measured by maximum wind speeds. Elementary considerations show that this limit increases with the amount of greenhouse gases in the atmosphere, but the magnitude of the increase is unknown, owing to large uncertainties about feedbacks in the climate system. However, for each one-degree rise in sea surface temperature, a 1 to 3 per cent drop in the minimum sustainable central pressure and a 5 to 7 per cent increase in the maximum wind speed of hurricanes may be expected (Emanuel, 1987). This result plays an important role in insurance loss modelling, discussed below.

Several researchers have undertaken a series of scenario integrations with the present and a doubled CO_2 concentration, to investigate whether the suggestions by Emanuel (1987) hold in a more realistic context. In the case of fixed or climatologically prescribed clouds, a doubling of CO_2 leads to an increase in the number of tropical storms, while in the case of predictive clouds the doubling of CO_2 leads to a decrease in the number of storms (Broccoli and Manabe, 1990). Haarsma et al. (1992), on the other hand, found a larger number of hurricane-type vortices in double-CO_2 experiments. More recently, Bengtsson et al. (1997) employed two 5-year integration's with a very high-resolution limited-area model (horizontal resolution 125 km). They showed that the low-level vorticity in the hurricane genesis regions is generally reduced compared to the present climate, while the vertical tropospheric wind shear is slightly increased. There is a clear indication of the development of an eye structure in the limited-area model run. The result of the double CO_2 run clearly shows a reduction in the number of storms by as much as 37 per cent. The main reasons for this can be found in changes in the large-scale circulation; that is, a weakening of the Hadley circulation, and a more intense warming of the upper tropical troposphere. The similarity of effects indicates that the warm ENSO signal may provide a useful analogue for climate change studies. What remains unclear is the likely effect that global warming would have on ENSO itself.

Bengtsson et al. (1997) point out that it is not possible at the present time to draw any final conclusions from the double-CO_2 experiment. Climate model simulations produce very different responses of hurricane frequency to climate change and should not be considered reliable at present, though in principle it should be possible to use such models to predict general levels of storm activity. There is no evidence that the regions susceptible to hurricanes would undergo any net expansion or contraction. Very few storms approach their limiting intensity (most reach only 55 per cent of potential), and the processes responsible for keeping storm intensities below their limiting value are poorly understood and not likely to be well simulated by GCMs. While the upper bound on intensity would increase, little is known about how the average intensity of storms might change.

182

It must be noted that neither basic theory nor numerical climate simulation is well enough advanced to predict how tropical cyclone frequency might change with changing climate, and the two give conflicting results concerning the change of tropical cyclone frequency on a doubling of atmospheric CO_2. There is no evidence that regions susceptible to tropical cyclogenesis would expand (Emanuel, 1997). The above result can be expected to be sensitive to the physical parameterisation, which in turn is open to some criticism due to the lack of appropriately detailed empirical knowledge, for example of cloud and radiation processes. Furthermore, the SST changes were obtained from another experiment using a less advanced atmospheric model than in this study and with limited capability to reproduce coupled ocean–atmosphere modes. However, the similarity between the warming pattern to that observed for warm ENSO events (having fewer hurricanes, at least in some regions) lends credibility to the present result. Progress in confronting the important relationship between tropical cyclone activity and climate cannot be made unless there are fundamental advances in understanding the basic physics of hurricanes.

7.3 Impacts and vulnerability

Population has dramatically increased during the past three decades in portions of the US Gulf of Mexico and Atlantic coasts, as well as in Mexico and most of the Caribbean island nations, with attendant increases in residential and industrial vulnerability to hurricanes (Sheets, 1985). In the USA, there are about 45 million people living in the hurricane-prone coastal counties from Texas to Maine, with continued rapid growth in the sunbelt states (Figure 7.4). In areas with the highest growth rate from Texas to Maine, population has increased by about 15 per cent (5 million people) over the 1980–93 period. Of the 154 US hurricanes originating in the Atlantic since 1900, fifty-five have struck Florida. Behind Florida in the frequency of occurrence of direct and indirect hurricane-force winds and surges up to 1.5 m are Texas, Louisiana, North Carolina and South Carolina, in that order.

Many of the 45 million people living in the coastal areas exposed to hurricanes have moved there during the past 25 years. Over 85 per cent of new residents (1980–93) have never experienced a direct hurricane hit. During this period about one-fifth of the sixty-two direct hits by hurricanes of category 3 or higher occurred. In contrast, approximately 50 per cent of the costliest (> $25 million in damage) storms were during this period (Hatheway, 1996). Thus the damage is clearly reflective of demographic trends and property at risk and not increases in hurricane activity. For the top ten counties (by total dollar amounts of insured property and a 5 per cent annual chance exceedance in any given year), Hatheway (1996) calculated damage potentials of 10 to 15 per cent for category 3, and 20 to 30 per cent for category 4 storms. These ranges are estimates and may fluctuate depending upon whether the event has both severe winds and storm surge (e.g. Hugo), or is primarily a surge (Opal) or a wind event (Andrew). The value of

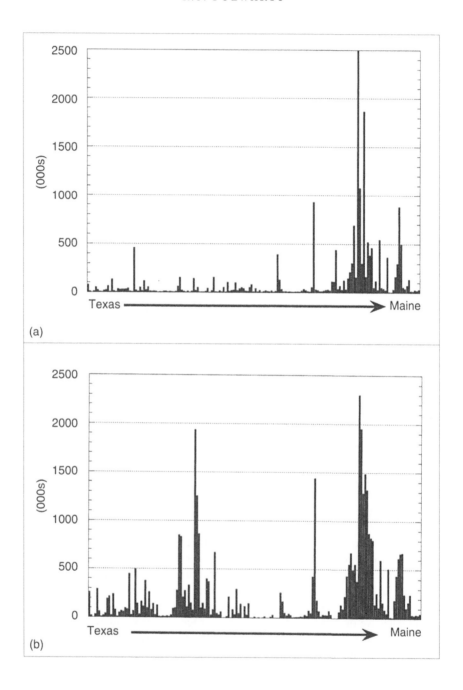

(a)

(b)

insured property at risk in these coastal counties (not counting flood insurance) is now above $3 trillion (Sheets, 1994).

Infrastructure has not kept pace with growth. In most coastal regions, it now takes much longer to evacuate a threatened area than it did a decade ago. For Florida alone, 2 million people are estimated to live in the category 1 zone identified by the National Hurricane Center as the area that would be inundated with water from a category 1 hurricane (74–95 mph, 119–153 kph).

Assessments of societal vulnerability and resilience inextricably link the impact of natural variations to trends in resource use and development, the dynamics of regional political economy, and the range of practical choices available (Pulwarty and Riebsame, 1997). Major structural developments are the luxuries of developed areas (e.g. freshwater storage, seawalls, strapped roofing, etc.), while in developing areas the direct effects of wind and rain may have greater livelihood consequences. However, as described above, damages in developed areas have increased dramatically with increased coastal development even as storm-related mortality has decreased (Figure 7.5). The means through which federal agencies, communities and industries at risk integrate the needed information into their disaster preparedness and development plans is not always uniform or compatible. Important in this assessment is the perception of the relative importance of hurricane risk and its acceptance by government agencies, the media and the private sector. For instance, it is well documented that the desired response to hurricane watches and warnings is heavily dependent on how this information is carried through by the news media (Baker, 1980; Sheets, 1985; Felts and Smith, 1997).

7.3.1 Climate change and property-loss insurance

The low hurricane activity since 1960 left insurers relatively free of underwriting large hurricane losses for a long period. Prior to Hurricane Andrew, the largest insured hurricane loss in the USA, $3.5 billion, was due to Hurricane Hugo in 1989. Before Hugo, the worst damage inflicted by a hurricane was just over $1 billion from Hurricane Alicia in 1983. Hurricane Andrew served as a wake-up call to the insurance industry, with over $20 billion worth of insured losses (Clark, 1997).

The property-casualty insurance industry has estimated that its insured property

---

Figure 7.4a (opposite) US coastal county population by county and state for 1930. *Source*: Pielke and Pielke, 1997; the US Census Bureau. Reprinted with permission from Springer-Verlag Publishers, Heidelberg

Figure 7.4b (opposite) US coastal county population by county and state for 1990. *Source*: Pielke and Pielke, 1997; the US Census Bureau. Reprinted with permission from Springer-Verlag Publishers, Heidelberg

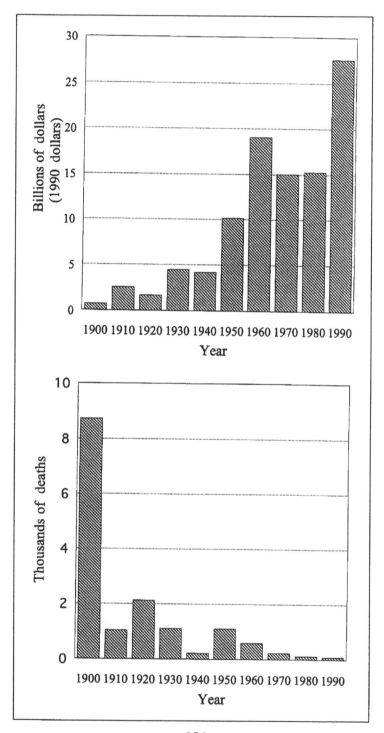

exposure in residential and commercial coastal counties in the eighteen Gulf and Atlantic coast states increased from $1.13 trillion in 1980 to $3.1 trillion in 1993. This includes a 50 per cent (after inflation) increase since 1988, according to the Insurance Institute for Property Loss Reduction. This represents increasing property values as well as the greater numbers and value of properties insured.

A major concern for damage estimation is that the plausible range of insured losses in several regions and the likely impact of such losses on affected insurers and other financial institutions have not been accurately estimated (Berz, 1993). In addition, insurance companies need information in a form that can be directly input to their decision-making processes (Michaels *et al.* 1997). Variations among study goals, methodology, frequency and length limit comparative studies of property-loss vulnerability along the entire exposed coasts of the USA.

The cost of Hurricane Andrew exceeded the property insurance premiums paid over the 1970–92 period. Increased premiums result in cancelled or reduced coverage, further creating problems for financial investors. Since Andrew, eleven insurance companies have become insolvent and some forty companies are either pulling out or greatly curtailing their insuring of property in Florida. It is estimated that a major hurricane striking both Miami and Fort Lauderdale would result in claims of $60 billion, a significant per centage of the $160 billion that the insurance industry reserves for catastrophes (Clark, 1997). Based on 1993 insured property values, Table 7.3 shows estimated probabilities of hurricane losses for the entire Gulf and east coasts. The loss estimates are derived using a hurricane computer simulation model and a database of US insured-property values developed by Applied Insurance Research (AIR). The estimation technique is discussed by Clark (1997). Because of limited historical loss data, problems of data reliability and changes in property values and repair costs, standard actuarial techniques that employ historical losses to project future losses are not appropriate for estimating potential hurricane losses. Applied Insurance Research was the first company to develop an alternative methodology based on Monte Carlo simulation.

The loss estimates above show the probabilities of experiencing losses of different magnitudes from a single hurricane in any given year. The probabilities are also expressed in the table as return periods. For example, $7.8 billion, the loss listed in the return period of 10 years, represents the 90th per centile of the hurricane loss distribution. This amount is likely to be exceeded 10 per cent of the time, or one year out of ten, on average. It is estimated that a 1 in 100-year insured loss amount is just over $30 billion. These figures are not static and continue to change (usually increasing) as coastal properties and their values change. It is important to keep in mind that these numbers include only insured

---

Figure 7.5 (opposite) Comparison of (a) property losses and (b) fatalities from US hurricanes during the various decades of the twentieth century.

Source: Jamieson and Drury (1997). Reprinted with permission from Springer-Verlag Publishers, Heidelberg

Table 7.3 US hurricane loss potential

Insured property loss ($ billions)	Estimated probability of exceeding	Estimated average return interval (years)
7.8	0.100	10
13.2	0.050	20
23.6	0.020	50
30.7	0.010	100
34.5	0.005	200
50.9	0.002	500
51.5	0.001	1,000

Source: Clark (1997). Reprinted with permission from Springer-Verlag Publishers, Heidelberg

damage to building and contents, that is, only a portion of losses. No time element (i.e. coverage that pays for additional living and business interruption expenses or non-insured losses) is included. No loss to infrastructure is estimated. The southeast region dominates the country-wide hurricane-loss potential.

An important value of the AIR model lies in its ability to project potential losses in areas that have not experienced hurricanes in the recent past. The 1940s were the worst decade with respect to potential damage-producing hurricanes. AIR has estimated that if the hurricanes that made landfall in this time period were to recur today, the insured losses would exceed $60 billion over the decade (Table 7.4). A repeat of the 1950s would average $30 billion, while the low-incidence period of the 1970s and 1980s would average about $10 billion in insured losses per decade.

Table 7.5 shows the estimated 1990 insured losses for three major hurricanes affecting the USA. It also shows the estimated losses when the maximum wind speed is increased by 5, 10 and 15 per cent corresponding (approximately) to 1, 2 and 3°C sea-surface temperature rises, respectively. While the precise results vary by storm, the data show that if the maximum wind speeds of these hurricanes were only 15 per cent higher, insured wind losses from these hurricanes would have more than doubled. For example, if the wind speeds of Hurricane Hugo, which caused estimated insured losses of $3.7 billion in 1989, had been higher by only 15 per cent, the expected loss would have been closer to $8.5 billion. Note that storm-surge losses would also have been greater since surge heights also depend on the central pressure. The AIR results indicate that if the average and/or minimum central pressures of hurricanes do, in fact, fall by as little as 1 or 2 per cent (see Emanuel, 1987), the storms will have greatly enhanced destructive power.

The observed, variable nature of hurricane landfall frequency in the past century indicates that the loss potential in the future from the onset of a period of more frequent and/or intense hurricanes could increase dramatically from that experienced in recent decades. The level of risk to life and property is related as much to actions taken by human beings as it is to the level of the underlying natural risk. Roth (1997) underscores the importance of loss mitigation by

Table 7.4 Estimated insured losses today from historical hurricanes of the 1940s

Year	Saffir–Simpson	Landfall	Estimated industry loss ($ billions)
1941	3	Texas	0.442
1941	1	Florida	0.024
1942	3	Texas	0.700
1943	2	Texas	0.447
1944	2	North Carolina	4.400
1944	4	Florida	8.070
1945	3	Texas	0.143
1945	2	Florida	8.173
1947	3	Georgia	0.333
1947	2	Florida	21.850
1948	4	Florida	2.117
1948	4	Florida	0.133
1949	2	Florida	4.700
1949	3	Texas	2.020
1950	3	Florida	2.600
1950	3	Florida	4.500
1950	2	Alabama	0.124

Source: Clark (1997). Reprinted with permission from Springer-Verlag Publishers, Heidelberg

Table 7.5 Loss potential in future hurricanes for increases in maximum wind speed of 0%, 5%, 10% and 15%

Storm	SS	Estimated losses in 1990 dollars (000s)			
		0%	5%	10%	15%
Hugo (1989)	4	3,658,887	4,902,705	6,514,172	8,542,428
Alicia (1983)	3	2,435,589	3,382,775	4,312,884	5,685,853
Camille (1969)	5	3,086,201	4,120,733	5,438,332	7,095,008

Source: Clark (1997). Reprinted with permission from Springer-Verlag Publishers, Heidelberg

Note: SS is the Saffir–Simpson Intensity Scale

advocating a proactive strategy that promotes prudent loss-reduction actions, such as improved and better enforced building codes, the implementation of more rational land-use policies, and greater efforts to inform the public of the levels of risks to natural disasters in different regions.

7.3.2 Environmental impacts

Hurricanes wreak havoc on society and nature alike. It is estimated that Hurricane Hugo, which struck the environs of Charleston, South Carolina, in 1988, resulted in a greater loss of timber than was due to the Yellowstone fires and the Mount St Helens eruption combined. For Hurricane Andrew, the loss of human life and property is well documented. However, the environmental effects are less well known. For south Dade county alone, 33 per cent of reefs were destroyed in Biscayne National Park and up to 70 per cent of hardwoods lost. The northernmost Keys were stripped of vegetation, while in the mangroves, defoliation and wood damages killed large old stands on the shoreline. In Louisiana, Andrew stripped sand from 70 per cent of the barrier islands, exposing old coastal marshes and smothering 80 per cent of oyster reefs with 0.3–0.9 m thick blankets of sediment (USGS, 1993). Barrier islands in Louisiana protect productive estuarine and wetlands environments that support a $10 billion per year fishing industry. The variability of shoreline types, barrier islands, mangroves, etc. makes potential impacts difficult to predict.

Mangrove ecosystems are especially vulnerable because they thrive in the intertidal zone of tropical regions where hurricanes originate and are most frequent. Because mangroves are found at the land–sea interface where hurricanes are often most intense, these coastal forests are subject to damage from both high wind and surge. Mangroves are also an important habitat for many colonial and migratory birds and other wildlife, along with being nursery grounds for fisheries. The fate of mangrove habitat as influenced by hurricanes may be threatened in the future under a global warming environment that might yield more intense storms than have previously been observed. Doyle and Girod (1997) develop simulation models of hurricane abiotics and mangrove community dynamics to evaluate the effects of hurricanes on mangrove habitat across the south Florida landscape. They show that hindcast simulations of actual hurricane tracks and conditions seem to account for the structural composition of modern-day mangrove forests across south Florida. A recurrence interval of major storms every 30 years over the past century is the major factor controlling mangrove ecosystem dynamics in south Florida. Model results of climate-change scenarios indicate that future mangrove forests are likely to be diminished in stature and perhaps include a higher proportion of red mangroves. This modelling approach offers the ability to assess the effects of decadal and longer-time-scale changes in hurricane behaviour on community structure and distribution of important plant associations such as the fate of mangrove habitat.

7.3.3 Hazard mitigation and costs

Mitigation has been called the most neglected aspect of emergency management (FEMA, 1996). Mitigation generally includes structural approaches such as reinforcing buildings against winds, and non-structural approaches, for example zoning that locates communities away from vulnerable shores (i.e. retreat). Analysing the economic value of potential responses to the effects of climate change lies at the heart of decision-making under conditions of long-term uncertainty (Yohe, 1991). For instance, several researchers have produced variable projections of the costs of sea-level rise. Impact estimates for sea-level rise in the USA, based on lost future property values, range from $39 to $350 billion, while costs of protection is estimated as much as $120 billion (Broadus, 1994). (See West and Dowlatabadi, Chapter 8, this volume, for a review of sea-level rise and storm damage costs in the USA).

As the 1993 Midwest event and floods have demonstrated, the country is not taking maximum advantage of available mechanisms to ameliorate the effects of disasters (FEMA, 1996). In 1978 it was estimated that better construction could reduce the combined hazards cost (hurricanes, floods, earthquakes, etc.) by 25 per cent. Nationally the National Flood Insurance Program (NFIP) estimates that $569 million a year in flood damage prevention is attributable to community compliance with its regulation. More recently, the cost of lack of compliance with building codes has been estimated to have accounted for as much as a quarter of the $20 billion insurance bill for Hurricane Andrew (Kunreuther, 1993).

Coastal management programmes occur on the federal, state and local scale (Beatley *et al.*, 1995). It should be noted that the benefits of State efforts such as beach nourishment versus retreat are difficult to ascertain because most projects have been in existence only a short time (Powell, 1996). Certainly the lack of precision in assessing risks makes it easier for short-term benefits of economic development to dominate long-term costs. It is fair to say that structural solutions will continue to be pursued as more immediately economical over other strategies, such as retreat, simply because large investments have already been made on the beach front. In addition, as Wiener (1997) points out, in cost–benefit studies of losses due to flood-plain (or coastal) hazards and mitigation strategies, the opportunities lost by inland regions as a result of induced development on the shore are real but difficult to take into account. Indeed, the trade-off in costs of mitigation versus the benefits of living on the coast during less hazardous periods is not well quantified (R. Pielke, personal communication, 1996). It is also clear that the distributional consequences of projected changes or of the proposed policies are not well understood. For instance, in the aftermath of Hurricane Hugo (1989), Rubin and Popkin (1990) observed that the major obstacle to relief and reconstruction on the South Carolina coast was bureaucratic blindness to the needs of poor, often illiterate, people who lacked insurance and other support systems. This component of vulnerability is discussed in detail in Bender (1994), Blaikie *et al.* (1994) and Pulwarty and Riebsame (1997).

7.4 Social learning, coastal policy and the 'retreat from retreat'

Learning is of strategic importance in the decades-long process of responding to climatic change and variability, whether global or regional. Differing views as to what constitutes vulnerability have had profound influence on the study of the impacts of natural hazards and the choice of response strategies. Policy needs, public values and differing scientific perspectives are all elements in problem definition (Bradbury, 1989). Problem definition can be viewed as a process in which the objects on which focus is to be placed are denoted (i.e. what is 'wrong') and the context and directions under which analysis is pursued, determined (Jamieson, 1992; Schon, 1983). It directs the lessons which are drawn and the responses to be undertaken. The notion that many solutions to reducing social and environmental vulnerabilities, cognisant of social and economic time frames, are available but remain unused is not new (see Ascher and Healy, 1990; Pulwarty and Riebsame, 1997). However, society is always adapting incrementally to a variety of changes in diverse ways.

In an interesting comparison, Nigg (1993) showed that Charleston, South Carolina, clearly at risk to Hugo owing to coastal proximity, responded with measures within its mitigation portfolio to successfully minimise impacts on lifeline infrastructure, while Charlotte, further inland, simply opted 'to bear the loss' (that is, no preparatory actions took place). Charlotte had presumed that since its inland location had been enough to offer it protection from previous 'coastal' storms, the same would happen on this occasion. This illustrates not only differences in perception and expectation but a fundamental lesson, common to rare but severe events, that communities develop protective measures based on 'past experience'; that is, future hazard policies are based on mitigating a problem in the past. Pulwarty *et al.* (1997) review the social learning literature in the context of water management on the decadal time scale in the western USA, with similar conclusions.

Reduced death rates, where they occur, are due almost entirely to improved warning systems and preparation. However, the results of hurricane impacts, including the ability to recover, can be said to be 'prefigured' by developments in resource use and local to regional policy (Hewitt, 1983; Dove and Khan 1995). Griggs and Gilchrist (1983), in reviewing high-risk situations on the Gulf and Atlantic coasts of the USA concluded that 'people want to live in the sun and be able to look at the ocean, realtors and developers want to make money, and local governments want more tax dollars'. Indeed, the aforementioned demographic and economic pressures have transformed the sparsely developed eastern shorelines of the USA of the 1940s and 1950s into 'cities on the beach' (Platt *et al.*, 1991). In addition, as Platt *et al.* observe, 'For every "volunteer" resident in a high risk coastal location ("sun and surf") there are thousands who have no alternatives because their livelihoods are tied to jobs in oil refineries, export enclaves, tourism, fishing, coastal farms and plantations.'

The key requirements, actions and experiences that have significantly influenced federal-level disaster mitigation polices and programmes fall into three broad categories (Rubin, 1996): (1) organisational/institutional milestones: enabling legislation, programme developments in the Federal Emergency Management Agency (FEMA), executive orders; (2) disaster event milestones: major catastrophic events that become defining events nationally; and (3) mitigation assessment milestones: Congressional studies and reports, disaster evaluations. These directly or indirectly influence policy-makers and legislators responsible for emergency management.

Significant events, responses and milestones in learning are shown in Table 7.6 (Platt, 1994; Rubin, 1996). Before the mid-1960s, Congress tended to respond directly to coastal disasters by enacting remedial legislation, usually authorising structural shore protection projects. However, after Hurricane Betsy resulted in losses of $2 billion on the Gulf coasts, Congress passed the Southeast Hurricane Disaster Relief Act of 1965, which called for study of alternative methods of federal disaster insurance. The follow-up National Flood Insurance Act of 1968 marked a clear departure from the long-standing reliance on purely structural approaches to coastal hazard reduction. The Coastal Zone Management Act (1972) provided for the Coastal Zone Management Program to fund eligible coastal states to develop and administer their own coastal programme, which must address specified national concerns including coastal hazards. By the 1980s, however, researchers were questioning the wisdom of providing flood insurance and other forms of federal assistance; that is, subsidising rebuilding on coastal barriers.

Changes in perspective can be seen by comparing the evolution of alternative views, influencing factors and approaches to coastal hazards management in transition from the 1970s to the 1980s (Platt, 1994). They illustrate a change from: (1) the belief that people and institutions are committed to removing known risks from life and fail to do so only where the risk is highly uncertain, to (2) the increasing realisation that vulnerability is also constructed from an open-ended development process that determines the ways in which a hazard is likely to constitute a disaster (Hewitt, 1983; Bohle et al., 1994; Blaikie et al., 1994).

As Platt (1994) notes, the 1970s were characterised by (1) high property values of coastal real estate; (2) cultural preferences to build as close to the water as possible despite hurricane, erosion or flood risk; (3) prevalent faith in conquering nature rather than adjusting to it; and (4) generous assistance from the federal government to construct shoreline protection efforts. The 1980s, on the other hand, were characterised by (1) decreasing federal funding for shore protection and higher levels of non-federal participation; (2) public interest in aquatic and coastal habitats; (3) judicial support for public regulatory programmes despite reduction of private-property values; and (4) predictions of sea-level rise.

In 1990, revisions to the South Carolina Beachfront Management Act created somewhat stronger restrictions for erosion control devices but represented a political retreat from retreat (Beatley, 1992; Beatley et al., 1995). Opposition to the rebuilding restrictions following Hugo was intense, especially by beach-front property owners. Several 'takings' decisions (see *Lucas v. South Carolina Coastal Council*) suggested that the state's financial liability, in cases where the dead-zone restrictions prevented all reasonable use of a parcel, could exceed $100 million. These political dynamics led to a substantial softening of the law, completely eliminating the dead zone (i.e. the no-construction zone) and allowing development to occur even further seaward under certain conditions.

The evolution of incremental learning so far has left the USA with no

Table 7.6 Significant US coastal disaster events, responses and milestones in learning

Disaster or event	Response and/or milestone
1900, Galveston Hurricane	Galveston seawall
1938, 'Great Northeast Hurricane'	Structural protection
1954–5, Six hurricanes – NE	P.L. 84–71 (Structural Protection)
1962, 'Ash Wednesday Storm'	P.L. 87–874 (Structural Protection)
1965, Hurricane Betsy	P.L. 89–339 (Department of Housing and Urban Development to study flood insurance)
1972, Tropical Storm Agnes	Flood Disaster Protection Act of 1973 – P.L. 93–234 (strengthened NFIA and added erosion as insurable hazard)
1979	Federal Emergency Management Agency (FEMA) created
1979–80, Hurricanes Frederic and David	Sheaffer and Roland Report, 1981, Coastal Barrier Resources Act of 1982 – P.L. 97–348
1985–6, Great Lakes coastal flooding	Michigan Home Moving Program
North and South Carolina coastal erosion	Upton–Jones Amendment to NFIA: P.L. 100–242 sec. 544
	SC Beachfront Management Committee Report 1987
	SC Beachfront Management Act 1982
1989	Stafford Act. Defines the process that releases federal disaster assistance (except for crop losses)
1990, Hurricane Hugo	National Research Council (NRC) Erosion Report
	H.R. 1236 adopted NRC erosion recommendations – 1991 (S. 1650 died in committee after US Supreme Court decision in *Lucas v. SC Coastal Council*, 1992)
1991	Government Accounting Office (GAO) report: federal, state and local responses to disasters need improvements
1992, Hurricane Andrew Winter storms – northeast (H.R. 62)	Coastal Amendments to NFIP reintroduced – 1993
1994	NFIA Reform Act

Source: Adapted from Platt (1994)

Note: References to citations can be provided by the author comprehensive national policy in response to coastal hazards. Platt (1994) lists the barriers to the development of such a policy as follows:

- privatism: owners are presumed to be entitled to use their land as they wish subject only to reasonable constraints in the public interest;
- localism: land use planning is considered a local government prerogative;
- spatial and temporal variations of hurricane frequency, intensity and duration;
- fragmentation of authority over coasts (federal, state, local, private);
- intermingling of public and private benefits from shoreline protection projects;
- intergovernmental cost-sharing of hazard mitigation projects;
- fear of 'takings' issues;
- conflict between economic and environmental objectives in coastal management; and
- circumstances of political representation in Congress.

In an insightful comparison, Birkland (1997) concludes that the creation of a congressionally mandated national hurricane policy will be difficult. He shows that at the national level, hurricanes are considered a land-use problem about which the federal government can currently do little, while earthquakes are considered a scientific problem about which the federal government can do more. As Burby (1996) has shown, the case for induced vulnerability due to federal relief may not be clear-cut. In one study he showed that communities that have received federal disaster aid were the most likely to invest in hazard-mitigation measures, while communities that had not experienced disastrous losses nor received disaster-relief funds that forced them to comply with various mitigation measures were the least likely. Communities that had suffered disasters but had been denied federal relief funds fell in the middle.

In fact, the lesson of 30 years of federal effort to foster hazard mitigation is that nothing short of a strong federal mandate forcing states to implement mitigation measures will succeed. However, *Lucas v. South Carolina* is a stern reminder of the limitations of applying this lesson without decentralised controls on implementation.

The National Mitigation Strategy published by FEMA in 1995 calls for a reduction of all losses caused by natural hazards within fifteen years through mitigation as the primary strategy. This certainly indicates learning in the form of rhetoric if not yet implementation. As pointed out by Rubin (1996), constraints to hazard mitigation have been identified many times, but there have been few implementation and evaluation analyses. Implementation is a superior indication of what is readily useful by policy-makers (Brunner and Ascher, 1992). A predictive model is important but not sufficient for policy use because projecting the probable consequences of action alternatives is only one task in a rational decision process. The other tasks include clarifying policy goals, evaluating action

195

alternatives and reconciling differences through politics. Projections therefore must be objective in the sense of bringing all relevant information to bear on a problem, and this includes adaptation to changes in goals, alternatives, political realities and scientific understanding. The development planning field can potentially make an important contribution by suggesting factors that foster institutional capacity for undertaking adaptive learning before disaster occurs (Berke *et al.*, 1993; Gunderson *et al.*, 1995).

7.5 Conclusion: Lessons and recommendations

Dewey (1899) proposed that humanity exists in a hazardous natural world which results in human insecurity. This has led to an appreciation of the fact that the impacts of environmentally initiated hazards like floods, tornadoes, etc. are not completely independent of society, since these perils are defined, reshaped and redirected by human action (Mileti *et al.*, 1995). In 1987, a US National Research Council committee estimated that the International Decade for Natural Disaster Reduction's (IDNDR) efforts would result in a 50 per cent reduction in losses within 10 years (NRC, 1994). There has been much activity in processing and distributing scientific findings on natural hazards' occurrence and risk in the first half of the IDNDR. However, by the middle of the decade, losses were significantly greater than in any previous 5-year period (White, 1994). While the IDNDR resolution stipulates that substantial knowledge exists on technical solutions to disaster reduction, less work has been done on understanding why these solutions have not been implemented (Berke *et al.*, 1993). National and international relief and development agencies have increasingly recognised that local organisational and individual capacity to use these solutions is inadequate (OFDA, 1990; OAS, 1991).

Historically, disaster planning has been a fairly static process. Recent research indicating the quasi-cyclic nature of tropical cyclone formation, when combined with human-induced changes to the environment, requires disaster planners to consider the problem of how to create dynamic disaster plans. However, experience from the Caribbean Disaster Mitigation Project's efforts in the Caribbean shows that while the technical challenges to incorporating variability in the disaster planning process are achievable, the political and economic implications are extensive, and relatively more difficult to overcome (Watson and Vermeiren, 1997).

It is not enough to say that given local control, provisional shelter can be transformed into acceptable safe housing or that low-income settlements can evolve into stable communities (Bender, 1992). Where possible, all-hazard insurance and government reinsurance as risk-spreading mechanisms will have to be used (Kunreuther, 1995). However, most of the risk-based principles of insurability (Kunreuther, 1995; Roth, 1997) may not be met in poorer areas. In fact, a perfect forecast of decadal-scale hurricane risk may lead insurers to reduce their financial exposure by not providing coverage for the most vulnerable regions (Kunreuther, 1993). This leads to the recommendation that the match rate for disaster assistance to states with effective mitigation and preparedness should be increased and that insurance or loans must be used to cover more of the disaster costs of state and local governments (see FEMA, 1996). Disaster insurance offers two clear benefits over direct federal relief (see also Kunreuther, 1996); (1) it provides greater assurance to disaster-stricken governments that losses will be covered, and (2) insurance can mandate risk-reduction measures by limiting participation to those states that comply with such measures.

In recent years, there have been indications that the recent period of relatively low Atlantic tropical cyclone activity may be ending. The 1995 and 1996 hurricane seasons were quite active, with several storms affecting the USA, Mexico, Central America and the islands of the Caribbean. It is not yet possible to determine whether this increase in tropical storm activity represents a change towards more frequent storm activity, as was prevalent during the earlier decades of this century.

Among the lessons and conclusions to be drawn from this review are:

- Improvements in scientific knowledge about hurricanes, and technological and communication advances have resulted in a large decline in the number of casualties resulting from these powerful storms.
- Changes in the level of Atlantic hurricane activity spanning several years to decades have occurred in the observational record.
- Most of the largest losses during the past 50 years in the Caribbean have resulted from freshwater-induced floods, mud slides and landslides.
- The historical record can provide useful lessons about the impacts of particularly severe events. During the month of October 1870 three hurricanes claimed the lives of over 24,000 people. Four storms in the twentieth century have each resulted in fatalities exceeding 8,000 people.
- Understanding decadal trends in hurricane activity may be critically dependent on understanding the broader issues of decadal variations of the major ocean circulation.
- Modelling studies indicate that the upper bounds on hurricane intensity will increase under global warming scenarios but little is known about how the average intensity would change. However, large-scale atmospheric circulation changes resulting in weakening of the Haley circulation and warming of the upper troposphere may result in a significant reduction in the number of hurricanes.
- Ecosystem dynamics, including structural composition and distribution, are linked to the occurrence of low-frequency hurricane events. Natural history may be inextricably tied to decadal-scale variations in hurricane impact.
- Recent large losses from landfalling hurricanes in the USA are due primarily to exponential growth in coastal populations and the valuation of coastal property.
- The relationship between hurricane intensity and damage is non-linear.

Damages in a particular location can be doubled by relatively small changes in intensity.

- Vulnerability to hurricanes may be defined in terms of risk, exposure and capacity to recover.
- Changes in vulnerability can occur without changes in hurricane frequency.
- 'Voluntary choice' does not provide a sole explanatory model for why people are located in risk-prone areas.
- There is a need to document the long-term social, economic and political trends that increase vulnerability.
- Crises can provide an opportunity to introduce innovations in mitigation. The success of programmes is inextricably tied to the people, place, values, institutions and environments concerned.
- Insurability must be coupled to mitigation strategies, particularly where private insurance goals and social objectives may be similar.
- Distributional effects of hazards and of mitigation policies are not well understood.
- There is a strong need to increase not only vertical but horizontal communication, education and co-ordination across sectors, agencies and local groups.

These lessons are especially immediate as programmes for hazard reduction and sustainable development, such as the IDNDR, are being designed, implemented and evaluated. Uncertainty is unavoidable. Thus, the cycle of preparedness, response and recovery must be framed within strategies for longer-term mitigation. The Intergovernmental Panel on Climate Change (IPCC) in 1996 called for a broad framework for the analysis, planning and management of coastal zones in the context of climate change, recognising the co-evolution of natural and social systems. However, a strong need also exists for incorporating mechanisms for evaluating programmes throughout their planning and implementation stages and in the larger contexts of disaster management. Scientific information and knowledge for understanding and reducing vulnerability exists (see Table 7.7), however, as the history of US coastal policy suggests, the mobilisation of political will to undertake needed actions remains a central concern.

The methods of contextual inquiry into vulnerability, advocated here, find a reflection in the adaptive management efforts from the ecological literature (Gunderson *et al.*, 1995) and prototyping studies from the policy sciences arena (Lasswell, 1963). These, together, provide promising ways of linking case-study and experimental planning approaches to increase practical learning at socially relevant scales. The issue remains one of how to better integrate natural-hazard management concerns into the development planning process (Watson and Vermeiren, 1997). What needs to be analysed in each case is how the structures of society, as rooted in individuals, environment, institutions and values, determine the ways in which a hazard is likely to constitute a disaster and to be explicit about the uncertainties of such assessments in a changing environment.

Table 7.7 First steps in coastal zone management

---

1 Revamp the National Flood Insurance Program.
 • Direct FEMA to identify and map non-flood related erosion zones.
 • Mandate erosion management standards.

2 Improve disaster assistance.
 • Require states and localities to adopt mitigation measures as a condition of disaster assistance.
 • Review and, if necessary, revise the criteria used by the president to declare disasters.

3 Strengthen coastal zone management.
 • Mandate stronger risk-reduction measures when the Coastal Zone Management Act (CZMA) is reauthorised.
 • Consider implementing a coastal hazards management programme.

4 Promote public education.
 • Authorise and fund education programmes to foster greater knowledge about coastal erosion, sea-level rise, flooding risks, and other threats.

5 Require increased state and local contributions to beach-nourishment programmes.
 • Redistribute the costs more evenly between the federal government (currently paying 65 per cent) and the states.

Source: Recommendations from OTA (1993)

References

Alexander, D. (1991) Natural disasters: A framework for research and teaching. *Disasters* 15: 209–223.

Anderson, M.B. (1995) 'Vulnerability to disaster and sustainable development: A general framework', in M. Munasinghe and C. Clarke, eds, *Disaster Prevention for Sustainable Development: Economic and Policy Issues*, New York: IDNDR and World Bank, 41–60.

Ascher, W., and Healy, R. (1990) *Natural Resource Policymaking in Developing Countries*, Durham, NC: Duke University Press.

Baker, E.J. (1980) 'Coping with hurricane evacuation difficulties', in E.J. Baker, ed., *Hurricanes and Coastal Storms*, Gainesville: Florida Sea Grant College, 13–18.

Beatley, T. (1992) *Hurricane Hugo and Shoreline Retreat*, Final Report to National Science Foundation, Washington, DC.

Beatley, T., Brower, D., and Schwab, A.K. (1995) *An Introduction to Coastal Zone Management*, Washington, DC: Island Press.

Bender, S.O. (1992) 'Disaster prevention and mitigation in Latin America and the Caribbean: Notes on the decade of the 1990s', in Y. Aysan and I. Davis, eds, *Disasters and the Small Dwelling: Perspectives for the UN IDNDR*, London: James & James Science.

Bender, S.O. (1994) The sustaining nature of the disaster–development linkage. *Ecodecision*, April: 50–52.

Bengtsson, L., Botzet, M., and Esch, M. (1997) 'Numerical simulation of intense tropical storms', in H. Diaz and R.S. Pulwarty, eds, *Hurricanes: Climate and Socioeconomic Impacts*, Heidelberg: Springer-Verlag, 67–92.

199

Berke, P., Kartez, J., and Wenger, D. (1993) Recovery after disaster: Achieving sustainable development, mitigation and equity. *Disasters* 17: 93–109.

Berz, G. (1993) 'Global warming and the insurance industry', in R. Bras, ed., *The World at Risk: Natural Hazards and Climate Change*, New York: American Institute of Physics, 217–223.

Birkland, T. (1997) Factors inhibiting a National Hurricane Policy. *Coastal Management* 25, 387–403.

Blaikie, P., Cannon, T., Davis, I., and Wisner, B. (1994) *At Risk: Natural Hazards, People's Vulnerability, and Disasters*, London: Routledge.

Bohle, H.G., Downing, T.E., and Watts, M. (1994) Climate change and social vulnerability. *Global Environmental Change* 4: 37–48.

Bradbury, J. (1989) The policy implications of differing concepts of risk. *Science, Technology and Human Values* 14: 380–399.

Brewer, P., Broecker, W., Jenkins, W., Rhines, P., Rooth, C., Swift, J., Takahashi, J., and Williams, R. (1983) A climatic freshening of the deep Atlantic North of 50°N over the past twenty years. *Science* 222: 1237–1239.

Broadus, J. (1994) The human dimensions of global change in coastal areas. Paper read at Ocean/Land/Atmosphere Interactions in the Intertropical Americas, meeting at Inter-American Institute for Global Change Research, Panama City, Panama, 7–10 February.

Broccoli, A., and Manabe, S. (1990) Can existing climate models be used to study anthropogenic changes in tropical cyclone climate? *Geophysical Research Letters* 17: 1917–1920.

Brunner, R., and Ascher, W. (1992) Science and social responsibility. *Policy Sciences* 25: 295–331.

Burby, R. (1996) Is federal disaster relief at odds with mitigation? *Forum for Applied Research and Public Policy* 11, 3: 109–113.

Burpee, R. (1972) The origin and structure of easterly waves. *Journal of Atmospheric Science* 29: 77–90.

Burton, I., and Kates, R.W. (1964) The floodplain and the seashore. *Geographical Review* 54: 366–385.

Burton, K., Kates, R.W., and White, G.F. (1993) *The Environment as Hazard*, 2nd edition, New York: Guilford Press.

Clark, K. (1992) Predicting global warming's impact. *Contingencies*, May/June.

Clark, K. (1997) 'Current and potential impact of hurricane variability on the insurance industry', in H. Diaz and R.S. Pulwarty, eds, *Hurricanes: Climate and Socioeconomic Impacts*, Heidelberg: Springer-Verlag, 273–284.

Dewey, J. (1899) 'The school and society', in J. Boydston, ed., *The Middle Works: 1899–1924*, vol. 1, Carbondale, IL: Southern Illinois University Press.

Diaz, H., and Pulwarty, R.S., eds (1997) *Hurricanes: Climate and Socioeconomic Impacts*, Heidelberg: Springer-Verlag.

Dove, M., and Khan, M. (1995) Competing constructions of calamity: The April 1991 Bangladesh cyclone. *Population and Environment* 16: 445–471.

Doyle, T., and Girod, G. (1997) 'The frequency and intensity of Atlantic hurricanes and their influence on the structure of south Florida mangrove communities', in H. Diaz and R.S. Pulwarty, eds, *Hurricanes: Climate and Socioeconomic Impacts*, Heidelberg: Springer-Verlag, 109–120.

Emanuel, K. (1987) The dependence of hurricane intensity on climate. *Nature* 326: 483–485.

Emanuel, K. (1997) 'Climate variations and hurricane activity: Some theoretical issues', in H. Diaz and R.S. Pulwarty, eds, *Hurricanes: Climate and Socioeconomic Impacts*, Heidelberg: Springer-Verlag, 55–66.

Federal Emergency Management Agency (FEMA) (1996) *Recommendations and Actions: Create Results-Oriented Incentives to Reduce the Costs of Disaster*, Washington, DC: FEMA.

Felts, A., and Smith, D. (1997) 'Communicating climate research to policy-makers', in H. Diaz and R.S. Pulwarty, eds, *Hurricanes: Climate and Socioeconomic Impacts*, Heidelberg: Springer-Verlag, 233–250.

Friedman, D.G. (1977) Assessment of the magnitude of the hurricane hazard. Paper in proceedings *11th Conference on Hurricanes and Tropical Meteorology*, 13–16 December, Miami Beach, FL, 294–301.

Glantz, M.H., and Price, M.F. (1994) 'Summary of discussion sessions', in M.H. Glantz, ed. *The Role of Regional Organisations in the Context of Climate Change*, Berlin: Springer-Verlag, 33–56.

Goldenberg, S., and Shapiro, L. (1996) Physical mechanisms for the association of El Niño and West African rainfall with major Atlantic hurricane activity. *Journal of Climate* 9: 1169–1187.

Gray, W.M. (1984) Atlantic seasonal hurricane frequency, Part 1: El Niño and 30 millibar quasi-biennial oscillation influences. *Monthly Weather Review* 112: 1649–1668.

Gray, W.M., Landsea, C.W., Mielke, P.W., and Berry, K.J. (1992) Predicting Atlantic seasonal hurricane activity 6–11 months in advance. *Weather and Forecasting* 7: 440–455.

Gray, W.M., Landsea, C.W., Mielke, P.W., and Berry, K.J. (1993) Predicting Atlantic basin seasonal tropical cyclone activity by 1 August. *Weather and Forecasting* 8: 73–86.

Gray, W.M., Landsea, C.W., Mielke, P.W., and Berry, K.J. (1994) Predicting Atlantic basin seasonal tropical cyclone activity by 1 June. *Weather and Forecasting* 9: 103–115.

Gray, W.M., Sheaffer, J.D., and Landsea, C.W. (1997) 'Climate trends associated with multidecadal variability of Atlantic hurricane activity', in H. Diaz and R.S. Pulwarty, eds, *Hurricanes: Climate and Socioeconomic Impacts*, Heidelberg: Springer-Verlag, 15–54.

Griggs, G.B., and Gilchrist, J.A. (1983) *Geological Hazards, Resources and Environmental Planning*, 2nd edition, Belmont, CA: Wadsworth.

Gunderson, S., Holling, C., and Light, S. (1995) *Barriers and Bridges to the Renewal of Ecosystems and Institutions*, New York: Columbia University Press.

Haarsma, J., Mitchell, J., and Senior, C. (1992) Tropical disturbances in a GCM. *Climate Dynamics* 8: 247–257.

Hastenrath, S. (1990) Decadal-scale changes of the circulation in the tropical Atlantic sector associated with Sahel drought. *International Journal of Climatology* 10: 459–472.

Hatheway, D.J. (1996) Correlation of hurricane magnitude, percentage chance of exceedance and damage potential for coastal counties from Texas to Maine. Paper in *Coast to Coast: 20 Years of Progress. Proceedings Twentieth Annual Conference of the Association of State Floodplain Managers*, San Diego, CA, 10–14 June, 101–106.

Hebert, P.J., and Taylor, G. (1979a) Everything you always wanted to know about hurricanes, Part I. *Weatherwise* 32: 60–67.

Hebert, P.J., and Taylor, G. (1979b) Everything you always wanted to know about hurricanes, Part II. *Weatherwise* 32: 100–107.

Hebert, P.J., Jarrell, J., and Mayfield, M. (1993) *The Deadliest, Costliest and Most Intense United States Hurricanes of This Century (and Other Frequently Requested Hurricane Facts)*, NWS NHC-31, NOAA Technical Memorandum. Washington, DC: NOAA.

Hewitt, K. (1983) 'The idea of calamity in a technocratic age', in K. Hewitt, ed., *Interpretations of Calamity from the Viewpoint of Human Ecology*, Boston: Allen & Unwin.

Houghton, J.T., Meira Filho, L.G., Callander, B.A., Harris, N., Kattenberg, A., and Maskell, K., eds (1996) *Climate Change 1995: The Science of Climate Change*, Cambridge: Cambridge University Press.

Jamieson, D. (1992) Ethics, public policy and global warming. *Science, Technology and Human Values* 17: 139–153.

Jamieson, G., and Drury, C. (1997) 'Hurricane mitigation efforts at the U.S. Federal Emergency Management Agency', in H. Diaz and R.S. Pulwarty, eds, *Hurricanes: Climate and Socioeconomic Impacts*, Heidelberg: Springer-Verlag, 251–260.

Jarrell, J., Hebert, P., and Mayfield, M. (1992) *Hurricane Experience Levels of Coastal County Populations from Texas to Maine*, NWS NHC-46, Washington, DC: NOAA.

Karl, T.R., Knight, R.W., Easterling, D.R., and Quayle, R.G. (1995) Trends in U.S. climate during the Twentieth Century. *Consequences* 1: 3–12.

Kunreuther, H. (1993) Workshop Presentation. Paper read at 18th Annual Hazards Research and Applications Workshop, Boulder, CO, July.

Kunreuther, H. (1995) 'The role of insurance in reducing losses from natural hazards', in M. Munasinghe and C. Clarke, eds, *Disaster Prevention for Sustainable Development: Economic and Policy Issues*, Washington, DC: IDNDR and World Bank, 87–102.

Kunreuther, H. (1996) Mitigating disaster losses through insurance. *Journal of Risk and Uncertainty* 12: 171–187.

Landsea, C.W., and Gray, W.M. (1992) The strong association between western Sahel monsoon rainfall and intense Atlantic hurricanes. *Journal of Climate* 5: 435–453.

Landsea, C., Gray, W., Mielke, P., and Berry, K. (1994) Seasonal forecasting of Atlantic hurricane activity. *Weather* 49: 273–284.

Landsea, C., Nicholls, N., Gray, B., and Avile, L. (1996) Downward trends in the frequency of intense Atlantic hurricanes during the past five decades. *Geophysical Research Letters* 23: 1697–1700.

Lasswell, H.D. (1963) 'Experimentation, prototyping, intervention', in *The Future of Political Science*, New York: Atherton Press, 95–123.

Mather, J., and Sdasyuk, G. (1991) *Global Change: Geographical Approaches*, Tucson: University of Arizona Press.

Michaels, A., Malmquist, D., Knap, A. and Close, A. (1997) Climate science and insurance risk, *Nature* 389, 225–227.

Mileti, D., Darlington, J., Forrest, B., Passerini, E., and Myers, M.F. (1995) Toward an integration of natural hazards and sustainability. *Environmental Professional* 17: 117–126.

National Research Council (NRC) (1994) *Facing the Challenge: The U.S. National Report to the IDNDR World Conference on Natural Hazard Reduction*, Yokohama and Washington, DC: NRC.

Nigg, J. (1993) 'Societal response to global climate change: Prospects for natural hazards reduction', in R. Bras, ed., *The World at Risk: Natural Hazards and Climate Change*, New York: American Institute of Physics, 289–294.

Office of Foreign Disaster Assistance (OFDA) (1990) *FY 1989 Annual Report*, Washington, DC: US Office of Foreign Disaster Assistance, Agency of International Development.

Office of Technology Assessment (OTA) (1993) 'Coasts', in Office of Technology Assessment, ed., *Preparing for an Uncertain Climate*, Washington, DC: US Government Printing Office, 153–207.

Organization of American States (OAS) (1991) *Primer on Natural Hazard Managment in Integrated Regional Development Planning*, Washington, DC: Department of Regional Development and Environment, Organization of American States.

Pielke, R.A. Jr. (1990) *The Hurricane*, London: Routledge.

Pielke, R.A. Jr. and Landsea, C. (forthcoming) Normalized hurricane damages in the United States: 1925–1995. *Weather and forecasting*.

Pielke, R.A. Jr. and Pielke, R. (1997) 'Vulnerability to hurricanes along the U.S. Atlantic and Gulf coasts', in H. Diaz and R.S. Pulwarty, eds, *Hurricanes: Climate and Socioeconomic Impacts*, Heidelberg: Springer-Verlag, 147–184.

Platt, R.H. (1994) Evolution of coastal hazards policies in the United States. *Coastal Management* 22: 265–284.

Platt, R.H., Beatley, T., and Miller, H.C. (1991) The folly at Folly Beach and other failings of U.S. coastal erosion policy. *Environment* 33: 7–32.

Popkin, R. (1990) 'The history and politics of disaster management in the United States', in A. Kirby, ed., *Nothing to Fear: Risks and Hazards in American Society*, Tucson: University of Arizona, 101–129.

Powell, M.S. (1996) Evaluation of the costs associated with managing Delaware's Atlantic Ocean coast through a policy of retreat. Paper in *Coast to Coast: 20 Years of Progress. Proceedings Twentieth Annual Conference of the Association of State Floodplain Managers*, at San Diego, CA, 10–14 June, 101–106.

Pulwarty, R., and Riebsame, W. (1997) 'The political ecology of vulnerability to hurricane-related hazards', in H. Diaz and R.S. Pulwarty, eds, *Hurricanes: Climate and Socioeconomic Impacts*, Heidelberg: Springer-Verlag, 185–214.

Pulwarty, R., Lee, K.N., and Miles, E.L. (1997) *Social Learning in Adaptation to Climate Variations and Climatic Change: Lessons from the Western United States*, Boulder, CO: NOAA.

Roth, R. (1997) 'Insurable risks, regulation and the changing insurance environment', in H. Diaz and R.S. Pulwarty, eds, *Hurricanes: Climate and Socioeconomic Impacts*, Heidelberg: Springer-Verlag, 261–272.

Rubin, C. (1996) 'Review of literature on federal hazard mitigation efforts (1979–1995). Paper in *Coast to Coast: 20 Years of Progress. Proceedings Twentieth Annual Conference of the Association of State Floodplain Managers*, at San Diego, CA, 10–14 June, 14–24.

Rubin, C., and Popkin, R. (1990) *Disaster Recovery after Hurricane Hugo in South Carolina*, Working Paper no. 69, Boulder, CO: Natural Hazards Research and Applications Information Center.

Schon, D. (1983) *The Reflective Practitioner: How Professionals Think in Action*, New York: Basic Books.

Shapiro, L.J. (1982a) Hurricane climatic fluctuations, Part I: Patterns and cycles. *Monthly Weather Review* 110: 1007–1013.

Shapiro, L.J. (1982b) Hurricane climatic fluctuations, Part II: Relation to large-scale circulation. *Monthly Weather Review* 110: 1014–1023.

Sheets, R. (1985) The National Weather Service hurricane probability programme. *Bulletin of the American Meteorological Society* 66: 4–13.

Sheets, R. (1994) The Natural Disaster Protection Act of 1993. Statement to the U.S. House of Representatives Committee on Public Works and Transportation Subcommittee on Water Resources and Environment, Washington, DC: House of Representatives.

Simpson, R., and Riehl, H. (1981) *The Hurricane and its Impact*, Baton Rouge: Louisiana State University Press.

US Geological Survey (USGS) (1993) *Hurricane Impacts on the Coastal Environment, USGS Fact Sheet*, St Petersburg, FL: USGS Center for Coastal Geology.

Walsh, R., and Reading, A. (1991) Historical changes in tropical cyclone frequency within the Caribbean since 1550. *Würzburger Geographische Arbeiten* 80: 199–240.

Watson, C., and Vermeiren, J. (1997) 'Incorporating variability in the disaster planning process', in H. Diaz and R.S. Pulwarty, eds, *Hurricanes: Climate and Socioeconomic Impacts*, Heidelberg: Springer-Verlag, 215–232.

Weaver, A., Aura, S., and Myers, P. (1994) Inter-decadal variability in an idealized model of the North Atlantic. *Journal of Geophysical Research* 99: 12423–12441.

White, G.F. (1945) *Changes in Urban Occupancy of Flood Plains in the United States*, Department of Geography Research Paper 7, Chicago: University of Chicago Press.

White, G.F. (1994) A perspective on reducing losses from natural hazards. *Bulletin of the American Meteorological Society* 75: 1237–1240.

Wiener, J. (1997) Research opportunities in aid of federal policy. *Policy Sciences* 26, 321–345.

Wu, G., and Lau, N. (1992) A GCM simulation of the relationship between tropical storm formation and ENSO. *Monthly Weather Review* 120: 958–977.

Yohe, G. (1991) Uncertainty, climate change and the economic value of information. *Policy Sciences* 24: 245–269.

8

ON ASSESSING THE ECONOMIC IMPACTS OF SEA-LEVEL RISE ON DEVELOPED COASTS

J. Jason West and Hadi Dowlatabadi

8.1 Introduction – past impact assessments

Of all the potential impacts of climate change, the impacts of sea-level rise on coastal regions have been among the easiest to visualise. Consequently, images of sea-level rise and the inundation of vulnerable regions have figured prominently in discussions of climate change in the popular press, as well as in assessments of the economic, social and environmental impacts of climate change. This chapter addresses perhaps the most apparent and widely discussed effect of sea-level rise: its impacts on developed coastal regions. Other important impacts of sea-level rise, not addressed in this chapter, include the intrusion of salt water into rivers and groundwater and losses of or changes in estuaries and coastal wetlands.

The first generation of quantified economic assessments of sea-level rise impacts followed the methods of Schneider and Chen (1980). In these early studies (see also Milliman *et al.*, 1989), the economic impacts from inundation were assessed using a 'colouring book' approach: a scenario of future sea-level rise was overlaid on topographical maps of coastal regions, and the economic impact was taken as the present market value of the inundated land and property. Yohe (1990) employed these same methods to assess economic vulnerability in the USA, but discussed the shortcomings of using vulnerability as a measure of damages. Such assessments generated large economic impacts and sensitised the public to possible impacts of climate change.

Subsequent assessments introduced a recognition that sea-level rise is a gradual process during which human adaptation is likely. In the second generation of assessments, this adaptation came in the form of structural protection: analysts recognised that many wealthy nations already engineer their coastlines to protect developed coastal regions or have the financial resources and technical capability to put such protection measures in place (Titus, 1986; Smith and Tirpak, 1989; Den Elzen and Rotmans, 1992; Fankhauser, 1995). Such structural protection includes hard protection through the construction of sea walls or bulkheads, and

205

soft protection through beach replenishment and dune-building. Where protective measures are assumed, the economic impact of sea-level rise is estimated as simply the cost of protection in these regions. Consideration of these factors and lower estimates of sea-level rise reduced assessed impacts to roughly one-half previous values (Titus *et al.*, 1991). The recent reports of the Intergovernmental Panel on Climate Change Working Groups II (Watson *et al.*, 1996) and III (Bruce *et al.*, 1996) are based on these second-generation methods.

The third, and most recent, generation of assessments features the adaptation of the market to a gradually eroding shoreline, where structural protection is not pursued. In estimating the economic damages from the loss of land to erosion, Yohe (1991) recognised that when an ocean-front lot is inundated or lost to erosion, the next lot inland becomes an ocean-front lot, and its value increases. The values of other inland lots also increase to reflect their increased proximity to the sea. The economic loss to the community is therefore not the value of an ocean-front lot, but is instead that of a comparable inland lot. Likewise, Yohe *et al.* (1995, 1996) propose that since erosion is a gradual process, the market will reduce the value of developed property which is at risk of inundation. With foresight, the market can anticipate when structural damage will occur, and can appropriately depreciate the value of the structure by the time that inundation occurs. Thus, zero is proposed as a lower bound for the social loss from structural damage when the market has perfect foresight. These third-generation methods are useful both for estimating damages where there is no structural protection, and for estimating the value of such protection. Figure 8.1 illustrates these three generations of impact assessment.

In addition to these methodological advances, improved methods of predicting the sea-level rise caused by climate change have decreased the estimates of sea-level rise from 1–3 m (in 2100, following a business-as-usual emissions scenario) to the most recent estimates of 20 to 86 cm (Warrick *et al.*, 1996; see also Titus and Narayanan, 1995). As a result of both lower estimates of sea-level rise and methodological changes emphasising adaptation, estimates of the economic impacts of sea-level rise in the USA have decreased by an order of magnitude between the first generation and the third generation (Yohe *et al.*, 1996) (see Table 8.1).

In this chapter, we evaluate critically the methods which have been most recently developed and employed to assess the economic impacts of sea-level rise, as well as their underlying assumptions, in an effort to identify potential sources of bias in national estimates.[1] Foremost among our insights is that while previous work has focused on the impacts of sea-level rise directly, storms are the events by which damages become manifest in many coastal regions. In assessing the economic impacts of sea-level rise, then, a critical question is whether the economic impacts of storms in coastal regions will be at all influenced by climate change and sea-level rise.

The remainder of this chapter is organised as follows. First, we consider the past methods of assessing economic impacts in light of the experience, practices

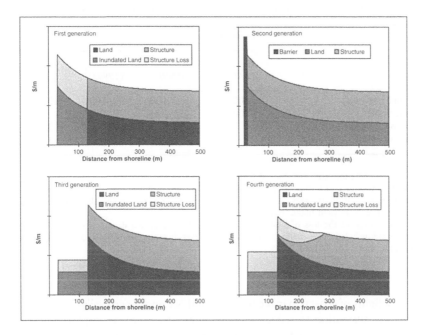

Figure 8.1 The four panels depict the profile of land and structure values in a hypothetical community, as a function of distance from the shoreline. In each
case, there is a beach of 30 m, and the social damages from 100 m of erosion
are shown using each of the four generations of methods. In the first
generation, social damages were simply taken as the present market value
of land and structures in the area inundated. In the second generation,
analysts recognised that barriers could be engineered to protect against
sea-level rise. Where structural protection is pursued, social damages could
be reduced to the cost of the barrier (given here by its area). Third- and
fourth-generation methods focused on damages where structural protection
is not pursued. The third-generation methods recognised market adaptation.
Following erosion, the land value profile shifts inland and the social loss is
given by the value of inland property, not ocean-front property. Likewise,
because there is foresight of damage, the value of structures at risk could be
discounted in the market, and the social loss is reduced. In the fourth-
generation methods outlined in this study, the loss of land is estimated
following third-generation methods. However, structural damages increase
because storms cause damage before structure values can be fully discounted.
In addition, storms cause damage further inland, and create the potential for
repeated damages where structures are rebuilt.

Table 8.1 Assessments of economic damages of sea-level rise to 2100

	Generation			
	1 (1980)	2 (1990)	3 (1996)	4 (present)
Sea-level rise (cm)	~200	50–100	50–100	40
United States[a]	~300	138–321	20–45	?
Duck, NC (variable SLR)[b]	7.5	1.7	2.3[c]	3.2[d]
Duck, NC (SLR = 40 cm)[b]	2.0	0.9	1.2[c]	3.2[d]

Notes
a Undiscounted cumulative damages in US\$1990 × 10^9
b Social damges in US\$1990 × 10^6, discounted at 4%.
c The third generation damages exceed those of the second generation because different methods of projecting erosion yield higher estimates of eroded land
d A combination of the loss of land (\$1.9 million) and the expected damage to structures when storms are included (\$1.3 million). The assumed market discount rate is 8%, and individuals have discount rates from 6.8% to 16.6%

and paradigms of coastal management. Second, we analyse critically the potential for market adaptation to reduce economic impacts. Third, we explore the relevance of extreme events (storms) to the assessment of economic impacts, and the ability of markets to adapt. We then illustrate the relevance of storms and of human agency in estimating impacts by presenting the results of a coastal hazard model. We conclude by summarising the potential sources of bias in past methods of impact assessment.

8.2 Coastal management experience

Sea-level rise induced by global climate change is expected to accelerate the erosion of shorelines in coastal areas. But this change due to sea-level rise will take place in a coastal environment which is historically dynamic. The erosion attributed to sea-level rise, therefore, is superimposed on existing processes of erosion or accretion of shorelines. These processes include local changes in elevation (from subsidence or uplift), transport of sediment along the shore (and offshore), and changes in the supply of sediment (for example, due to the management of rivers). These processes can cause significant differences in the rate of erosion over rather small spatial scales (a few kilometres) and vary significantly on time scales of decades (Fenster and Dolan, 1993). Although it is clearly important internationally, erosion due to sea-level rise in many regions will be only one component (perhaps a small one) of the erosion that will be present. Further, while it is common to simply superimpose the erosion due to sea-level rise on to the erosion from other processes, if global sea-level rise influences those existing processes, the resulting erosion could be non-additive.

The prediction of coastal erosion due to sea-level rise is most difficult on low-lying, unstable coastlines, particularly deltas and sandy barrier islands. Such features are maintained through a balance of geomorphic forces and, because of their low elevations, they are particularly vulnerable to changes in sea-level. For example, the deltaic islands of Bangladesh appear and disappear depending on the supply of sediment from the Ganges and Brahmaputra Rivers, as well as the erosion induced by riverine floods and coastal storms. The barrier islands of the US east coast, where not maintained artificially, move slowly towards the mainland as erosion on the seaward beach is matched by deposition in the backwater lagoon. Past methods of assessing impacts have assumed that barrier islands will simply submerge, given a sufficiently large sea-level rise. In fact, the processes which have created and maintained barrier islands can be expected to continue to maintain the island in a dynamic equilibrium, though the changes through time are difficult to predict. Although such a dynamic environment presents challenges for permanent development, it is inaccurate to assume that such low-lying islands will simply disappear with higher sea levels.

Some past assessments of national economic impacts of climate change have tended to ignore the historical rates of erosion, imposing only the erosion due to sea-level rise (induced by climate) on coastal regions. Other assessments have included only the past local changes in sea-level (e.g., Yohe *et al.*, 1996), and not the observed rates of erosion. Instead, local assessments of sea-level rise damages should be estimated as the difference between damages with sea-level rise and without, where both include estimates of future erosion by existing processes. Further, in producing national estimates of damages, some aggregation of historical erosion rates is necessary, while erosion rates may vary significantly within each length of coast considered. Locally, accounting for actual rates of erosion may yield different results, depending on historical erosion rates and projected development along the coast, and it is important to consider whether the methods of aggregation used in national estimates bias the results. Bias may be introduced, for example, if communities tend to be developed where historical erosion rates are low.

Because coastal regions have historically experienced erosion and storm damage, humans have responded by developing strategies for managing and adapting to coastal hazards. The additional hazard posed by a global sea-level rise should therefore be viewed in the context of these evolving methods and paradigms of coastal management. Foremost among these adaptation measures is structural protection, including 'hard' protection (construction of sea walls and bulkheads) and 'soft' protection (beach nourishment and dune-building) (NRC, 1987). Second- and third-generation estimates of the economic damages due to sea-level rise have identified regions where structural protection is economically justified, and have assumed that protection measures would be employed in these regions. The economic damages of sea-level rise were taken as simply the cost of constructing and maintaining protection. (Where protection would be used even without a climate-induced sea-level rise, damages reflected the addition cost of protection.) As an example, Titus *et al.* (1991) assumed that all developed coastal land would be protected, while Fankhauser (1995) explicitly analysed the optimal strategy between protection and retreat.

However, the efficiency of structural measures as a long-term solution can be questioned on technical grounds. Soft protection is limited by the cost of delivered sand (determined by the availability of sand locally), while measures to build and maintain dunes may be unsustainable in maintaining barrier islands (Dolan, 1972). Structural policies may also encounter political difficulty, as only a small portion of the community (the ocean-front region) benefits directly, while environmentalists and others may oppose the loss of amenity from the beach. In short, structural protection is not always seen as good long-term coastal management (Howard *et al.*, 1985; Fischer, 1990; Titus *et al.*, 1991), and decisions over coastal management practices will be driven more by local politics by than cost efficiency. As a result, protection may not be implemented in all areas where it is cost-efficient, and may be used where it is not cost-efficient.

Currently, coastal management practices in the USA appear to be undergoing a change from an emphasis on structural policies to an emphasis on adaptation to the changing coastal environment. This recent emphasis on non-structural management has included setback regulations (often expressed in terms of historical erosion rates) and incentives to relocate endangered structures (NRC, 1990; Klarin and Hershman, 1990). Very little work has been done to evaluate these policies economically, or to evaluate their potential to change estimates of the economic damages attributable to sea-level rise (West, 1994). In particular, while non-structural policies can help preserve a natural beach for public use, it is not clear that mandated setback regulations lead to more economically efficient outcomes than if investors individually made informed choices to build, though with the same information about erosion that is available to the regulators.

Where structural protection is employed, investors will most likely perceive that the risk of building is reduced, and thus investment in areas prone to risk may be increased. The net result is that structural management, employed without concurrent limitations on investment, may actually increase the vulnerability of the community to economic damages. This increased investment will also have the effect of increasing the political pressure for continued structural protection. Similarly, once non-structural policies are implemented, they may not be easily reversed, owing to opposition from those already displaced. As a result, present decisions regarding structural or non-structural policy options may cause a future bifurcation: decisions to support one policy today will engender conditions for increased public support for that policy in the future.

Governments also influence incentives to invest in coastal regions (and thus the vulnerability to damages) through building codes, zoning regulations, infrastructure expenditures, tax incentives and disincentives, and land acquisition programmes (Klarin and Hershman, 1990). In the USA, the federal government has heavily influenced investment in coastal areas through the provision of federal flood insurance. Aside from pooling financial risks for investors, federal insurance has, during much of its history, subsidised construction in areas at risk of flooding (both coastal and riverine); from 1969 until 1985, the US National Flood Insurance Program paid 31 per cent more in compensation for damages (not including operating costs) than was earned in premiums (Reilly *et al.*, 1993).

8.3 Market adaptation and economic impacts

As mentioned previously, the economic impacts which are attributed to sea-level rise should be found as the difference in impacts with and without sea-level rise, where all other physical conditions are the same. But while many impact assessments assume that coastal developments are static, coastal communities will continue to develop in the future. Further, the course of that future development will depend on the time history of sea-level rise and extreme events.

In an economic assessment, then, one should consider two possible futures, one with sea-level rise and one without. As time progresses in each of these futures, the decisions that individual investors and institutions will make will be different. As an extreme example, a decision to build a home or business on a particular lot may be impossible in the future that includes sea-level rise because that lot may be under water. In each of these futures, the community may be able to gain a certain utility from the land and developed property, may engage in certain productive activities, and may experience certain damages due to erosion or other hazards. It is the difference in utility and damages, once all these decisions and the adaptation of the community and market are included, which measures the economic impacts attributable to sea-level rise (see also Gibbs, 1984).

Viewed in this context, the decisions made by individual investors in response to the risks they perceive are important in estimating the future vulnerability of the community to damages. These include decisions to maintain or improve existing property, as well as decisions to build on vacant land or replace existing structures. The economic impacts due to sea-level rise should therefore be found as the difference between futures with and without sea-level rise, where other physical conditions and the decision rules used by investors (and other adaptive agents) are kept the same. Such a decision rule can be framed to reflect expected utility and risk aversion. Is the utility the investor expects to derive from the investment, given the risk perceived, greater than if that investment were made elsewhere? Decisions to build therefore generate utility for the investor, though at the cost of vulnerability to damages. Collectively, these decisions of individuals seeking to satisfy their varied risk aversion combine to determine both the vulnerability of the community to damages and the total utility gained within the community (Schelling, 1978). Furthermore, because utility is believed to provide incentives for private decisions, it is necessary to consider economic impacts in terms of the utility gained by society, rather than just the economic damages, and whether that total utility is influenced by the presence of sea-level rise. Previous studies have generally focused on the economic damages, irrespective of changes in utility.

The third-generation methods emphasise market adaptation, as the market value of endangered property is reduced gradually in response to the perceived risk of erosion. In fact, property values on the coast of Lake Erie are observed to decrease significantly with the expected remaining lifetime of the structure due to erosion (Kriesel et al., 1993). Yohe et al. (1995, 1996) suggest that in the limit, where the market has perfect foresight, the value of the structure is zero at the moment it is inundated. Thus the lower bound on the social loss due to structural damage is zero. Likewise, Yohe (1991) posits that when ocean-front land is eroded, the social loss is the value of an 'equivalent' area of land far inland, since the value of proximity to the coast is transferred to other lots.

However, such reductions in property values must occur through transactions in the market. At each of these transactions, the structure must be inspected to maintain minimum safety requirements and satisfy lending and insuring institutions. In general, structures are built to strict specifications and have physical lifetimes longer than their assumed economic life. Because structural integrity must be maintained, the value of the property cannot decline indefinitely. Thus if structures are serviceable at the time of their destruction, the social loss is equal to the investment necessary to replace a lost stream of benefits over the expected remaining useful service lifetime[2] of the structure – not the expected time before the structure is destroyed (by erosion) in the coastal environment (i.e., the structure's market value). Hence the economic loss to society should be measured by the service lifetime of the structure, not its economic lifetime.

Further, we have cast the economic problem in terms of utility rather than simply damages. While damages include the loss of serviceable structures and the loss of inundated land, the change in utility reflects how well the resources of the community are utilised. These resources include structures and the amenity of the shore. Sea level rise can affect the total utility derived from structures by destroying property sooner and allowing for reinvestment further inland. If investors are wise in their investments and have the capital to finance development, this inland construction can be designed such that overall utility is increased – otherwise the newly constructed homes may be less desirable and a loss of utility may be attributed to sea-level rise. Sea level rise may also affect the utility gained from land and the amenity of the shore, reflected in the premium investors pay for proximity to the shore. Consider that many coastal communities have lot boundaries which are oriented to allow a large number of ocean-front properties. Where erosion is accelerated, ocean-front properties will be destroyed sooner and the orientation of lots further inland will probably be less efficient in capitalising on the proximity of property to the coast.[3] Furthermore, erosion could reduce the amenity of the shore for public use (e.g., the width of the beach). Thus sea-level rise is expected to cause a loss of utility from land.

Finally, estimates of the economic impacts of sea-level rise consider time scales of a century or more. Over this time scale the demand for housing or recreation near the coast may change substantially. Consider, for example, that the rapid development of coastal resorts in the USA since the 1940s has been the product of a wealthier middle class with greater mobility, as well as the prominence of a beach lifestyle in popular culture. Over the same period, increased mobility has caused resorts in the UK to decline, while Britons now enjoy coastal resorts elsewhere. While such changes in demand are not entirely predictable, they are clearly important for future economic impacts, and the resulting uncertainty in future economic impacts should be acknowledged.

8.4 The relevance of storms

As shorelines erode in coastal regions, particularly where there is a sandy beach (as along most of the eastern US coast), existing property becomes more vulnerable to damage by storms. Often these storms are what cause damage to structures rather than erosion directly. Storms also cause damage where there is no erosion, or where structural protection is used to halt erosion. Given that the damage from storms is widespread in coastal regions, it is important to consider whether we should expect that damage will be different in futures with and without climate change and sea-level rise.

Clearly, if the changing climate over the next century affects the frequency or severity of storms in coastal regions, estimates of the economic impacts attributed to climate change will be affected. Given that extreme tropical and extratropical storms are driven by certain features of the climate (e.g. tropical sea surface temperatures), it is not unreasonable to expect that anthropogenic influences on climate will affect the frequency or severity of storms, both globally and in particular coastal regions. However, current modelling efforts do not show any clear indication of whether tropical and extratropical storms can be expected to become more or less frequent or severe (Kattenberg et al., 1996).[4]

Even where there is no secular trend in the frequency or severity of extreme events, there are good reasons to believe that the damages caused by storms will differ in futures with and without sea-level rise. In general, we expect that the damage from storms will increase as the sea-level rises and the shoreline erodes for three primary reasons: (1) the stage elevation (tidal surge) of all storms increases with sea-level; (2) the erosion of the shoreline decreases the distance between the water and structures, increasing vulnerability; and (3) the erosion of the dune decreases the protection the dune offers.[5] If more damage is estimated to result with sea-level rise than without, or if that damage occurs sooner, the increase in discounted damages could be uniquely attributed to sea-level rise.

213

While third-generation methods propose that investors may have foresight of when damage will occur, the presence of extreme events, which are both episodic and unpredictable, makes any expectations of when damage will occur highly uncertain. Without reliable foresight of damage, investors may not take precautionary measures to avert damage. In addition to structural damage, extreme events can also lead to the loss of the contents of structures (which could be salvaged if erosion were the only risk). Extreme events could also lead to a loss of life (which is most common in less industrialised nations), although the loss of life can be reduced considerably through effective early warning.

However, recall that where there is sea-level rise, investors will make different decisions as compared with where there is no sea-level rise, according to the risk they perceive. It is possible that the future course of these decisions could actually make the community *less* vulnerable to damage where sea-level is rising. Furthermore, storms actually create the conditions under which investment decisions are made. That is, by damaging structures, storms require that there be more investment decisions in terms of repairing storm damage or completely reconstructing where the damage is complete. In addition to causing damage, then, storms also provide opportunities for new decisions about reinvestment, where such decisions would otherwise not be made. Not only will the damage by storms differ in futures with and without sea-level rise, the decisions made following storm damage will be different – and those decisions may prove wise or foolish depending on when damage occurs again. These decisions to reconstruct following storm damage will generate different sets of benefits for investors in futures with and without sea-level rise, and will determine the vulnerability of the community to subsequent damage from storms or erosion.

Finally, consider that because storms are episodic, they will be critical to the cognitive element of both private and public adaptation. As climate changes over the next century, storms will be the salient events which will heighten public concern. For example, the occurrence of a large storm may generate beliefs (true or false) about secular changes in sea-level and storm regimes. The acute public concern following a large storm will strongly influence private decision-making (and thus market adaptation), the policies and premiums of insurance companies, and the formation of public policy (Ives and Furuseth, 1988; Lave and Lave, 1991).[6] The episodic and unpredictable nature of storms therefore confuses the interpretation of observed events, and creates the conditions under which individuals and institutions may maladapt or superadapt to the hazard.[7]

8.5 Illustrative results of a coastal hazard model

We have modelled the effect of storms on estimates of the economic impacts of sea-level rise following the methods outlined in this chapter, while our methods and results are described in greater detail elsewhere (West *et al.*, 1997). We employ a model of coastal change which includes the erosion of the shoreline and dune, and models the potential for storm damage as a function which increases with erosion. In the economic model, we project the development of the community in a manner which is responsive to the time histories of sea-level rise and storm damage in each simulation. This community-scale adaptation to changing physical conditions is modelled by considering the decisions of private investors to construct (or reconstruct) structures on particular lots, based on the risk each investor perceives. The economic impacts attributed to sea-level rise are estimated as the difference between model simulations with and without sea-level rise, where the human decision rules and the time history of storms are the same. Although we have argued that changes in future streams of utility are important for economic impacts, only damages are estimated here.

Table 8.1 (page 208) lists the decreasing estimates of sea-level rise and economic damages in the USA through the past three generations of impact assessments. We have employed these same methods, but assuming no structural protection is pursued, using economic data from Duck, NC, a moderately developed community on the east coast of the USA. The third-generation results shown are based on the methods of Yohe et al. (1996) for their 'perfect foresight' case. Here, since the market has perfect foresight of when damage will occur, the structural damage due to erosion is the lower bound of zero, and all damage shown is due to the loss of land to erosion.

Our fourth-generation estimate of economic impacts includes $1.9 million from the loss of eroded land, which is higher than when previous methods are employed because we model explicitly the movement of the dune with erosion. In addition, rather than estimating the social damages of lost structures using the market value at the time of damage, we use the remaining serviceable lifetime of the structure. The damages due to sea-level rise are therefore positive and, without considering storm damage, amount to $1.2 million (attributed to sea-level rise).

Introducing storms into the model, we simulate storm damage randomly based on return period, and repeat many simulations which differ only in the stochastic occurrence of storms. The expected cost of damages to structures from both storms and erosion is $1.3 million, which is only slightly higher than if erosion were the only cause of structural damage. More importantly, however, this expected result exists within a wide range of possible damages from $6.8 million to −$7.7 million (the extremes from fifty simulations). This wide range of results is generated only by the timing and severity of storms, and illustrates the two competing effects of storms. First, the presence of storms hastens the damage to structures and generally causes more damage over the time horizon considered. But storms also create opportunities for reinvestment, which may differ if sea-level rise is present. For example, a storm early in the simulation may cause roughly the same damage with and without sea-level rise. Following this storm, more structures may be rebuilt without sea-level rise because those investors perceive a lower risk (since the shoreline will have eroded less). Without sea-level rise, then, more structures will be vulnerable to future damage, and thus the damages without sea-level rise could actually exceed the damages with sea-level rise. These results only report economic damages from the loss of land and the damage to structures by storms and erosion. As we suggested earlier, however, economic assessments should also address changes in the utility gained from property, which will also be influenced by sea-level rise, storms, and private investor decisions through time.

8.6 Summary and conclusions

This chapter has revealed possible sources of bias in the methods which have been used to produce national estimates of the economic impacts of sea-level rise on developed coastlines. Those sources of bias which may be present in previous national estimates include (our assessments of bias are shown in parentheses):

- the use of erosion from climate-induced sea-level rise, irrespective of the historical local sea-level rise and local rates of erosion or accretion (from other processes), which can vary significantly in space and time (indeterminate bias);
- the use of average historical sea-level rise or erosion rates over coasts, where developed communities may have systematically chosen areas with low erosion (indeterminate bias);
- the assumption that barrier islands will simply be submerged (over-estimate);
- the assumption that structural protection will be pursued in all areas where it is cost-effective (and will not be pursued where it is not cost-effective), while such decisions have important political elements (underestimate);
- the perceptions of security created by structural protection, which may increase investment in coastal regions (underestimate).

In addition, the effect of non-structural policies and other government initiatives (including insurance provision), and the projected future of these policies, on estimates of the economic impacts of sea-level rise have not been systematically addressed.

Potential sources of bias in the third-generation methods, which emphasise market adaptation, are:

- Damages should be assessed using the service lifetime of a structure, which is generally longer than the economic lifetime (underestimate).
- Assessments should consider changes in utility in addition to simply changes in damages which are attributed to sea-level rise (underestimate).
- Past methods have sometimes used static pictures of communities, rather than considering how communities will continue to develop over the next century. This future development will be responsive to market pressures

and the time histories of coastal hazards (indeterminate bias, but increasing the uncertainty of estimates).

Finally, we presented a case for the relevance of storms in economic assessments. Here, we argued that the potential for damage by storms increases with sea-level rise. The results of previous methods may be biased for the following reasons:

- Storms may damage structures sooner than expected by erosion alone (underestimate).
- Storms reduce the foresight investors have, so damage to the contents of structures and loss of life may result, and some of this damage may be attributable to sea-level rise (underestimate).
- Storms also create opportunities for decisions about reinvestment, which will influence utility and the vulnerability to damage in the future (overestimate).
- Extreme events will, in general, strongly influence public perceptions of risk, which will drive individual investor decisions, insurance premiums and public policy (indeterminate bias).

We propose that these potential sources of bias be explored systematically through selectively chosen case studies, as we have demonstrated in this chapter. Such case studies should be dynamic in nature, and should evaluate the economic impacts of sea-level rise as the difference in utility (not just damages) in simulations with and without sea-level rise, where both other physical conditions (erosion and storms) and the decision rules which govern human behaviour are the same. The results from these detailed local studies should then be compared to the results when the past methods are used. Through such detailed case studies, as well as an evaluation of the aggregation procedures and assumptions currently used, the extent and nature of bias in current national assessments can be determined.

Acknowledgments

We gratefully acknowledge the comments and advice of Anand Patwardhan, Mitchell Small, Tom Downing, Lester Lave, John Miller, Granger Morgan, James Risbey, Richard Tol and Gary Yohe. This research was funded through grants from the National Science Foundation (BCS-9218045), a co-operative agreement between NSF and Carnegie Mellon University (SBR-9521914), the Electric Power Research Institute (RP-3441–14), the US Department of Energy (DE-FG02–95ER62105) and an NSF Graduate Research Fellowship. The views expressed are those of the authors, as are any remaining errors in the analysis and presentation.

Notes

1 See Pulwarty (Chapter 7, this volume) for a discussion of hurricane risk in the eastern USA.
2 The service lifetime of the structure is based on when the structure becomes obsolete, and can be shorter than the structure's physical lifetime.
3 Consider, for example, that the erosion of oceanfront properties would often leave a road as the development feature nearest the beach. Thus the proximity of homes to the beach may be reduced and the coastal amenity may be decreased.
4 See Dorland *et al.* (Chapter 10, this volume) for observations on mid-latitude storms in Europe.
5 Even where structural protection is provided, we expect greater storm damage with sea level rise than without (for these same reasons), although protection reduces the overall damage.
6 Conversely, relatively quiescent conditions from the 1950s to the 1980s on the east coast of the USA may have contributed to rapid coastal development.
7 With maladaption, individuals or institutions may overadapt or underadapt, either of which will lead to economic losses. With super-adaptation, adaptation is planned so that the individual or community is actually better off than if there were no hazard.

References

Bruce, J.P., Lee, H., and Haites, E.F. (1996) *Climate Change 1995: Economic and Social Dimensions of Climate Change*, Contribution of Working Group II to the Second Assessment Report of the Intergovernmental Panel on Climate Change, Cambridge: Cambridge University Press.

Den Elzen, M., and Rotmans, J. (1992) The socioeconomic impact of sea-level rise on the Netherlands: A study of possible scenarios. *Climatic Change* 20, 3: 169–195.

Dolan, R. (1972) Barrier dune system along the Outer Banks of North Carolina: A reappraisal. *Science* 176: 286–288.

Fankhauser, S. (1995) Protection versus retreat: The economic costs of sea-level rise. *Environment and Planning A* 27: 299–319.

Fenster, M.S., and Dolan, R. (1993) Historical shoreline trends along the Outer Banks, North Carolina: Processes and responses. *Journal of Coastal Research* 9, 1: 172–188.

Fischer, D.W. (1990) Public policy aspects of beach erosion control. *American Journal of Economics and Sociology* 49, 2: 185–197.

Gibbs, M.J. (1984) 'Economic analysis of sea level rise: Methods and results', in M.C. Barth and J.G. Titus, eds, *Greenhouse Effect and Sea Level Rise: A Challenge for This Generation*, New York: Van Nostrand, 215–251.

Howard, J.D., Kauffman, W., and Pilkey, O.H. (1985) Strategy for beach preservation proposed. *Geotimes* 30: 15–19.

Ives, S.M., and Furuseth, O.J. (1988) Community response to coastal erosion: The view from two North Carolina beach areas. *Ocean and Shoreline Management* 11: 177–193.

Kattenberg, A., Giorgi, F., Grassi, H., Meehl, G.A., Mitchell, J.F.B., Stouffer, R.J., Tokioka, T., Weaver, A.J., and Wigley, T.M.L. (1996) 'Climate models: Projection of future climate', in J.T. Houghton, L.G. Meira Filho, B.A. Callander, N. Harris, A. Kattenberg and K. Maskell, eds, *Climate Change 1995: Impacts, Adaptation and Mitigation of Climate Change*, Cambridge: Cambridge University Press, 285–358.

Klarin, P., and Hershman, M. (1990) Response of coastal zone management programs to sea level rise in the United States. *Coastal Management* 18: 143–165.

Kriesel, W., Randall, A., and Lichtkoppler, F. (1993) Estimating the benefits of shore erosion protection in Ohio's Lake Erie housing market. *Water Resources Research* 29, 4: 795–801.

Lave, T.R., and Lave, L.B. (1991) Public perception of the risks of floods: Implications for communication. *Risk Analysis* 11, 2: 255–267.

Milliman, J.D., Broadus, J.M., and Gable, F. (1989) Environmental and economic implications of rising sea level and subsiding deltas: The Nile and Bengal examples. *Ambio* 18, 6: 340–346.

National Research Council (NRC) (1987) *Responding to Changes in Sea Level : Engineering Implications*, Washington, DC: National Academy Press.

National Research Council (NRC) (1990) *Managing Coastal Erosion*, Washington, DC: National Academy Press.

Reilly, F.V., Leikin, H.L., and Hayes, T.L. (1993) *National Flood Insurance Program Flood Insurance Rate Review*, Washington, DC: Federal Insurance Administration.

Schelling, T.C. (1978) *Micromotives and Macrobehavior*, New York: W.W. Norton.

Schneider, S., and Chen, R. (1980) Carbon dioxide warming and coastline flooding: Physical factors and climatic impact. *Annual Review of Energy* 5: 107–140.

Smith, J., and Tirpak, D. (1989) *Potential Effects of Global Climate Change on the United States, Appendix B: Sea Level Rise*, Washington DC: Environmental Protection Agency.

Titus, J.G. (1986) *Effects of Changes in Stratospheric Ozone and Global Climate*, vol. 4: *Sea Level Rise*, Washington DC: Environmental Protection Agency.

Titus, J.G., and Narayanan, V.K., eds (1995) *The Probability of Sea Level Rise*, Washington, DC: Environmental Protection Agency.

Titus, J.G., Park, R.A., Leatherman, S.P., Weggel, J.R., Greene, M.S., Mausel, P.W., Brown, S., Gaunt, C., Trehan, M., and Yohe, G. (1991) Greenhouse effect and sea level rise: The cost of holding back the sea. *Coastal Management* 19, 2: 171–204.

Warrick, R.A., Le Provost, C., Meier, M.F., Oerlemans, J., and Woodworth, P.L. (1996) 'Changes in sea level', in J.T. Houghton, L.G. Meira Filho, B.A. Callander, N. Harris, A. Kattenberg and K. Maskell, eds, *Climate Change 1995: The Science of Climate Change*, Cambridge: Cambridge University Press, 359–405.

Watson, R.T., Zinyowera, M.C., Moss, R.H., and Dokken, D.J. (1996) *Climate Change 1995: Impacts, Adaptations and Mitigation of Climate Change: Scientific-Technical Analyses*. Contribution of Working Group II to the Second Assessment Report of the Intergovernmental Panel on Climate Change, Cambridge: Cambridge University Press.

West, J.J. (1994) *Evaluation of Non-structural Coastal Management under Uncertain Physical Conditions: A Case Study at Duck, N.C.* Pittsburgh: Department of Civil and Environmental Engineering, Carnegie Mellon University.

West, J.J., Dowlatabadi, H., Patwardhan, A., and Small, M.J. (1997) Human agency, storms, and the economic impacts of sea level rise. Pittsburgh: Department of Engineering and Public Policy, Carnegie Mellon University (draft).

Yohe, G. (1990) The cost of not holding back the sea: Toward a national sample of economic vulnerability. *Coastal Management* 18: 403–432.

Yohe, G., Neumann, J., and Ameden, H. (1995) Assessing the economic cost of greenhouse induced sea level rise: Methods and applications in support of a national survey. *Journal of Environmental Economics and Management* 29: S-78–S-97.

Yohe, G., Neumann, J., Marshall, P., and Ameden, H. (1996) The economic cost of greenhouse induced sea level rise for developed property in the United States. *Climatic Change* 32, 4: 387–410.

Yohe, G.W. (1991) The cost of not holding back the sea: economic vulnerability. *Ocean and Shoreline Management* 15: 233–255.

9

TROPICAL CYCLONES IN THE SOUTHWEST PACIFIC: IMPACTS ON PACIFIC ISLAND COUNTRIES WITH PARTICULAR REFERENCE TO FIJI

Alexander A. Olsthoorn, W. John Maunder and Richard S.J. Tol

9.1 Introduction

Tropical cyclones constitute one of the most frightening and dangerous natural phenomena. They form over the open sea at latitudes between 5 and 30° latitude in both hemispheres. The causes of the formation of cyclones are not fully explained, but essential conditions include: (1) the sea at its warmest temperature of the year; (2) a deep unstable layer of moist, warm air; (3) an existing low-pressure (usually weak) circulation; and (4) the air moving on a curved path. The conditions for the genesis of tropical cyclones prevail in the North Atlantic Ocean, the Indian Ocean and the Pacific Ocean, in the areas where most of the world's small island states are located.

The damage caused by tropical cyclones is of varying nature, depending on the physical and social conditions in the regions affected:

- High winds may damage building structures either directly, or indirectly through fallen trees and flying debris. High winds affect also crops and forestry. Over land, tropical cyclones lose their extreme strength; severe damage therefore is usually limited to coastal areas.
- Associated heavy rainfall and resulting floods create even more damage.
- The death toll in several tropical cyclones has been enormous, owing to storm surges caused by coinciding high tides and wind forces driving sea water towards a densely populated coastal zone.[1]

Many of the countries vulnerable to tropical cyclones belong to the group of African, Pacific and Caribbean (APC) countries, and have relations with the European Union through the Lomé IV Convention. Under that convention, Europe provides emergency assistance in case of natural disasters. Also, the Convention agrees that some of the APC countries are threatened by adverse global environmental trends.

The case of tropical cyclones in the southwest Pacific and Fiji – one of the group of island ACP states specifically recognised in the Lomé Convention – was selected as a research topic for several reasons:

- The many low-lying islands in the area are obviously threatened by any significant sea-level rise.
- Some Pacific island states are among the poorest countries of the world and therefore particularly vulnerable; of particular interest is the comparison between the situation in the Pacific and the situation in highly developed Europe.
- Knowledge about this topic was considered to be minimal in comparison with the situation in the Caribbean and the Gulf of Bengal.
- Primary data were known to be available, in particular with respect to meteorological data.
- The area encompasses many (small) countries, which, together, represent a political force in international climate negotiations.

The objectives of this case study were an assessment of the 'vulnerability' of Fiji to tropical cyclones, or, more specifically, the vulnerabilities of various socio-economic sectors in Fiji, and the subsequent identification of (non-Fiji-specific) policy options for reduction of these risks against the background of a possible climate change. Resources did not allow an in-depth study collecting original data and specific analysis. The findings are based on information published earlier.

Section 9.2 goes into the past, present and future climatology of tropical cyclones in the southwest Pacific. Section 9.3 presents some general features of small island states in the southwest Pacific. Section 9.4 continues with the vulnerability of Fiji to tropical cyclones. Section 9.5 discusses policy options for damage mitigation. Section 9.6 presents conclusions.

9.2 The occurrence of tropical cyclones in the southwest Pacific

9.2.1 Geography, seasonality and early studies

The conditions for genesis of tropical cyclones are confined in space and in time (to the hurricane season). However, clear spatial and temporal boundaries are difficult to define. Tropical cyclones may, occasionally, occur slightly outside the 'usual' period. For instance, Hurricane Bebe occurred in October 1972, Hurricane Hannah in May 1972, The Son of Isaac on 30 May 1982 and Hurricane Ida in May/June 1972. Indeed, provided the essential conditions exist, a damaging cyclone can be expected at any time of the year.

Another aspect to be taken into account is the fact that although on average a country may have relatively infrequent cyclones, they may occur in rapid succession. For example, Fiji was struck by three cyclones in the 1977/78 season: Anne (25–26 December), Bob (4–5 January) and Ernie (18–19 February). Peni struck Fiji on 2–5 January 1980, followed by Tia on 24 March and Wally on 3–5 April.

9.2.2 Early studies

The knowledge of tropical cyclones on the part of Polynesians and others who roamed the tropical Pacific Ocean long before the arrival of Europeans is reflected in myths and legends. The British explorer Captain James Cook does not appear to have gathered any information about the tropical South Pacific cyclones, probably because he spent most of his time in the tropics during the southern winter. However, European whalers, traders and missionaries who followed Cook soon found that the South Pacific Ocean was not free of the violent storms that had been known in the Caribbean since Columbus's day (Kerr, 1976).

Thomas Dobson (1853) was the first to gather information systematically and to attempt to understand and explain the characteristics and behaviour of tropical storms in the South Pacific. He described twenty-four tropical cyclones. His descriptions are of considerable historical interest, but, considering the small amount of data and the lack of synoptic weather charts, his conclusions are of limited value. He did, however, establish that the storms are cyclones, or 'revolving storms', of which the circulation is clockwise (in the southern hemisphere). Knipping (1893) extended Dobson's list to 120. The classical works of Visher (1925) and Visher and Hodge (1925) covered a total of 259 tropical storms in the South Pacific.

In Visher's time and up to the Second World War, there was insufficient information on weather charts to permit accurate derivation of the tracks of tropical cyclones. The situation improved immediately before and during the war as meteorological services expanded greatly to meet the needs of international aviation and military operations. Hutchings (1953) provides the tracks and statistics of forty-three tropical cyclones in the period 1940–51 inclusive. Tropical cyclones were included in Hutchings's list only if a wind of at least Beaufort force 9 was actually reported at some station or ship within the storm's circulation at some time during its life north of 30° S. This objectivity had been lacking in earlier studies.

Hutchings's average number of 3.6 cyclones per year is low because a number of tropical storms slipped through the network without any weather station reporting a wind of force 9. Gabites (1956) extended the work to 1956. Gabites

(1963) discussed various aspects of tropical cyclones in the South Pacific and, in particular, gave a clear demonstration of the wide variety of track types that occur.

Charts of the tracks of tropical cyclones and depressions in the southwest Pacific for each of the seasons 1947/48 through to 1961/62 were published by Giovannelli and Robert (1964). The Australian Bureau of Meteorology, with the 1957/58 season, began to publish annual reports on tropical cyclones in the Australian region (see, for example, Newman and Bath, 1959).

The comprehensive chapter by Kerr (1976) aimed to consolidate the earlier work and extended it to 1969. Revell (1981) further extended it to 1978/79. An important difference between the analysis method and presentation by Revell and that of Kerr was that satellite information available from 1969 onward allowed the intensity of cyclones in terms of probable maximum wind speed to be estimated with a considerable degree of confidence. Accordingly, analyses of cyclones of different intensity classes were made.

Revell (1981) noted that in the 1969/79 decade the number of tropical cyclones per season varied from a maximum of fifteen in 1971/72 to a minimum of five in 1974/75, with a decadal average of 9.1. Monthly variations were shown to be much the same as in the preceding four decades, the peak occurring in January or February. However, the main difference in the monthly distribution between the 1969/70 to 1978/79 decade and the earlier ones was the greater frequency of cyclones near the end of the season.

Further work on tropical cyclones in the southwest Pacific has been carried out by Thompson et al. (1992). A total of 107 tropical cyclones affected the southwest Pacific region during the decade 1979/89. This is an 18 per cent increase in cyclones as compared with the previous decade.

Maunder (1995) updated the work of the previous studies. He concludes that, in spite of these careful studies, a reliable and comprehensive long-term history of tropical cyclones in the South Pacific is lacking. This is due to the vastness of the area, which, until recently, meant that several tropical cyclones may well have occurred but were not observed. Since satellite imagery became available, in 1970, the information base has improved, and since about 1980, when geostationary satellites came into operation, meteorologists have been able to obtain wide-coverage data of the South Pacific at regular 6-hour intervals, and all tropical cyclones have been observed. Table 9.1 summarises the data, compiled by Maunder (1995), from the 1969/70 cyclone season to the 1993/94 season. In this period 238 tropical cyclones occurred.

9.2.3 Future occurrence of tropical cyclones

The key question is how climate change will manifest itself in the occurrence of tropical cyclones. Will the incidence increase? Will they become more severe? Will the tropical cyclone season change? And will tropical cyclones occur in geographical areas previously not affected?

Table 9.1 Incidence and intensity of tropical cyclones in the South Pacific, 1969/70–1993/94

Season	Major cyclone	Cyclone	Storm	Gale	Annual average
Pentade averages					
69/70–73/74	0.8	2.8	2.6	3.8	10.0
74/75–78/79	0.6	2.4	3.0	2.2	8.2
79/80–83/84	2.0	3.0	3.0	3.4	11.2
84/85–88/89	0.6	4.2	3.2	2.6	10.2
89/90–93/94	0.8	2.6	2.6	0.4	8.0
Twenty-five-season average					
1969/70–1993/94	1.0	3.0	2.9	2.5	9.5

Source: Maunder (1995)

Note: Intensity is defined as: Major cyclone: sustained wind speeds over Beaufort force 12; cyclone: Beaufort force 12; storm: Beaufort force 10 or 11; gale: Beaufort force 8 to 9

Statistical analysis (Tol *et al.*, 1995) of the recorded history of the intensity and frequency of tropical cyclones in the South Pacific revealed an apparent small average increase in the number of the tropical cyclones in the southwest Pacific of 0.07 per year during the period 1939/40–1993/94. The upward trend is only partly explained, statistically, by increased changes of detection of tropical cyclones since 1940. The uncertainty about the 'real' trend is large, however.

Thus statistical trend analysis does not indicate a clear change. Physical modelling of the climate has not resulted in clear answers either, although it is relevant to note the conclusion from the Third International Workshop on Tropical Cyclones held in November 1993 in Mexico (Lighthill *et al.*, 1994), which stated that the 'present evidence suggests that the effect of global warming on either the frequency or severity of tropical cyclones will be minor'. This finding is confirmed by the Second Assessment Report of the IPCC (Kattenberg *et al.*, 1996). The symposium emphasised that there were many thermal and dynamic parameters which brought about tropical cyclone formation and that sea surface temperature was only one of them. Further, the conference noted that warmer sea surface temperatures alone will not produce tropical cyclones, as other parameters, such as weak vertical shear, need to be present for cyclogenesis to take place. The meeting also said that very few tropical cyclones appear to reach their potential intensity, implying that other influences play an important part in the overall development of a tropical cyclone.

9.3 Small island states

9.3.1 Geography

Many of the island countries in the vast Pacific are very small, in terms of population and non-sea surface. These islands are grouped in fourteen Pacific Island countries. By far the largest country, by both population and geographical size, is Papua New Guinea (over 4 million inhabitants, 462,234 km^2). The smallest country is Tokelau (1,600 inhabitants in a 12 km^2 area). Table 9.2 summarises basic data for seven island micro-states.

The economies of these countries exhibit the typical characteristics of small economies (Streeten, 1993): little diversified economic structure, large foreign trade ratio, large share of primary products in exports, diseconomies of scale, little room for macro-economic policy, and substantial remittances from their inhabitants earning a living abroad. Small countries may exhibit more flexibility in economic policy, as a small population is an obvious advantage in quickly reaching a consensus. It should be noted, however, that in a small country the role of specific individuals is more important.

With respect to the environment, there is a similar imbalance in environmental pollution from these countries compared with their environmental importance, in particular with respect to unique ecosystems, such as coral reefs.

Small countries are also more vulnerable than large countries. A storm, a flood or an earthquake may easily affect the whole country, and relatively more people and assets will be affected.[2] The damage caused by one tropical cyclone can easily exceed the GNP of the small countries, and set back development for years.

9.3.2 Economics and development

Economic policies and the wavering development of island micro states since the 1960s are discussed by Connell (1992). It is now possible to see the importance of several constraints and conflicting developments jeopardising the development

Table 9.2 Characteristics of the small island states in the Pacific

Country	Population (mid-1988)	Area (sq km)	GNP per capita (US$ 1988)
Fiji	732,000	18,272	1,520
Kiribati	67,000	690	650
Solomon Islands	303,000	28,530	630
Tonga	97,000	700	830
Tuvalu	9,000	26	570 (1980)
Western Samoa	159,000	2,935	640
Vanuatu	147,000	11,880	840

Source: Connell (1992)

efforts in the 1960s and 1970s. Agricultural policy often resulted in the introduction or expansion of the plantation system, which, *vis-à-vis* a growing population and food demand, hampered land reform and the possibilities for indigenous food production. Imports of food have been substituted for various indigenous crops and fruits, previously grown in subsistence gardening and farming. Nowadays, however, many developing countries have adopted a policy aimed at integrating their economies into larger markets, and several countries are associated to some extent with the European Union. In the South Pacific, small island states such as Fiji joined SPARTECA (South Pacific Regional Trade and Economic Co-operation Agreement), which opened markets in Australia and New Zealand to these island states. Essentially, such a policy can be successful only when aimed at exploiting a country's comparative advantages, such as a relative abundance of natural resources and cheap labour.

Perhaps the most important comparative advantage of island micro-states is that the natural environment can provide the possibility for extensive tourism. In the 1960s, in the Pacific, tourism was not considered an attractive option for economic development, in contrast to the general opinion in the Caribbean. This divergence might be associated with the different geographical 'hinterlands' of the Caribbean and the Pacific. In addition, the traditional character of societies in the Pacific resulted in a prudent attitude towards outside cultural influences (Butler, 1992). Nowadays the importance of the tourist industry is increasing in the Pacific, particularly in Fiji, Samoa and Tahiti.

A comparative 'advantage' is the small-country bias in development aid (Connell, 1992). The reason for this 'advantage' is partly the disproportionately large weight in international politics (possibly associated with voting power in the UN system), and the strategic military/environmental locations of many islands (Anthony, 1990).

9.3.3 Fiji

Fiji comprises the Fiji archipelago plus an island (Ceva-i-ra) and a small island group (Rotuma). The group consists of over three hundred islands and reefs. Most of the islands are of volcanic origin. Vitu Levi, Vanua Levu and Taveuni are the larger islands, where over 80 per cent of the Fiji population live. Many of the other islands are of low elevation, and are particularly vulnerable to tropical cyclones owing to the hurricane-induced storm surges. These surges can reach heights of several metres. For example, at Funanfuti, the surge caused by Cyclone Bebe in 1972 was about 4 m. Two years later, surges from the cyclone Meli reached 6 m in a few places on the island of Nayau (Fiji). Porter (1994) reviewed vulnerability in the context of climate change and sea-level rise, while Nunn *et al.* (1994) studied the possible impacts of storm surges on the Yasawa group.

Fiji is a country with a comparatively favourable natural resource endowment, a sound rural economy, a well developed entrepreneurial class, good social and physical infrastructure, and established structures of public administration (Hill and Tabor, 1993). It is a middle-income developing country with a per capita income of about US$ 1,900 in 1991. Table 9.3 and Figure 9.1 illustrate development from 1970 to 1990 of the Fijian economy towards a more diversified structure. Agriculture is the main economic sector; in terms of employment, it represents 80 per cent of the total.

Over the years, more than 50 per cent of the export trade of Fiji has been directed to Australia, New Zealand and the UK. The increasing diversity of the

Table 9.3 Export shares (%) by major commodity in Fiji, 1970–90

	1970	*1980*	*1990*
Sugar and molasses	65.6	81.1	41.1
Gold	6.8	5.4	11.3
Fish	—	3.7	8.1
Coconut products	11.6	3.1	0.9
Timber	0.5	1.8	5.7
Garments	—	n.a.	19.1
Other	15.5	4.9	13.7

Source: Chandra (1993)

Note: Total 1990 export of garments was F$113 million (in 1990 values, US$75 million)

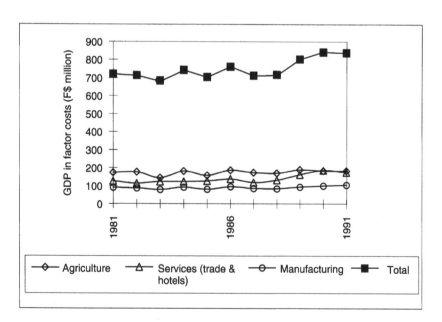

Figure 9.1 Development of Fiji's GDP 1981–91
Source: Treadgold (1992)

Fijian economy (Table 9.3) is the result of economic reform in late 1987. Not shown in Table 9.3 is the growth of the tourist industry in recent years (Hill and Tabor, 1993). Tourism earns about F$300 million per year (Treadgold, 1992), which represents 20 per cent of the country's foreign exchange earnings (Tiedemann, 1991). About half of the tourists are from Australia and New Zealand. The number of visitor arrivals increased from 198,935 in 1981 to 278,966 in 1990.

In 1991, the share of subsistence farming in the output of agriculture, forestry and fishing amounted to 28 per cent (Treadgold, 1992). The subsistence share of the economy is particularly large on the small peripheral islands, which have insufficient communication and infrastructure to develop a cash economy.

9.4 A thirty-year history of tropical cyclone disasters on Fiji

This section assesses the risk from tropical cyclones on Fiji. The assessment is done by briefly examining a thirty-year history of tropical cyclone disasters. Systematic overviews of the events (impacts, responses) have not been found, except for the information from the Fiji Meteorological Service series of information sheets, as compiled by Maunder (1995). Vulnerability is discussed sector by sector, starting with public safety, and subsequently dealing with various economic sectors. The emphasis is on the socio-economic aspects of the vulnerability.

1965: Thousands of coconut trees were blown down or severely damaged, breadfruit trees destroyed, houses flattened and food crops ruined. Driving sea spray blackened and killed vegetation not stripped away by the wind. It was estimated that copra production for the year at Vitu Levi would be halved, a loss of some F$75,000, and that the coconut plantations would take 5 or more years to fully recover. The loss of copra production from the much larger plantations at Taveuni was estimated at over F$500,000 (1965 $).

1972: Food crop damage in Fiji caused by hurricane Bebe was 100 per cent for crops in the Yasawas, 60 per cent in the Rewa-Suva area, 56 per cent in Rotuma and 44 per cent in the Lautoka-Ba, Tavua, Nadi, Nadroga, Navosa and Ra areas.

1975: Hurricane Val destroyed 21 per cent (3,524 acres – 1,426 ha) of coconut in the Lau group; the cost of replanting was estimated at F$141,000. In Tuvuca, 10 per cent of the coconut trees were uprooted. Heavy damage was also caused in Nayau and Cicia.

1977: Cyclone Anne set back copra production in Taveuni by 25 per cent. Anne badly damaged 30 per cent of the coconuts on Yacata Island, 40–45 per cent of coconut trees and 90–100 per cent banana trees on Yanuca Island, and 50 per cent of coconut trees on Naqulevu Island.

1978: Cyclone Bob caused widespread damage in the Western, Northern and Central Divisions of Fiji. It twisted the crowns of coconut trees in Rotuma and blew down palms, especially between Oinafa and Lpta villages. Sixty-five per cent of citrus, 85 per cent of bananas and breadfruit, and 80 per cent of tapioca and dalo were damaged. However, 75 per cent of dalo-ni-tana and 50 per cent of yams survived.

1979: Cyclone Meli added to the destruction caused by Fay. In Vanuabalavu, Meli destroyed the 50 per cent of root crops spared by Fay. In Vatulele, all root crops were damaged, 1,255 breadfruit trees were uprooted and 3,500 coconut palms (20 per cent) were knocked down. Breadfruit, bananas and cassava were totally destroyed in Cicia.

1981: Cyclone Arthur damaged 50–70 per cent of food crops in Nadi, 50–70 per cent food crops in Nadroga and Navosa and 70–80 per cent of root crops in the Yasawas.

1985: Within a few days two tropical cyclones, Eric and Nigel, hit Viti Levu, the most populous island of the group. Twenty-seven people were killed, fifteen deaths being caused by falling walls and roofs, and flying debris. Extensive damage was done to schools, electrical installations, the telephone system, Nadi International airport facilities and equipment, hospitals and health centres, and the Fiji Sugar Corporation's plant and equipment. Fortunately, however, the major part of the annual sugar crop had already been harvested and crushed.

In the event of a tropical cyclone, rural areas experience damage which is disproportionately severe, as livestock and crops, both subsistence, are particularly vulnerable. Table 9.4 presents as an illustration, data for damage done to crops. Damage to livestock made up about a quarter of the crop losses. As the people lost their food resources, relief food rations were issued. In total, a population of 275,747 – about a third of the total population of Fiji – is reported to have obtained food relief at a net cost of F$3.2 million (Chung, 1988).[3]

The total damage done by these 1985 tropical cyclones has been estimated at F$80–90 million, of which F$66 million was at the expense of the insurance industry (of which F$50 million was covered by reinsurance overseas). The events triggered a restructuring of the insurance market in Fiji (Rokovada and Vrolijks, 1993). Eight out of twelve insurance companies retreated from cyclone damage insurance; the remaining four introduced engineering codes and guidelines for insurance. Dwellings can now be insured only after an 'engineering certificate' has been granted. According to the insurers, this policy has considerably improved housing standards, in particular in urban areas.

1990: Three people were drowned in the floods from Rae. Only minor damage to crops and vegetation occurred. In November of the same year, the tropical

Table 9.4 Summary of crop damage caused by tropical cyclones Eric and Nigel in the Western and Central Divisions, 1985

	Area damaged (ha)	Estimated value (F$)
Cassava	1,349	1,653,512
Other root crops	1,477	2,877,150
Tree crops	13,370	1,464,301
Vegetables	505	1,226,452
Maize	223	64,650
Peanuts	4	3,242
Pulses	178	177,930
Rice	256	244,230
Pineapples	20	17,000
Cocoa	5	3,800
Yaqona	66	677,800
Total	17,453	8,410,067

Source: Chung (1988)

cyclone Sina caused no loss of human life. However, Fiji sustained FS$27.5 million worth of damage, nearly 80 per cent of which was accounted for by losses to sugar cane, pine forests and agriculture. Tonga, Niue and the southern Cook Islands sustained minor damage.

1993: Tropical cyclone Kina (Prasad, 1993) caused the greatest ever financial loss due to a tropical cyclone: F$170 million (US$110 million). Twenty-three people were killed, several others were reported missing, and there was uncounted loss of livestock. This tropical cyclone caused widespread heavy rainfall, which, in combination with other factors, including high tides and heavy seas, caused the worst flooding experienced in over sixty years, and accounted for much of the loss. A total of 5,544 people, mostly in rural areas, qualified for government housing assistance and funds of F$6.4 million were paid out to them. Major disruptions were caused to transportation as seven bridges broke under the pressure of torrential currents and accumulated debris in flooded rivers, while landslides cut roads in rural areas. The costs of replacement and reconstruction of these bridges exceeded F$28 million. The Fiji Public Works Department, the Fiji Posts and Telecommunications and the Fiji Electricity Authority expended F$12 million, F$2.7 million and F$7 million, respectively, on the reconstruction of roads, bridges, water and sewerage, telecommunications and power supply lines. Schools were also affected. Reconstruction costs amounted to about F$5 million in total (Rokovada and Vrolijks, 1993).

Sugar crop damage was about 150,000 tons of cane, mostly constituting young crops which did not grow owing to loss of nutrients, rain damage and washouts. The damage to cane mills and equipment used in sugar-cane-growing amounted to approximately F$15 million. Most of the sugar companies' assets were insured; an increase in premiums was expected. The damage to the agricultural sector, excluding the sugar industry, amounted to F$40 million, of which F$17 million was estimated to be crop damage. Consequently, market prices of food (e.g. cassava, kumala, vegetables, fruits) increased dramatically. Immediately after the cyclone, the government imposed for a few months a ban on the export of all food crops (except for crops grown under contract). A very large total of 115,000 people – out of a total population of 750,000 – received food aid for a period up to 3 months, costing the government more than FS$10.5 million. Damage to forestry operations accounted to nearly F$1 million (Fiji Pine Ltd, 1994).

The path of the cyclone stayed well clear of the main tourism areas. There were some difficulties with road transport and some disruption of power and telecommunication. Statistics of visitor arrivals in Fiji showed no dip which could be attributed to the cyclone events. The most significant impact seemed to be an increase in insurance premiums.

The insured damage of this single tropical cyclone amounted to F$50 million, of which F$42 million was reinsured on the international market.

The rehabilitation and relief effort was largely funded through redeployment of government funds, particularly those relating to the capital work programme. The total government budget redeployed amounted to F$40 million, which represented 32 per cent of the total 1993 capital budget.

Table 9.5 summarises the impacts of the tropical cyclones mentioned above. An earlier study of the windstorm risks in Fiji (Tiedemann, 1991) estimated the return period of a storm causing damage of F$10 million at 5 years.

It is noted that the kind of damage caused by tropical cyclones is different in each case. In the case of the hurricanes Eric and Nigel the level of damage caused to the pine plantations was surprisingly low (Chung, 1988); in the case of Kina, much of the damage was related to floods associated with the event.

The perspective from which to evaluate these figures is that the population of Fiji is about 750,000, while its gross domestic product (GDP) is about US$1,500 million (F$2,250 million). Gross investment in Fiji was in the range of F$250–300 million per year in the period 1982–9 (Treadgold, 1992).

Table 9.5 Summary data of recent tropical cyclone disasters in Fiji

Impact	1985	1990	1993
Casualties (death)	27	3	23
Direct damage (F$ million)	80–90	18.5	170
Insured damage (F$ million)	66	unknown	50
Government expenditures on food relief (F$ million)	3.2	unknown	10

Translated to the Dutch situation, the number of fatal casualties would be about 500, which would be enormous for the Netherlands.

9.5 Tropical cyclone risks in Fiji

Conceptually, risk assessment involves combining three sets of information: assessment of the evolution of the tropical cyclone activity; assessment of the development of the element-at-risk (e.g. population or economic sector); and the (future) vulnerabilities of these elements. There are no firm indications that the cyclone activity will change significantly (see Section 9.3).

9.5.1 Rural communities

The prospects for an amelioration of the Fiji rural economy are not promising. This is partly due to rigidities in (traditional) land ownership (Overton, 1993). In addition, the outlook for the developments of the world market for sugar, Fiji's main agricultural export commodity, is uncertain. Only a modest expansion of the sugar industry, stemming mainly from improvements in sugar yields, is anticipated (Treadgold, 1992). Prospects are that vulnerability of the rural people will not diminish, unless targeted policies are pursued (see next section). The need for food aid, by up to 25 per cent of the population when disasters occur, demonstrates the magnitude of the risk of loss of livelihood of the poor rural population. The majority of the people of Fiji live in rural areas, while the rural poor people are the most at risk (Anderson, 1990; Albala-Bertrand, 1993).

9.5.2 Public safety

In Fiji, the risk of falling victim to a hurricane is high. From Table 9.5, the estimated mortality risk from tropical cyclones for Fiji is of the order of magnitude of 1 in 100,000 per year. In the Netherlands the risk of falling victim to drowning from a dike failure is 1 in 10,000,000 per year. The vulnerability of the transport, power supply and communications infrastructure was shown in the event of Cyclone Kina. In particular, landslides caused by heavy rainfall proved to be a major threat to infrastructure.

9.5.3 Risk at the macro-economic level

Figure 9.1 gives a short time series (1981–90) of the performance of the Fiji economy (Treadgold, 1992). One might expect that the time series would reveal a macro-economic effect of the 1985 disasters. However, at least by visual inspection, no influence is seen. There are many other influences on macro-economic performance, among them civil disturbances, labour disputes, economic reform and droughts (Treadgold, 1992). Of course, macro-economic indicators (e.g., GDP) do not directly relate to disaster losses (see Tol and

Leek, Chapter 12, this volume). Fiji does not occur on the list of fifty countries earmarked by 'economic proneness to disasters' in the period 1970–89 by the UNDRO-DHA.

9.5.4 Agriculture

The effects of strong winds on plants vary with the types of plant and circumstances. Crops differ in their vulnerability (e.g., coconut trees and sugar cane are less vulnerable than banana trees[4]). The amount of shelter is obviously an important factor, while soil depth determines the anchorage of plants. The stage of growth is important; plants tend to be particularly vulnerable in the flowering stage. Finally, the health of the crop influences its vulnerability. The force of the wind is a critical factor; however, other phenomena which accompany tropical cyclones may be as devastating, such as heavy rainfall, salt sprays, and storm surges in coastal plains.

The contribution of agriculture (sugar cane, other cash crops and subsistence farming) to the GDP in 1990 was about 15 per cent, of which 30 per cent was from sugar cane. Crop damage from the 1985 hurricanes and the 1993 hurricane Kina was reported to have cost F$8 million and F$14 million, respectively. This is on the order of 5 per cent of agricultural production.

9.5.5 Plantation forestry

Silviculture in Fiji dates back to the beginning of the 1970s with planting of *Pinus caribes* in the framework of the pine scheme. By 1993 the plantation estate covered about 37,000 ha. Forestry, and its associated wood products industry (logging and wood chips), is expected to become a major industrial sector in Fiji. Over the period 1981–91 the contribution of forestry (excluding timber activities) to the GDP almost doubled from 0.9 to 1.6 per cent, and is expected to expand further (see Table 9.3). Trials are under way to establish fast-growing pulpwood (eucalyptus and acacia) plantations. Export earnings of Fiji's Tropic Wood Industry Ltd totalled F$27.2 million in 1993 (Fiji Pine Ltd, 1994).

Forestry is a sector particularly vulnerable to severe storms, and this may be an important factor in the future. The annual reports of the Fiji Pine Commission[5] mention tropical cyclone losses in several years. These losses amounted to about 5 per cent of the stands; in 1990, for instance, the tropical cyclones Rae and Sina caused uprooting and propping of about 4,200 ha. These losses were not insured as insurance was considered to be too expensive (Fiji Pine Commission, 1987). Damage to plantations caused by Cyclone Kina in 1993 totalled F$943,906 and planting operations had to be suspended to enable rehabilitation operations (Fiji Pine Ltd, 1994). It was the main cost factor leading to Fiji Pine Ltd's loss of F$460,350 in 1993. These tropical cyclone losses surpass the losses from fire damage, and they are considered to constitute the limiting factor to forestry's potential economic performance (Waterloo, 1994).

9.5.6 Tourism

Tropical cyclones and associated surges may erode beaches and top trees along the coasts, thus destroying the assets which are important to market tourism (Dixon, 1991). It is obvious that any sea-level rise combined with tropical cyclone hazards puts beaches at greater risk in Fiji (Nunn *et al.*, 1994). Protection of beaches may remove their very attractiveness to tourists; in Western Samoa this has led to a decision not to protect after a risk analysis indicated a low return period for tropical cyclones in that area (Hecker, 1994). Little information is available on how the existing hazards of tropical cyclones constitute a psychological barrier to tourists considering visiting tropical islands. For example, with respect to the aftermath of the 1989 Hurricane Gilbert, which had significantly damaged tourist resorts on the peninsula of Yucatán (Mexico), it was noted that 90 days following the hurricane, hotels were open for the new season and tourists appeared unaware of the previous disaster (Dixon, 1991). In Fiji in February 1993, after Kina, a dip in visitor arrivals was noted (Rokovoda and Vrolijks, 1993). However, seasonal effects may fully explain this phenomenon, and from March 1993 onwards, visitor arrivals were higher than in the previous year. Kina did not pass over the tourist resort area. The most significant impact on tourism of tropical cyclones may well be the costs of insurance; many of the smaller resorts had insured their properties overseas, and insurance premiums were raised recently. Local insurers required upgrading of standards. Probably the main threat is tourist areas becoming known, rightly or wrongly, as being dangerous. The timing of the hurricane season relative to the tourism season is relevant to this issue. The World Tourism Organization (WTO) stresses in its manual for disaster management the importance of good communications (public relations), to avoid giving potential tourists erroneous impressions (WTO, 1995).

9.5.7 Tropical cyclones and property insurance

In Fiji it is possible to insure against the hurricane risk of damage to buildings. The total premium of non-life insurance amounted to about 1.5 per cent of Fiji's 1986 GDP (US$28 per capita). This compares with rates typically in the range of 5 to 6 per cent in industrialised countries (OECD–DAC, 1995) and less than 1 per cent in developing countries like Bolivia (0.45 per cent in 1983), India (0.49 per cent in 1986), Indonesia (0.49 per cent in 1986) and the Philippines (0.58 per cent in 1986) (Outreville, 1992).

In 1985 Fiji was hit by four tropical cyclones (Eric, Nigel, Gavin and Hina) within 2 months. The insured losses amounted to F$66 million, which is large compared to the F$10 million premium gathered for fire insurance (which includes tropical cyclone damage). The return period of the Eric type of tropical cyclone may be less than 10 years. The event was far from a possible maximum loss (PML) scenario (Tiedemann, 1991). As noted above, these events triggered

Table 9.6 Summary data of the impacts on Fiji of major tropical cyclones, such as in 1985 and 1993

Vulnerable sectors	Stock at risk (1989–93)	Impact
Public safety	Population 750,000	25 killed
Infrastructure	Roads, bridges, telecommunication equipment, public buildings (e.g. schools)	Broken lifelines Interruption of services
Subsistence farming	Value F$51.4 million (25% of total agriculture)	100,000 people requiring food aid for a few weeks
Government budget	Capital expenditure, F$79 million (1990)	F$5–10 million, especially for food relief
Tourism	Foreign exchange earnings, F$338 million	Nuisance to tourists
Agriculture	F$400 million at factor cost	F$10 million
Forestry (Pine woods)	34,000 ha, 1990 export lumber F$15 million (2.5% of total exports)	1990: 10% of stands written off; F$4.4 million damage

a restructuring of Fiji's insurance sector: eight out of twelve firms retreated from cyclone damage insurance. The remaining firms tightened the building and construction standards needed in order to obtain insurance. According to Roko-vada and Vrolijks (1993), this initiative paid off: it is still possible to buy tropical cyclone insurance cover for buildings in Fiji, unlike Western Samoa. Recently, the reinsurance cost for calamity insurance has increased by 100 to 300 per cent for insurance operators in Fiji (May, 1994). This is not the result of the Kina event, but rather the aftermath of the occurrence of major disasters all over the world, notably Hurricane Andrew and the Mississippi floods of 1993, which have forced adjustments by the international reinsurance markets.

The impacts of a severe tropical cyclone on Fiji are summarised in Table 9.6. It appears that most of the impacts are in rural areas, where subsistence farming is important. Moreover, the food relief which is necessary after a tropical cyclone disaster, and which usually leads to some abuse, tends to increase the dependency of the local people on aid agencies. In other words, the socio-economic vulnerability of the local communities increases.

9.6 Policy responses

This discussion is organised by the type of policy responses or coping mechanisms, with the nature of the actors (e.g., local communities or a national government) as a second factor. Measures, or policy options, in the context of the area of natural disasters are often grouped into warning, preparedness, mitigation and prevention. The list of options here is certainly not comprehensive, but it does give an overview of the types of activities needed in less developed countries which are vulnerable to tropical cyclones. It should also be noted that for many of these countries their current vulnerability will be sufficient reason to adopt such policies, irrespective of possible climate change.

9.6.1 *Warning*

Before the advent of modern technological warning systems, in some communities a few people experienced with weather, the sea and fauna were considered to be able to predict tropical cyclones to some extent. The ancient Fijians believed that when a breadfruit tree was heavily laden with young fruit, there would be a hurricane.

Since then, the availability of satellite imagery and modern meteorological technology has made it possible to predict more accurately the movement and intensity of tropical cyclones (Obasi, 1994). This, in combination with improved radio communications, is a real advance in comparison with 'traditional' warnings. The accuracy of forecasts is very important for their effectiveness, to the extent to which people believe in the warnings and react accordingly.

In the Fiji context, the improvement of the meteorological techniques in weather forecasting is predominantly an international development. The World Meteorological Organization (WMO), through its Tropical Cyclone Programme, seeks to mitigate the effects of these storms through participation of those nations at risk in developing effective tropical cyclone warning systems (WMO, 1977, 1992, 1994a, b). The programme notes that effective communication of forecasts to the public should be attuned to local situations and local cultures.

The above relates to weather forecasting; at another level, warning relates to climate forecasting. The context of this activity is also international; Fiji can support the development of the research through taking part in intergovernmental organisations (e.g., the Intergovernmental Panel on Climate Change).

9.6.2 *Preparedness*

Loss of harvest and resulting food shortage is traditionally managed by food conservation and preservation techniques, diversification of crops and the sheltering of gardens from exposure to strong winds. In case of food shortages, food from local 'wild' plants and trees provides relief. Examples are the berries of the *Damadamu* tree (Nunn *et al.*, 1994), wild yams (*ibid.*) and the roots of *Cordyline fruticosa* (Overton, 1988).

The process of shifting from subsistence farming to growing cash crops, and the inclusion of the rural economy into the cash economy tend to decrease the importance of the indigenous methods for preparedness as the availability of cash makes it possible to buy manufactured foodstuffs which can be kept. Another development having the same result is that food relief programmes in the event of a cyclone tend to decrease the incentive to preserve emergency foodstuffs (Chung, 1988). The evolution of dependency is described in detail by Chung (1988). He notes that in Fiji plenty of food aid is supplied and, therefore, dependency on the government is taken for granted. Superfluous food relief is the result of inappropriate operational practices, the politicisation of relief activities and the general propensity to overreact when a cyclone occurs.

Preparedness relates to the capacity of enterprises to be able to react appropriately after the cyclone (e.g., in the case of forestry, the capacity to salvage damaged forests, at least in part). The actual substance of such contingency plans is dependent on the outcome of an economic risk analysis, evaluating costs and benefits of the possible measures to reduce post-disaster effects.

In Fiji, activities and decision-making in case of disaster are laid out in the National Disaster Management Plan. Emergency operations centres are operational at national, divisional and district level during an emergency. Disaster management councils at different government levels manage the operations.

Currently the UN carries out the activities making up the International Decade for Natural Disaster Reduction (IDNDR), which includes many programmes directed at increasing the efficacy of disaster management in all countries and among international organisations. Improved cyclone warnings are a major emphasis.

9.6.3 Mitigation

Risk-sharing and mutual assistance is one of the functions of the traditional social system between families as well as between villages or islands (Torry, 1978; Lessa, 1964). However, dignity and self-respect, which are important to South Pacific islanders, have limited the dependence on charity (Nunn et al., 1994)

Cyclone insurance is marketed in Fiji, although not in all sectors. In plantation forestry, for instance, it was considered too expensive (Fiji Pine Commission, 1991). Crop insurance is not available in the small island countries (Tiedemann, 1991; Amerasinghe, 1984). However, buildings are insured in Fiji. This market was restructured after 1985 (see above).

Actions to reassure the travelling public are often more significant than physical repairs. A single negative rumour can destroy the marketability of an area. Thus preparations for dealing with the media and tourists scheduled to arrive after the disaster event are critical.

Several international schemes have provided assistance to Fiji during times of disaster. Over the years, aid has been obtained from the United Nations Department of Humanitarian Affairs (UNDHA) (formerly the UN Disaster Relief Organization, UNDRO), from other UN organisations, from the European Union through the Lomé Convention, and through regional agreements, such as the South Pacific Bureau for Economic Co-operation. Individual countries, including neighbouring countries and in particular Australia and New Zealand, have also provided aid and grants. This aid is often partly earmarked, e.g. to reconstruct a bridge.

9.6.4 Prevention and protection

Traditionally, houses were built from local products and could be repaired easily, in contrast to modern houses built from concrete and corrugated iron sheet. In the case of the island of Nacula (one of the Yasawa islands) there is a tendency to return to building more traditional houses (from wood and grass) as these are cooler and easier to repair after being damaged by tropical cyclones (Nunn et al., 1994: 49). Another approach to prevent damage to houses is the observing of building codes. Often, simple safety techniques can be used. Obstacles to their use are that the villagers are not aware of these techniques, or lack either the tools to construct these devices from local building material or suitable local construction material. Appropriate aid from outside the villages is needed. In Fiji, a training programme for village carpenters has been carried out recently. The programme involves the construction of cyclone-resistant structures at training venues involving participants themselves, with the inspectors mainly engaged in the supervision and training aspects (Rokovada and Vrolijks, 1993).

In forestry, routes to prevention include revised planting practices and the breeding of more wind-resistant varieties (Fiji Pine Commission, 1990). The vulnerability of (market) agriculture can be decreased by developing and using varieties of crops which are more wind resistant (Tiedemann, 1991). Other strategies to increase resistance include (1) development of varieties of crops which are more resilient to salt spray, (2) increased growing of crops which are harvested in the cyclone-free season, and (3) improved use of windbreaks around plots of crops susceptible to wind damage. An obvious prevention measure in the tourism sector is to avoid the development of tourist resorts at low elevation and near sand beaches which are vulnerable to high winds and storm surges.

The government plays a major role in prevention. Prevention investments may be targeted merely at reducing physical vulnerability; the housing improvement programme is mentioned above. However, the government may also try to link vulnerability reduction with programmes for economic development. Many (Anderson, 1990; Kreimer and Munasinghe, 1991; Tisdell, 1993; Albala-Bertrand, 1993; May, 1994) advocate a greater role for risk analysis in the design of development programmes and in taking into account effects of development programmes on vulnerability in decision-making.

Under the Lomé Convention the EU specifically designates funds for developing rural areas in island APC states, including Fiji. Project appraisal should take account of increasing risks of climate change.

In the international negotiations in the Framework Convention on Climate Change (FCCC), the island states are collaborating in the Alliance of Small

Island States (AOSIS) in seeking to limit climate change and seeking assistance from the industrialized countries to adapt to the change that does occur.

In the southwest Pacific, a tropical cyclone disaster is a normal phenomenon. In the region's countries, individuals, local communities, economic sectors and governments have developed ways to cope with this risk. If the risk increases, it can be addressed by reinforcement of existing policies. New policies are more likely to be needed for the new risk imposed by sea-level rise (Nunn *et al.*, 1994).

9.7 Conclusions

Small island states are particularly vulnerable to climate change, as it may manifest itself as sea-level rise, increased occurrences of tropical cyclones and a combination of the two. Many of these countries are poor and, therefore, vulnerable.

In small island states in the southwest Pacific, major disasters from tropical cyclones occur with return periods typically of about 10 years. On Fiji, the largest island state, a typical major tropical cyclone causes the death of 25 people (out of a population of 750,000) and destroys 25 per cent of the standing crops, while relief and reconstruction consumes up to 10 per cent of the government's capital expenditures. In Fiji, tropical cyclones may constitute a limiting factor to the development of plantation forestry.

Statistical analysis of the documented history of the occurrence of tropical cyclones in the South Pacific shows a slight (0.07 per cent) increase in incidence. The outcomes of climate modelling are not conclusive about the future incidence of tropical cyclones, their intensities, the length of the hurricane season, nor the geographical area where tropical cyclones are active.

These countries already cope with tropical cyclone risks by pursuing various policies; as the nature of the risk will not change greatly as a result of climate change (in the near term), we cannot recommend new types of policy responses. However, new risks will evolve from the combination of sea-level rise and the existing incidence of tropical cyclones.

Acknowledgements

Critical comments by Tom Downing and Angela Liberatore on an earlier version of this chapter are gratefully acknowledged (see Olsthoorn *et al.*, 1996).

Notes

1 For example, in November 1970 and April 1991, tropical cyclones caused storm surges of 6–9 m in coastal areas of Bangladesh, resulting in 225,000 and 140,000 casualties, respectively.

2 In rankings of countries according to their proneness to disaster, small island countries figure prominently (see DHA-UNDRO information as quoted in OECD–DAC, 1995).
3 Chung (1988) also points out the unwanted side-effects of relief, in particular the evolution of dependency to relief, as aid expectations in the event of a disaster tend to reduce self-reliance.
4 In 1972 the tropical Cyclone Bebe destroyed banana plantations which had been planted to develop exports of this agricultrual product (de Haan, 1978). Apparently the tropical cyclone hazard proved to be a limiting factor for this type of agriculture.
5 Succeeded by Fiji Pine Limited.

References

Albala-Bertrand, J.M. (1993) *Political Economy of Large Natural Disasters*, Oxford: Clarendon Press.

Amerasinghe, A.R.B. (1984) *Crop Risk Management in the South Pacific*, Wellington: Bowring, Burgess, Marsh & McLellan.

Anderson, M.B. (1990) *Analyzing the Costs and Benefits of Natural Disaster Responses in the Context of Development*, Environment Working Paper no. 29, Washington, DC: World Bank.

Anthony, J.M. (1990) 'Conflict over natural resources in the Pacific', in L.T. Ghee and M.J. Valencia, eds, *Conflict over Natural Resources in S-E Asia and the Pacific*, Singapore: United Nations University Press, 182–245.

Butler, W. (1992) 'Tourism development in small islands: Past influences and future directions', in D.G. Lockhart, D. Drakakis-Smith and J. Schembri, eds, *The Development Process in Small Island States*, London: Routledge, 71–91.

Chandra, R. (1993) 'Breaking out of import-substitution industrialization: The case of Fiji', in D.G. Lockhart, D. Drakakis-Smith and J. Schembri, eds, *The Development Process in Small Island States*, London: Routledge, 205–227.

Chung, J. (1988) 'Tropical cyclones and disaster relief', in J. Overton, ed., *Rural Fiji*, Suva: Institute of Pacific Studies, University of the South Pacific, 85–96.

Connell, J. (1992) 'Island microstates: development, autonomy and the ties that bind', in D.G. Lockhart, D. Drakakis-Smith, and J. Schembri, eds, *The Development Process in Small Island States*, London: Routledge, 117–150.

de Haan, P.S. (1978) *Fiji*, Amsterdam: Koninklijk Instituut voor de Tropen (in Dutch).

Dixon, C. (1991) Yucatán after the wind: Human and environmental impact of Hurricane Gilbert in the central and eastern Yucatán peninsula. *GeoJournal* 23, 4: 337–345.

Dobson, T. (1853) *Australasian Cyclonology, or The Law of Storms in the South Pacific Ocean, and on the Coasts of Australia, Tasmania and New Zealand, etc.*, Hobart Town: [n.s.].

Fiji Pine Commission (1987) *Annual Report 1986*, Lautoka.

Fiji Pine Commission (1990) *Annual Report 1989*, Lautoka.

Fiji Pine Commission (1991) *Annual Report 1990*, Lautoka.

Fiji Pine Ltd (1994) *Annual Report 1993*, Lautoka.

Gabites, J.F. (1956) 'A survey of tropical cyclones in the South Pacific', in *Proceedings of the Tropical Cyclone Symposium*, Brisbane, December.

Gabites, J.F. (1963) '(i) Historical survey of tropical cyclones; (ii) The origin of tropical cyclones; (iii) Development and decay of tropical cyclones; (iv) The movement of tropical cyclones,' in *Proceedings of the Inter-regional Seminar on Tropical Cyclones*, Tokyo, 18–31 January 1962.

Giovannelli, J., and Robert, J. (1964) Quelques aspects des dépressions et cyclones tropicaux dans le Pacifique Sud-ouest. *Monographies de la Météorologie Nationale* 82.

Hecker, G. (1994) A review of the disaster related activities of the Asian Development Bank: An economic perspective. Paper read at World Conference on Natural Disaster Reduction, 23–27 May, at Yokohama.

Hill, A.E.H., and Tabor, S.R. (1993) Liberalization and diversification in a small island economy: Fiji since the 1987 coups, *World Development* 21, 5: 749–769.

Hutchings, J.W. (1953) *Tropical Cyclones in the Southwest Pacific*, no. 37, Wellington: NZ Meteorological Office.

Kattenberg, A., Giorgi, F., Grassi, H., Meehl, G.A., Mitchell, J.F.B., Stouffer, R.J., Tokioka, T., Weaver, A.J., and Wigley, T.M.L. (1996) 'Climate models: Projection of future climate', in J.T. Houghton, L.G. Meira Filho, B.A. Callander, N. Harris, A. Kattenberg and K. Maskell, eds, *Climate Change 1995: Impacts, Adaptation and Mitigation of Climate Change*, Cambridge: Cambridge University Press, 285–358.

Kerr, I.S. (1976) *Tropical Storms and Hurricanes in the Southwest Pacific, November 1939 to April 1969*, Misc. Publ. 148, Wellington: NZ Meteorological Service.

Knipping, E. (1893) *Die tropischen Orkane der Sudsee, Archives, Deutschen Seewarte*, Hamburg.

Kreimer, A., and Munasinghe, M. (1991) *Managing Natural Disasters and the Environment*, Washington, DC: World Bank.

Lessa, W. (1964) The social effects of typhoon Ophelia on Ulithi. *Micronesia* 1: 1–47.

Lighthill, J., Holland, G.J., Gray, W.M., Landsea, C., Emanuel, K., Craig, G., Evans, J., Kurihara, Y., and Guard, C.P. (1994) 'Global climate change and tropical cyclones', *Bulletin of the American Meteorological Society* 75: 2147–2157.

Maunder, W.J. (1995) *An Historical Overview Regarding the Intensity, Tracks and Frequency of Tropical Cyclones in the South Pacific during the last 100 years and an Analysis of any Changes in these Factors*, WMO/TD No. 692, Geneva: World Meteorological Organization.

May, P. (1994) *Natural Disaster Reduction in Pacific Island Countries: Report to the World Conference on Natural Disaster Reduction, 1994*, Emergency Management Australia.

Newman, B.W. and Bath, A.T. (1959) Occurrence of tropical depressions and cyclones in the northeast Australian region during the season 1957–58. *Australian Meteorological Magazine* 24: 35–64.

Nunn, P.D., Ravuvu, A.D., Balogh, E., Mimura, N., and Yamada, K. (1994) *Assessment of Coastal Vulnerability and Resilience to Sea-level Rise and Climate Change. Case Study: Yasawa Islands, Fiji*, Apia: South Pacific Regional Environment Programme.

Obasi, G.O.P. (1994) 'Natural disasters and sustainable development of small developing islands. Paper read at World Conference on Natural Disaster Reduction, May, Yokohama.

OECD–DAC (Organization for Economic Co-operation and Development – Development Assistance Committee) (1995) *Guidelines on Aid and Environment No. 7: Guidelines for Aid Agencies on Disaster Mitigation*, Paris: OECD.

Olsthoorn, A.A., Maunder, W.J., and Tol, R.S.J. (1996) 'Tropical cyclones in the southwest Pacific: Impacts on Pacific Island countries with particular reference to Fiji', in T.E. Downing, A.A. Olsthoorn and R.S.J. Tol, eds, *Climate Change and Extreme*

Events: Altered Risk, Socio-economic Impacts and Policy Responses, Amsterdam: Institute for Environmental Studies, Vrije Universiteit, 251–272.

Outreville, F.J. (1995) The relationship between insurance, financial developments and market structure in developing countries: An international cross-section study. *UNCTAD Review* 3: 53–61.

Overton, J. (1988) *Rural Fiji*, Suva: Institute of Pacific Studies, University of the South Pacific.

Overton, J. (1993) The limits of accumulation: Changing land tenure in Fiji. *Journal of Peasant Studies* 19, 2: 326–342.

Porter, J. (1994) *The Vulnerability of Fiji to Current Climatic Variability and Future Climate Change*, Sydney: Climatic Impacts Centre, Macquarie University.

Prasad, R. (1993) *Tropical Cyclone Kina: 26 December 1992 to 5 January 1993*, Tropical Cyclone Report 92/1, Suva: Fiji Meteorological Service.

Revell, C.G. (1981) *Tropical Cyclones in the Southwest Pacific: November 1969 to April 1979*, New Zealand Meteorological Service Miscellaneous Publication 170, Wellington: NZ Meteorological Service (Ministry of Transport).

Rokovoda, J., and Vrolijks, L. (1993) Case study Fiji: Disaster and development linkages. Paper read at South Pacific Workshop UN Disaster Management Training Programme, Apia, 29 November–4 December 1993.

Streeten, P. (1993) The special problems of small countries. *World Development* 21, 2: 197–202.

Thompson, C., Ready, S., and Zheng, X. (1992) *Tropical Cyclones in the Southwest Pacific: November 1979 to May 1989*, New Zealand Meteorological Service Miscellaneous Publication 170, Wellington: NZ Meteorological Service.

Tiedemann, H. (1991) *Mission Report: Disaster Management and Mitigation Possibilities and Insurance Aspects*, Suva: UNDRO South Pacific Programme.

Tisdell, C. (1993) Project appraisal, the environment and sustainability for small islands. *World Development* 21, 2: 213–219.

Tol, R.S.J., Dorland, C., and Olsthoorn, A.A. (1995) *Statistical Evidence on Genuine and Observational Trend in Tropical Cyclone Frequency in the Southwest Pacific*. Amsterdam: Institute for Environmental Studies.

Torry, W.I. (1978) Natural disasters, social structure and change in traditional societies. *Journal of Asian and African Studies* 13, 3–4: 167–183.

Treadgold, M. (1992) *The Economy of Fiji: Performance, Management and Prospect*, Canberra: Australian Government Publishing Service.

Visher, S.S. (1925) *Tropical Cyclones of the Pacific*, Bulletin 20, Honolulu: Bernice P. Bishop Museum.

Visher, S.S., and Hodge, D. (1925) *Australian Hurricanes and Related Storms, with an Appendix on Hurricanes in the South Pacific*, Bulletin no. 16, Melbourne: Commonwealth Bureau of Meteorology.

Waterloo, M.J. (1994) *Water and Nutrient Dynamics of Pinus Plantation Forests on Former Grassland Soils in Southwest Viti Levu, Fiji*, Amsterdam: Vrije Universiteit.

World Meteorological Organization (WMO) (1977) *The Use of Satellite Imagery in Tropical Cyclone Analysis*, Technical Note no. 153, Geneva: WMO.

World Meteorological Organization (WMO) (1992) *Human Response to Tropical Cyclone Aspects of Tropical Cyclones*, Geneva: WMO Tropical Cyclone Programme.

World Meteorological Organization (WMO) (1994a) *The Management of Meteorological*

Hazards to Reduce Disaster Risk, prepared by P. Burton, Y. Aysan and I. Davis, Geneva: WMO.

World Meteorological Organization (WMO) (1994b) *The Human Response to Tropical Cyclone Warnings and Their Content* Technical Document TCP-34, prepared by D. Wernly, Geneva: WMO Tropical Cyclone Programme.

World Trade Organization (WMO) (1995) *Handbook on Natural Disaster Reduction in Tourist Areas*, prepared by D.W. Schnares *et al.*, Madrid: World Tourism Organization.

10

IMPACTS OF WINDSTORMS IN THE NETHERLANDS: PRESENT RISK AND PROSPECTS FOR CLIMATE CHANGE

Cornelis Dorland, Richard S.J. Tol, Alexander A. Olsthoorn and Jean P. Palutikof

10.1 Introduction

In the late 1980s and early 1990s Western Europe was hit by a series of windstorms which had an unprecedented impact. Figures published by the reinsurance company Munich Re show a dramatic increase in the costs of severe storm events from 1960 onwards (Table 10.1).

The insurance industry was taken by surprise; windstorm losses of more than just a few billion US dollars had been thought unlikely (Berz and Conrad, 1993). The winter storm in 1987 and the series of winter storms in 1990, with estimated economic losses of US$3.7 and US$15 billion (valued at 1992 prices) respectively, proved different. The January 1990 winter storm alone caused economic losses of around US$6.8 billion, of which US$5.2 billion were insured (valued at 1992 prices) (Munich Re, 1993; Swiss Re, 1994; Dlugolecki, 1992). In these years, similarly dramatic storm and tropical cyclone events occurred in the USA and the question was raised whether these events were early signals of climate change.

Models used for estimating storm losses are developed and used by reinsurance companies, such as Swiss Re and Munich Re, insurance companies, such as Nationale Nederlanden (the Centre for Insurance Statistics in the Netherlands), and consultants, such as EQE International in the UK and Applied Insurance Research in the USA. The Munich Re model showed that if the January 1990 storm track had been slightly different, damages could easily have been doubled or trebled (Berz and Conrad, 1993).

This study was prompted by the assumption that storminess may indeed increase because of the enhanced greenhouse effect, and it analyses socio-economic implications of this assumption. The chapter assesses the current vulnerability of the Netherlands to severe storms and presents scenarios for future storm damage in

Table 10.1 Number of and losses due to major world-wide windstorm events between 1960 and 1992

	1960–9	1970–9	1980–9	1983–92	Rate of increase	
					1980s/1960s	1983–92/1960s
Number of windstorms	8.0	14.0	29.0	31.0	3.6	3.9
Economic loss (in US$ billion)	22.6	33.6	38.0	88.1	1.7	3.9
Insured loss (in US$ billion)	5.3	8.3	18.9	52.1	3.6	9.8

Source: Berz and Conrad (1993)

Notes: Major events cost over US$500 million; damages are valued at 1992 prices and are corrected for inflation only

northwestern Europe, with an emphasis on the Netherlands. The damage scenarios are based on an analysis of the impacts of recent storms and scenarios for future storm risks. To better understand the physical, social and economic impacts of a severe storm in northwest Europe, a case study on impacts of the Daria event in the Netherlands was carried out (Section 10.2).

Unfortunately, storm loss models built by insurance companies are not publicly available. Therefore, two alternative models which relate storm characteristics and storm losses in the Netherlands have been constructed (Dorland and Tol, 1995). The first is an elaborate, geographically explicit model built from empirical information about the patterns of storm intensity and damage from a series of storms (Section 10.3). The second model investigates the annual average of aggregate damage, focusing on storm frequency and intensity (Section 10.4). Using these models, the sensitivity of storm damage to variations in storm intensity is analysed by means of specific storm scenarios. Next, policy options to influence the vulnerability to storminess are addressed (Section 10.5). Finally, the conclusions are given in Section 10.6.

10.2 Daria over the Netherlands

This section describes the physical characteristics of the 1990 winter storms in Europe. Furthermore, the results of a case study on the impacts of the Daria event in the Netherlands are given (Dorland *et al.*, 1994).

10.2.1 Physical characteristics of the 1990 winter storms over Europe

The risks from storms were revealed once again in the first three months of 1990 when northwest Europe was hit by a series of eight severe storms. Europe had experienced stormy weather in the last two months of 1989. Deep lows developed over the Atlantic, and the Atlantic jet stream went further south than is normal for this time of year. At the time, the pressures measured in the mid-Icelandic low and in the mid-Atlantic were the lowest since recording began in 1873. Several mercury barometers spilled mercury because the pressures were outside the range of their recording capabilities. These meteorological circumstances were followed by a very stormy period which lasted for 5 weeks and comprised fifteen different storm events (McCallum and Norris, 1990; McCallum, 1990; Hammond, 1990; Munich Re, 1993). Eight of the storm tracks passed over Europe (see Table 10.2 and Figure 10.1). Note that the highest wind speeds are always measured south of the centre of the depression area.

The first deep depression (Daria) was halfway over the Atlantic Ocean in the evening of 24 January and came in over Ireland and northern England at noon on 25 January. It was travelling at high speed and the pressure in the centre of the low at that time was 968 mbar. A cold front reached the Dutch coast between 2 and 3 p.m., after the speed of movement of the depression had reduced. At the same time it started to rain. A severe storm came in over the Netherlands at 3 p.m. The highest wind speeds were measured in the central and southwestern part of the country. To the south of the depression centre (949 mbar at that time) a small secondary low-pressure area developed, hitting the northern part of the Netherlands. At midnight the storm had passed the Netherlands and moved over northern Germany and the Skagerrak to end in the south of Finland on 26 January at 12 p.m. On 25 and 26 January record wind gusts, causing major damage, were measured in the whole of northwest Europe. On 27 January and 28 January there was widespread flooding in Great Britain, caused by heavy rainfall during the 25 January storm (McCallum, 1990; McCallum and Norris, 1990; Wieringa, 1990; RWS, 1990).

The depressions in the next few weeks followed naturally from the first because

Table 10.2 Storms in Europe between 25 January and 1 March 1990

No.	Period (d/m)	Name	Region of largest impact
1	25–26/1	Daria	Western and northern-Europe (especially the UK and Benelux)
2	3–4/2	Herta	Western and central Europe (especially the Paris region, Luxembourg and central Germany)
3	7–8/2	Judith	Western and central Europe
4	11–12/2	Nana	Western Europe
5	13–14/2	Otille	Central Europe
6	14–15/2	Polly	Central Europe
7	25–27/2	Vivian	Western, central and northern Europe (especially central Europe)
8	28/2–1/3	Wiebke	Western, central and northern Europe (especially central Europe)

Source: Munich Re (1993); OECD (1992)

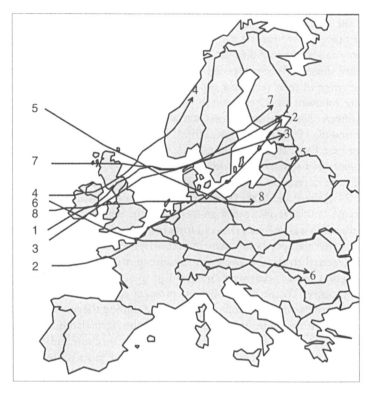

Figure 10.1 Tracks of the eight storms between 25 January and 2 March 1990 over
Europe. See also Table 10.2.
Source: Munich Re (1993)

the temperature gradients were amplified. The average 24-hour wind speeds of
these storm events were higher than the ones observed with Daria but the peak
gusts had lower intensities. Of the seven storm surges following Daria, the last
two (associated with Vivian and Wiebke) caused extensive damage along European coasts. Just before Vivian hit, a deep low formed over the Atlantic Ocean
on 25 February. On 26 February the depression (949 mbar) moved over
Ireland and northern England to the southwest coast of Norway through to
Finland. A severe storm was active over the North Sea in the morning. In the cold
front of this storm a new low had formed to the southeast of Iceland. The storm
(Wiebke) moved over Ireland with a pressure of 984 mbar on 28 February. From
Ireland it moved over central England, the North Sea and the Netherlands into
Germany, where it arrived on 1 March. Because a high pressure system formed
over the Azores the storm period ceased on 2 March in the afternoon.

10.2.2 *History of storms in the Netherlands*

The Royal Netherlands Meteorological Office (KNMI) classifies storms as heavy if the highest hour-average intensity is 10 on the Beaufort scale (Bft) or higher (Table 10.3). Between 1910 and 1990 the Netherlands was hit by twenty-nine severe storms (Table 10.4) (van Mourik, 1990). Eleven of these storms had an intensity of 11 Bft or higher and one, in September 1944, even had an intensity of 12 Bf. The peak gust intensities measured during the storm event of 27 August 1921 are the highest measured in this century. The peak gust intensities during the 25/26 January 1990 storm (Daria) are the second highest. The series of storms between January and March 1990 was very exceptional for the Netherlands. Between 1910 and 1989, only in 1928 was a storm period consisting of more than two separate storms succeeding each other in a short time span recorded in the Netherlands. Another characteristic of the 25/26 January storm event was the unusually extensive area in Europe affected. KNMI records show that the average return period of a storm like Daria in the Netherlands, depending on the location, is once in every 7 to 40 years, in the absence of any influence of climate change.

For the Netherlands, Daria was the most important storm event in early 1990. The storms Herta, Judith, Vivian and Wiebke also caused damage but the overall losses of these events were about ten times lower than the Daria losses. After Daria, the Dutch media paid relatively little attention to later events. Also, almost all insurance industry reports and scientific publications focus on Daria.

Table 10.3 The Beaufort scale

Bft	Description	Mean wind velocity at 10 m elevation				Wind pressure (kg m^{-2})
		m s^{-1}	kph	mph	knots	
0	Calm	0–0.2	0–1	0–1	0–1	0
1	Light air	0.3–1.5	1–5	1–3	1–3	0–0.1
2	Light breeze	1.6–3.3	6–11	4–7	4–6	0.2–0.6
3	Gentle breeze	3.4–5.4	12–19	8–12	7–10	0.7–1.8
4	Moderate breeze	5.5–7.9	20–28	13–18	11–15	1.9–3.9
5	Fresh breeze	8.0–10.7	29–38	19–24	16–21	4.0–7.2
6	Strong breeze	10.8–13.8	39–49	25–31	22–27	7.3–11.9
7	Near gale	13.9–17.1	50–61	32–38	28–33	12.0–18.3
8	Gale	17.2–20.7	62–74	39–46	34–40	18.4–26.8
9	Strong gale	20.8–24.4	75–88	47–54	41–47	26.9–37.3
10	Storm	24.5–28.4	89–102	55–63	48–55	37.4–50.5
11	Violent storm	28.5–32.6	103–117	64–72	56–63	50.6–66.5
12	Hurricane	32.7 +	118 +	73 +	64 +	66.6 +

Source: Munich Re (1993)

Table 10.4 Heavy winter storms over land in the Netherlands from 1910 to 1990

No	Year	Day/month	$\hat{V}_{max}$ Bf	Intensity $m\ s^{-1}$	$\bar{V}_{max}$ $m\ s^{-1}$
1	1911	30/9–1/10	11	30	38
2	1912	27/8	10	27	41
3	1914	28/12–29/12	11	32	42
4	1916	13/1	10	27	42
5	1921	6/11	11	32	45
6	1925	9/2–10/2	10	27	39
7	1926	9/10–10/10	10	26	35
8	1928	16/11–17/11	11	29	38
9	1928	23/11	10	25	38
10	1928	25/11	10	28	37
11	1938	4/10	10	26	38
12	1940	13/11–14/11	10	26	38
13	1943	7/4–8/4	11	31	35
14	1944	7/9	12	34	—
15	1949	1/3–2/3	11	29	39
16	1953	31/1–1/2	10	27	40
17	1960	20/1	10	26	41
18	1967	17/10	10	27	40
19	1972	13/11	11	29	42
20	1973	2/4	11	30	43
21	1976	2/1–3/1	11	30	41
22	1983	1/2	10	27	38
23	1983	27/11	10	27	40
24	1984	14/1	10	27	37
25	1987	16/10	10	27	41
26	1990	25/1–26/1	11	29	44
27	1990	3/2–4/2	10	24*	34*
28	1990	25/2–27/2	10	26	41
29	1990	28/2–1/3	10	24	35*

Source: Wieringa (1990)

Notes:
$\hat{V}_{max}$ is the highest measured hourly mean wind speed
$\bar{V}_{max}$ is the highest measured wind gust
* Estimated values from Swiss Re (1994)

10.2.3 Physical characteristics of Daria in the Netherlands

Rain commenced from the south/southeast on 25 January at 6 a.m. The air pressure dropped and temperature rose from 5.6 to 15.9°C in a few hours. A cold front came in between 2 and 3 p.m. and the storm intensity increased while the wind changed direction and at 7 p.m. came in from the southwest/west. Wind speeds of 9–10 Bf with gusts up to 11 Bf were measured. Over the whole of the western part of the Netherlands 10 Bf was measured while the highest wind speeds were measured in the provinces of Zealand, South Holland and Utrecht. A second depression area formed in the northern part of the Netherlands at 6 p.m. that day. This depression caused high wind speeds in the Leeuwarden area (the north), the middle, western and mid-southern part of the country. However, the highest wind gusts of that day, up to 44 m s^{-1} at Schiphol and 38 m s^{-1} in Huizen, were measured in the middle of the country. These peak gusts were the second highest measured in the Netherlands in this century (van Mourik, 1990). The temperature started to fall again after 7 p.m. and the air pressure began to rise. After 8 p.m. the storm subsided and it started to rain periodically. However, peak wind gusts up to 28 m s^{-1} were measured until midnight. The highest hourly mean wind speeds in Schiphol and Leeuwarden are given in Figure 10.2. The highest hourly mean potential wind speeds ($\bar{V}_p$)[1] in the Netherlands on 25 January are given in Figure 10.3 (Haverkamp and Schaap, 1990; Wieringa, 1990; RWS, 1990).

10.2.4 The socio-economic impacts of Daria in the Netherlands

A storm of force 9 Bf was forecast to hit the Netherlands on 25 January by the meteorologists of the Royal Netherlands Meteorological Office (KNMI) on the evening of Wednesday 24 January. The national news, however, did not warn the public of the coming storm event before the early morning of 25 January. The first warning was formulated as 'a period with severe storm' and was given with the normal weather broadcast at the end of the news bulletin (COT, 1990; Volkskrant, 1990a). This resulted in the public not being aware of the approaching storm. Also, many government institutions and economic sectors vulnerable to the storm were not aware of the threatening situation as the storm turned out to be much more severe than forecast. The lack of preparedness, however, was not only due to the poor forecasts and warnings. In the Netherlands, government bodies lack disaster plans and have little interest in storm impacts (COT, 1990). These circumstances were ideal for the Daria storm event to cause major impacts and paralyse public life for several hours. The impacts of the storm on Dutch society were extensive: twenty-one deaths, twenty-six severely injured people and very severe damages (DRC, 1990; Munich Re, 1993; Schraft et al., 1993).

The storm was at its peak as many people were on their way home from work. Trees fell on cars and houses. Roof tiles blew from houses, injuring people and damaging property. The railway system was blocked completely, owing to the

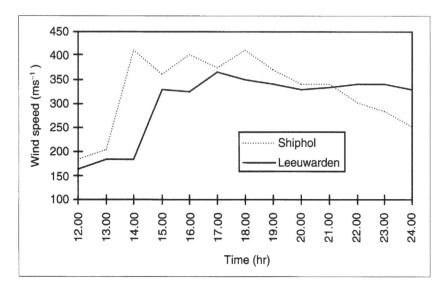

Figure 10.2 $\hat{V}_{max}$ (ms⁻¹) in Leeuwarden and Schiphol on 25 January 1990
Source: Wieringa (1990)

many broken electric wires and fallen trees. Travellers were not properly informed of the ongoing situation (NS, 1990; W. Assink (Nederlandse Spoorwegen (Netherlands Railways), Utrecht), personal communication, 1994; K. Wierda (Travellers Transport Organisation, ROVER, Amsterdam), personal communication, 1994; M.J.C.M. Smeets (Logistics Department, Nederlandse Spoorwegen, Utrecht), personal communication, 1994). Because Netherlands Railways was not prepared for relief, the Academic Medical Centre in Amsterdam, the Jaarbeurshallen (a large exposition and conference centre) in Utrecht, the Salvation Army and many friendly people helped stranded travellers by offering them shelter and food (COT, 1990; G&E, 1990a, b; Volkskrant, 1990b). More than 4,500 train travellers could not get home that evening. Also, road traffic was brought to a virtual standstill at the height of the storm, although less than 6 per cent of private car drivers initially reacted to calls from the Algemene Verkeersdienst (Public Traffic Service) to stop all traffic (Volkskrant, 1990b, c). The operational services (police and fire departments) were very busy in assisting travellers and undertaking cleaning-up operations. Of the media, the regional and local broadcasting stations especially appeared to play an important role in informing people on the condition of roads, available public transport and about people in distress. They independently assumed the task of informing people, without any governmental request (COT, 1990). In contrast to Netherlands Railways, the Dutch air transportation company reacted well and, although some people had to cope with long delays, chaos at air terminals was prevented

Figure 10.3 $\bar{V}_p$ (ms^{-1}) in the Netherlands on 25 January 1990
Source: Wieringa (1990)

(COT, 1990). Industrial capital goods and glasshouses were also damaged by the storm. Impacts on the environment were extensive. Forests were severely damaged and, because of the lack of disaster plans, there was chaos not only in the forests but also on the domestic timber market (Bussink, 1990; Ekkelboom, 1990; Volkskrant, 1990d). The dunes, a natural part of the Dutch coastal defence system, also suffered severely in some places (RWS, 1990). The shipping sector reacted adequately to the weather forecasts and therefore the impacts in this sector were very small (COT, 1990). From 5 p.m. onwards, the Home Office kept five men ready for emergency calls from other government offices. They did not have much work to do, as almost all other ministries were left unstaffed after office hours (COT, 1990).

Total damages are estimated at about Dfl. 2.6 billion (1990 values[2]), of which

Dfl. 1.5 billion (damage to private houses, business, industrial and commercial buildings and cars) was covered by the insurance industry (Munich Re, 1993; Schraft et al., 1993; Dorland et al., 1994). Other damage items were the costs of repair of the dunes (Dfl. 15 million) (RWS, 1990; Volkskrant, 1990e), damage to forests (Dfl. 13–15 million) (Ekkelboom, 1990) and damage to aeroplanes and buildings at Schiphol Airport (Dfl. 5 million) (G&E, 1990c). The economic losses due to transport delays are estimated at Dfl. 10 million (Dorland et al., 1994). Clearly, these damages or economic losses were minor with respect to the Dfl. 1.03 billion damage to private houses and Dfl. 0.38 billion damage to business, industrial and commercial buildings (CVS, 1994). Of the total losses in the Netherlands, it is estimated that 38 per cent were not insured (Dorland et al., 1994). In these figures income losses and other losses due to business interruptions are not included because data are not available. These losses can, however, be quite substantial in severe storms. Also, losses to government buildings and most infrastructure could not be included because data are not available. These damages are normally not insured.

Societal responses

A Crisis Onderzoeksteam (COT; Crisis Research Team) was formed after the event to evaluate the responses of the media, the transportation companies and the government. From their results and our own research it appears that in only a few sectors was any analysis, let alone action, undertaken. There are, however, quite a number of lessons that can be learned from the Daria storm event.

First, the role of the press, the radio and TV in disaster situations should be nationally planned, regulated and co-ordinated so as not to confuse people. Clear and uniform information should be widely broadcast. Warning of a storm in the weather forecast at the end of the news bulletin was shown not to be convincing. Therefore, an extreme-event forecast should be a separate item at the beginning of the news. A national disaster plan should be available for the press, radio and TV. People should also be made aware that warnings and advice must be taken seriously; it appears that Dutch people are not disciplined in following instructions given by the media during a disaster (COT, 1990).

Second, it turned out that municipal, provincial and national government offices were generally not interested and did not react to events. Emergency plans for these types of large-scale disasters were lacking (COT, 1990). However, according to the COT, the interest of government organisations in managing storm impacts has been increasing since 1990. The COT proposed that lessons could be learned from Belgium and the UK. In the UK government officials and the Queen spoke to the people through national television (COT, 1990).

Third, the Dutch railway company, which was criticised for providing inadequate information to stranded travellers, wrote an evaluation of internal and external communication during the event, and the responsibilities of employees (NS, 1990). Some recommendations from this report were implemented several months later. Informing travellers and internal communication within such a large company will, however, remain a difficult problem (W. Assink, M.J.C.M. Smeets, K. Wierda, personal communications, 1994). Apart from emergency plans for readily available alternative types of transportation (for example, buses taking over from trains to transport passengers), relatively little attention is paid to emergency plans where alternative types of transportation will not be available. Shelter and food for stranded passengers will continue to be mainly organised by local communities and organisations such as the Red Cross.

Fourth, the Bosschap (the forest organisation in the Netherlands) already had a contingency plan before the event but it was not well known within the branch organisations (Bosschap, 1993). The plan should be adopted and implemented by all forest owners to prevent chaos in the forests and on the timber market following storm events.

The aftermath

Some effects lasted only for minutes or hours (electricity cuts, delays for travellers), other effects lasted for several days or weeks (damage to houses and infrastructure). Only the effects on forests, dunes and the morbidity and mortality effects were more long-lasting. Forests remained closed to the public for several weeks. The Dutch Red Cross paid compensation to relatives of the dead and to severely injured people from funds made available for relief by the European Union (DRC, 1990). In 1993 the building codes were updated, so that buildings should withstand higher winds (Netherlands Standardisation Institute, NEN 6702, 1993). However, statistical evidence that the higher building standards lead to a decrease in damage is lacking (P.C. Staalduinen (Applied Natural Science Research Centre, Delft), personal communication, 1995; L.J. Buth (Nederlands Normalisatie Instituut, Delft), personal communication, 1995). Insurance companies have changed insurance conditions and slightly increased storm damage insurance premiums.

Five years later, hardly any traces of the events are left in Dutch society and, apart from changes into building codes, little action has been taken to prevent future storm disasters in the Netherlands. Mitchell *et al.* (1989) found the same result when studying the October 1987 windstorm impacts in the UK. Apparently, a developed society is not highly vulnerable and interest in disaster preparedness is low. However, (re)insurance companies see the increasing losses as threatening their profits, and even their existence. By studying and modelling past storm losses they try to improve their understanding of the damage aspects of storms.

10.3 A storm-damage model for the Netherlands

In this section the damage aspect of storm vulnerability is captured in a model of insured damage to houses and business (including industrial and commercial buildings) in the Netherlands. The relation between storm intensity, storm track and damage is modelled quantitatively by statistically correlating empirical damage data with synoptic meteorological data from five storms in the period 1987–92 (see Table 10.5).

The methodology to derive a storm-damage model from these data involves four steps. The first step is storm mapping, the second is to specify the model variables and to set a functional form for the model, the third step is to estimate the coefficients for the model variables and the last step is to evaluate the model results.

10.3.1 Storm mapping

For the storm mapping, the measured wind data are spatially disaggregated to the level of the impact data; that is, the two-digit postcode level. Wind data were

Table 10.5 Empirical data used for estimating the storm model

Damage data	Storm data	Other data
Source: Centre for Insurance Statistics (CVS)*	Source: The Dutch Met Office (KNMI)	Number of private houses by two-digit postcode, H
Claim periods: 10/14–20/1987; 1/24–28/1990; 2/24–3/2/1990; 12/17–26/1991; 11/24–29/1992	Measurement periods: 10/15–17/1987; 1/25–26/1990; 2/25–1/3/1990; 12/22–24/1991; 11/25–27/1992	Number of businesses by two-digit postcode, B Area of the postcode area (sq km), A
Stock-at-risk: policies on private houses valued at less than Dfl. 500,000 per two-digit postcode area. Number of two-digit postcode areas 90	Synoptic measurements at 30 locations (10 m height) in the Netherlands Maximum hourly mean wind speed data, $\bar{V}_{max}$	Average income in postcode areas (used as a proxy for type and value of houses), I
Numbers of claims per 1,000 policies per two-digit postcode area (for each storm) Corresponding claim averages over postcode area (Dfl.)	Maximum wind gust data $\hat{V}_{max}$ (in hourly periods) Duration of the storm (number of hours $\hat{V}_{max} > 20$ m s^{-1})	

Note: * The CVS database normally covers 80 per cent of all policies. These data were scaled upwards to cover 100% of the policies for this study. The deductibles were added to the CVS claim data to give total damage estimates

256

available from only thirty measurement stations in the Netherlands (see Figure 10.4). Storm maps were based on distance-related weighting spatial interpolation between the measurement sites. The interpolation of the point measurement values to average values for a 5 km × 5 km grid over the Netherlands was carried out with a geographical information system (GIS). Figure 10.4 also gives an example of the maximum hourly mean wind speed ($\bar{V}_{max}$) contour map of the 25 January 1990 storm event. The wind data so derived were again aggregated to the two-digit postcode level by (arithmetically) averaging over the grids in postcode areas. This results in values for maximum wind gust ($\hat{V}_{max}$), maximum hourly mean wind speed ($\bar{V}_{max}$) and storm duration (D) for every postcode area level (I) and every storm event (t).

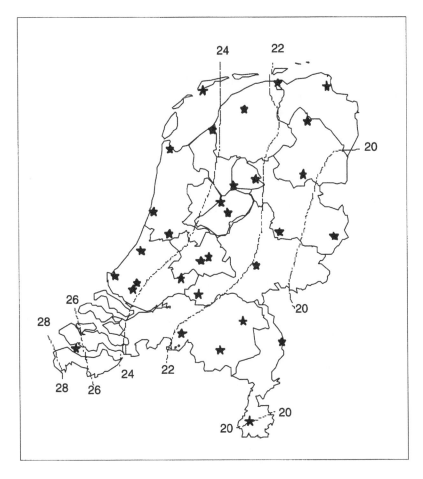

Figure 10.4 The $\bar{V}_{max}$ contour map of the 25 January 1990 storm event obtained with the distance-related weighting method for thirty wind measurement stations in the Netherlands

10.3.2 Specification of the model variables

The model variables relevant to storm damage are speed of onset, maximum gust wind speed, wind direction, variability of wind speed and wind direction, mean wind speed, storm, and types and amount of the stock at risk in the area affected by the storm. The spatial distribution of all these parameters over the affected area is also of great importance. The variables available for this storm damage model on a two-digit postcode area are:

- number of private houses (H) and number of businesses (B);
- the wind speed, either $\hat{V}_{max}$ or $\bar{V}_{max}$;
- the storm duration D (hours);
- the average income per household, as a proxy for the value of the houses; and
- the postcode area A (km^2).

The number of parameters in the model must be limited because there are only ninety observations (ninety two-digit postcode areas) per storm. More than three to four different parameters would easily lead to overfitting, resulting in a meaningless model.

Based on maximum likelihood statistics, a logarithmic functional form was preferred over the physically plausible square and cubic functional forms, resulting in:

$$\ln \text{TDH}_{i,t} = \ln H_{i,t} + \ln A_{i,t} + \gamma \hat{V}_{max:i,t} + c \tag{1}$$
$$\ln \text{TDB}_{i,t} = \text{In } B_{i,t} + \ln A_{i,t} + \gamma \hat{V}_{max:i,t} + c \tag{2}$$

where:

$\text{TDH}_{i,t}$ = total damage to houses (in Dutch guilders) in area i for storm event t;

$\text{TDB}_{i,t}$ = total damage to businesses (in Dutch guilders) in area i for storm event t;

$H_{i,t}$ = number of houses in area i for storm event t;

$B_{i,t}$ = number of businesses in area i for storm event t;

$A_{i,t}$ = postcode area i for storm event t (in km^2);

$\hat{V}_{max:i,t}$ = storm input data in area i for storm event t;

c = a constant;

α, β, γ = coefficients;

i = postcode area number (10, 11, . . . , 99); and

t = storm event number (1, 2, . . . , 5).

The ordered records of the wind speed data and the damage data were found to co-integrate (see Figures 10.5a and 10.5b), which indicates that the model can

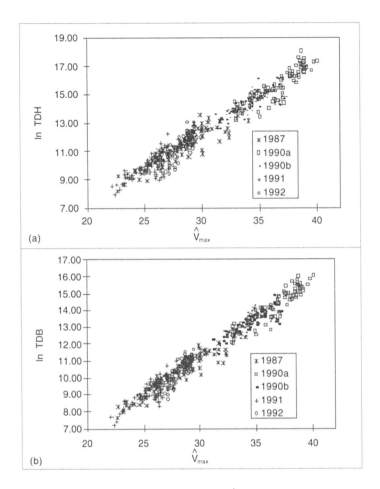

Figure 10.5a Logarithmic damage to houses versus $\hat{V}_{max:i}$ for all five storm events.

Figure 10.5b Logarithmic damage to houses and businesses versus $\hat{V}_{max:i}$ for all five storm events.

safely be applied. From a physical point of view, it is interesting that the indicated *exponential* relation performs better than a *cubic* relation, as it is known that wind power is proportional to wind speed to the third power. In wind models less elaborate than ours this is taken for granted. According to our analysis such reasoning is too simple.

Storm duration appears not to be significant. The average income, which was considered as a proxy for house value, is also not statistically significant and, therefore, not included in the model. Thus socio-economic vulnerability in storm risk reduces to only the number of houses per square kilometre.

10.3.3 *Estimating the coefficients of the model variables*

The coefficients were estimated by ordinary least-squares regression based on empirical data for the five storms. The resulting model is:

$$\ln \mathrm{TDH}_{i,t} = 0.93 \ln H_{i,t} + 0.26 \ln A_{i,t} + 0.50 \ \hat{V}_{\mathrm{max:},i,t} - 14.4 \qquad (3)$$

$$\ln \mathrm{TDB}_{i,t} = 0.58 \ln B_{i,t} + 0.42 \ln A_{i,t} + 0.46 \ \hat{V}_{\mathrm{max:},i,t} - 10 \qquad (4)$$

Several F-tests were performed to see whether:

- the model would be stable over areas with low and high terrain roughness (as suggested in the building codes (NEN 6702, 1993));
- the model would be stable over a split of the Netherlands into north versus south, or east versus west;
- the individual parameters are stable over sparsely and densely populated areas;
- the model fitted to four storms would be valid for the fifth storm; and
- the models for individual storms correspond to the model for all five storms.

The model parameters are constant over sparsely and densely populated areas and other regional divisions. With respect to models for individual storms, there is an increase in the overall uncertainty. However, more problematic is the instability of the wind speed between the different models. This indicates that results from the model should be treated with caution. Although damage proportional to the exponent of the wind speed is preferred over proportionality to wind speed squared or cubed, and the 'co-integration' test did not find structural deviations between the natural logarithm of damage and wind speed, the tests above suggested that there may be a better specification. More research is needed to address this. The parameter stability test showed that the damage curve for higher wind speeds should be a little flatter, so that the analyses for increasing wind speed in the next section probably slightly overstates the damage. Partly because of the limited number of observations, no ready improvements to the model could be found. Hence one and the same function will be used for all storms and the entire domain of the explanatory variables in the following. For further elaboration see Dorland and Tol (1995).

10.3.4 *Model evaluation*

Table 10.6 evaluates the model by comparing calculated and observed damage. The 95 per cent confidence interval of the calculated mean damage is indicated in parentheses, assuming the damage has a normal distribution.

Table 10.7 evaluates the model by presenting damage (and the 95 per cent confidence interval), assuming a storm which is similar to the Daria storm, except

Table 10.6 Modelled and observed wind damage to private houses and businesses in the
Netherlands (Dfl. million)

Year of storm:	1987	1990	1990	1991	1992
Event number:	1	2	3	4	5
Total damage to private houses					
Modelled	21	903	214	3.3	12
	(12)	(286)	(57)	(0.9)	(3.0)
Observed	10	977	232	3.3	22
Modelled − observed	11	−74	−18	0	−10
Total damage to businesses					
Modelled	6.8	202	59	1.3	4.3
	(4.1)	(66)	(17)	(0.4)	(1.2)
Observed	4.1	310	42	1.5	8.2
Modelled − observed	2.7	−108	17	−0.2	−3.9

Note: Twice the standard deviation of the model estimates is given in brackets

Table 10.7 Increased damage (in Dfl. million) from a Daria-like windstorm in the
Netherlands due to increased storm intensity

	Percentage increase in wind intensities					
	0%	2%	4%	6%	8%	10%
Houses						
Increase in damage	0	419	1,034	1,936	3,255	5,191
2 × standard deviation	309	550	867	1,315	1,962	
Businesses						
Increase in damage	0	85	205	375	617	962
2 × standard deviation	67	117	181	267	387	

for an increase in $\hat{V}_{\max; i,t}$. It is concluded that the exponential functions derived
for both private houses and businesses provide a reasonable model of storm
damage. Though the functions might not hold for every individual storm, they
give a reasonable estimation of the average damage due to a storm in a given year.
Better functions can be derived only by analysing more storm events and including more meteorological and physiographic (e.g., surface roughness) data. These
were not available for this research.

10.4 Scenarios for losses in the Netherlands and northwest Europe

Forecasting storm damage is difficult not only because of to the limited empirical basis upon which models can be built, but also because of the large uncertainties in long-term forecasting. Not only climatic changes and their impacts on extreme weather events, but also economic and demographic growth need to be forecast over a long period. Even harder to predict is the way in which insurance companies will deal with the changing behaviour of policy holders, which is itself hard to foresee. Consequently, the analyses presented here are to be interpreted as what-if scenarios and vulnerability studies, rather than forecasts.

Many studies of historic trends in storminess have been performed in the past 10 years by, among others, Hammond (1990), Schmidt and von Storch (1993), von Storch et al. (1993), Jones (1991), Palutikof et al. (1991, 1993), Schinke (1993), Isemer (1992) and Ward (1992). Lamb (1991) even produced a catalogue of severe storms for northwestern Europe which extends back to the mid-1500s. Palutikof and Downing (1994) found that the evidence for the twentieth century suggests that there is no long-term trend, either upwards or downwards, in a range of measures of storm frequency and severity over Europe. Where trends are observed, they can be explained by changes in observational or analytical practices. The relatively high frequency of storm occurrence and/or severity over Europe in the past few years is not yet found to be outside the bounds of natural variability. However, the extent of storm damage has increased over the past decades. According to Munich Re (1990) and Dlugolecki (1992), the reasons for the increase in the storm damage include an increase in the socio-economic exposure to extreme weather.

The influence of the enhanced greenhouse effect on storminess in Europe has been studied extensively by many researchers. Among others, Palutikof et al. (1993), Gates et al. (1992), Hall et al. (1994), von Storch et al. (1993), Balling et al. (1991) and Agee (1991) have analysed observational records of atmospheric pressure data sets and/or results from transient-response general circulation models (GCMs).

Palutikof and Downing (1994) reviewed these studies and found that many of them give conflicting results, and forecasting storms and storm tracks in detail is not possible. They developed scenarios for wind hazard based on the UK Meteorological Office transient simulation (UKTR) (Hadley Centre, 1992; Viner and Hulme, 1994; Murphy, 1992). This simulation experiment is similar to the UKHI experiment (Mitchell et al., 1990), but atmospheric radiative forcing is increased each year for a transient response (as compared to an equilibrium response). The model run for years 66–75 was chosen as broadly comparable to the UKHI scenarios scaled to 2050. The resulting increases in mean monthly wind speed are taken *a priori* to be similar in direction and magnitude to changes in storminess. In the absence of model results for maximum wind speeds, this is a reasonable first approximation. Figures 10.6a and 10.6b plot the changes in wind speeds for four seasons. The most notable features are that (1) wind speeds in July change little, or even decrease, throughout Europe, and (2) wind speeds increase across the whole of Europe in October, by 4–12 per cent. In January and April, the changes are less uniform, with areas of greater and of reduced wind speeds, and changes of the order of ± 5 per cent. Palutikof and Downing (1994) conclude that the mean wind speed could, as a result of climate change, increase by 1–9 per cent in winter over the next 75 years. Experiments from the Canadian Climate Centre GCM (Lambert, 1995) and the Max-Planck GCM (Lankeit *et al.*, 1996) point in the same direction. The best estimate of the increase is about 6 per cent in 75 years. This estimate is, of course, highly uncertain. The direction of the wind is expected to become more westerly. A change in the frequency of the storms could not be predicted. It is assumed that the maximum gust wind speed changes linearly with the mean wind speed.

In the following sections, scenarios for individual storm losses in the Netherlands in 2015, and scenarios for the average damage in the Netherlands and western Europe until 2065, partly based on the studies mentioned above, are given.

10.4.1 Individual storm losses in the Netherlands in 2015

Given a socio-economic scenario based on scenarios for future numbers of houses and businesses per postcode area, and a description of a future storm, given as $\hat{V}_{max}$ by postcode area, it is possible to estimate future damages resulting from a scenario storm.

For growth in the number of houses, long-term scenarios for population growth for Europe are used, based on work by the Central Planning Bureau in the Netherlands (CPB, 1992). The following assumptions are made: (1) the growth in the number of houses is equal to the population growth rate; (2) the growth in the number of businesses is equal to the economic growth rate; (3) the growth rates in the Netherlands are equal to the growth rates in Europe; and (4) the growth rates are equal for all postcode areas in the Netherlands. The names and the growth rates for these scenarios are given in Table 10.8. For the 'No Change' scenario it is assumed there will be no change in the number of houses and businesses. This 'No Change' scenario is used as a reference case only.

The future storm scenarios for this analysis are based on the January 1990 storm in the Netherlands (see Section 10.3.4), with an increase in the modelled wind speeds ($\hat{V}_{max}$) of between 0 and 10 per cent. In other words, the future storms are spatially identical with the January 1990 storm but more intense.

Table 10.9, and Figures 10.7a and 10.7b present the results of the analysis. No correction for changes in the vulnerability of buildings has been made in this analysis, because statistical evidence for changes in damage as a result of higher

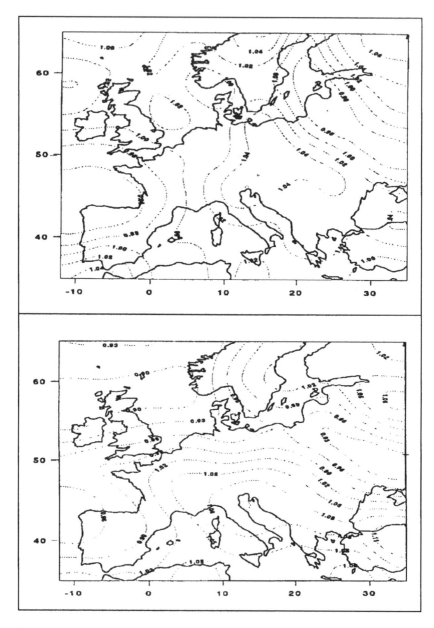

Figure 10.6a Change in mean wind speed (ratio) over Europe between the control and perturbed run decade for years 1966–75 of the UK transient experiment for January (top) and April (bottom)
Source: Palutikof and Downing (1994)

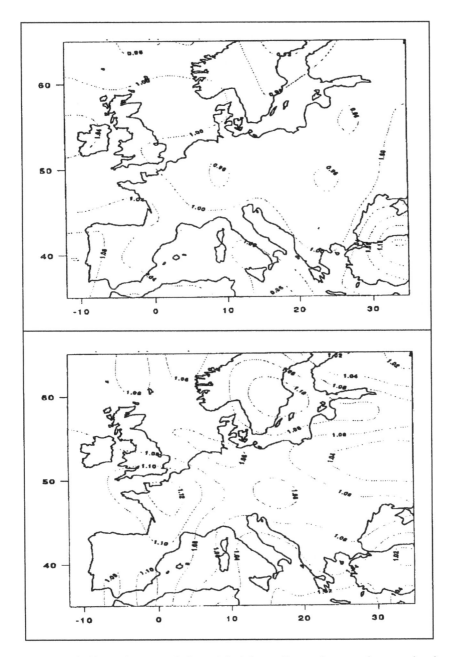

Figure 10.6b Change in mean wind speed (ratio) over Europe between the control and perturbed run decade for years 1966–75 from the UK transient experiment for July (top) and October (bottom).
Source: Palutikof and Downing (1994)

Table 10.8 Population and economic growth rate scenarios for Europe

No.	Name	Population growth rate in Europe	Economic growth rate in Europe
1	Global Shift	0.4%	1.9%
2	European Renaissance	0.3%	2.8%
3	Global Crisis	0.4%	2.0%
4	Balanced Growth	0.3%	3.2%

Source: CPB (1992)

Table 10.9 Mean wind damages to private houses and businesses in 2015 in the Netherlands for different socio-economic scenarios and a 0 to 10% increase in the wind intensity

Socio-economic scenario	Percentage increase in the wind intensity					
	0	2	4	6	8	10
Houses						
No Change	903	1,322	1,937	2,838	4,158	6,094
	(286)	(421)	(620)	(918)	(1,346)	(1,983)
Global Shift/Global Crisis	991	1,451	2,125	3,114	4,563	6,687
	(314)	(462)	(680)	(1,000)	(1,476)	(2,176)
European Renaissance/	968	1,418	2,077	3,043	445	6,534
Balanced Growth	(307)	(451)	(665)	(979)	(1,443)	(2,156)
Businesses						
No Change	202	287	407	578	820	1,164
	(66)	(94)	(135)	(192)	(275)	(393)
Global Shift	266	377	535	759	1,077	1,529
	(87)	(124)	(177)	(253)	(361)	(516)
Global Crisis	270	382	543	770	1,093	1,551
	(88)	(126)	(180)	(257)	(367)	(524)
European Renaissance	302	428	608	862	1,224	1,737
	(99)	(141)	(202)	(288)	(411)	(587)
Balanced growth	319	453	643	912	1,295	1,838
	(105)	(149)	(213)	(305)	(435)	(622)

Notes: Twice the standard deviation is given in parentheses. Figures are given in Dfl. million at 1990 values

building standards is lacking (L.J. Buth, personal communication, 1995). The coefficient of variation of damage is assumed to be independent of both the level of wind intensity and building density. This assumption was already tested and not rejected above.

The results clearly show that the damage risks are very sensitive to the intensities of future storms and that the damage under the different CPB growth scenarios are not statistically different. It should be noted that these damages are

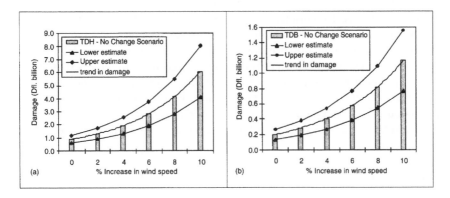

Figure 10.7 Scenario results for the mean damage to (a) houses and (b) businesses in 2015 in the Netherlands (1990 values)

only a part of the impact of a storm. Other effects of storms – that is, casualties and losses from business interruption – presumably will also increase. The damage that would be caused by the Daria event occurring in 2015 would, assuming an increase of 2 per cent in the wind speed over 25 years, be increased by around 40 per cent relative to 1990 levels. In the year 2065, assuming a 6 per cent increase of the wind speed in 75 years, the damage of the Daria event would increase around by 300 per cent.

10.4.2 Average damage in the Netherlands and western Europe

The previous section focused on specific storms with a geographically explicit wind and damage pattern. This allowed us to study vulnerability to variations in Daria-like storms. Daria is only one storm, however. Equally interesting is an analysis of annual mean storm damage, and how that may be affected by climate change. Obviously, the five storms analysed so far yield little information on the average. Given the inescapable problems with data quality and data availability, the analysis here has to step back from geographical explicitness and examine aggregate damage instead. Consequently, storm intensity must also be reduced to a single index.

An alternative storm model for the Netherlands

Table 10.4 displays the Dutch storm history of the twentieth century. Twenty-nine storms affected the country between 1910 and 1990; that is, on average 0.36 storms per year. This KNMI record contains data on both the storm frequency and the storm intensity, and is used to calibrate a storm model.

Annual mean maximum wind gust is calculated with the stochastic weather generator described by Tol (Chapter 13, this volume) and fitted to the data of Table 10.4. Figure 10.8 displays the observed and fitted storm density.

The database with which to estimate a storm damage function is rather poor. We have only the four observations on total damage to houses and business mentioned above (the 1991 event is not a storm, according to the KNMI scale). These damage data need to be related to the maximum wind gust. The damage model derived in the previous sections cannot be used without imposing the restriction that geographical storm patterns in the past are like Daria. Unfortunately, indicators of the stock at risk cannot be taken into account either. The models estimated are

$$\text{TDH} = 2.48 \times 10^{-7} \exp(0.48 \hat{V}_{max}); \qquad \text{TDB} = 1.49 \times 10^{-7} \\ \exp(0.46 \hat{V}_{max}) \qquad (5)$$

The parameters are of course highly uncertain, but the central estimates are reassuringly close to those found for the five storms together (see above). Figure 10.9 depicts the observed and modelled damages.

Annual mean storm damage in the Netherlands

Given the stochastic storm generator and storm damage model (valid for the early 1990s only), it is straightforward to calculate the expected storm damage in 1990. Table 10.10 displays the results. The expected damage is calculated from 100,000 replications of the year 1990. Over this sample, the maximum damage is

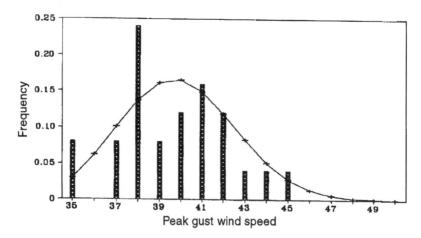

Figure 10.8 Observed (bar) and fitted (line) frequency of peak gust wind speed as a measure of storm intensity. Storm is defined as maximum peak gust wind speed in excess of 35 m s^{-1}. Observations are for 1910–90 for the Netherlands

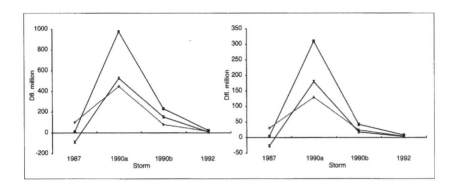

Figure 10.9 Damage to private houses (left) and businesses (right) as observed (filled squares) and as modelled (asterisks). The residuals (pluses) are also displayed

Table 10.10 Average and maximum storm damage in the Netherlands (Dfl. million at 1990 values)

Year	Average scenario			Maximum scenario
	Houses	Businesses	Total	Total
1990	34	10	44	17,000
2015	61	18	79	50,000
2065	219	59	278	414,000

Notes: Average and maximum from over 100,000 random realisations from the stochastic storm generator coupled to the storm damage model, both fitted to historical data. For 2015 there is a 2% increase in modal wind speed; no change in vulnerability. For 2065 there is a 6% increase in modal wind speed; no change in vulnerability

also extracted, and displayed in Table 10.10. Average total damage is Dfl. 44 million, maximum total damage is Dfl. 17 billion. These numbers are too uncertain to make it possible to derive a confidence interval.

The storm generator can readily be altered to produce a hypothetical future climate. When this is coupled to the damage model, the impact of such a change on the present socio-economic situation can be calculated. Sensitivity analyses reported by Dorland and Tol (1995) reveal that the estimates are highly uncertain. Table 10.10 displays the average and maximum storm damage for a 2 per cent (representative of the 2015 scenario) and 6 per cent (possible by 2065) increase in modal wind speed. For a 2 per cent increase in modal wind speed, wind damage increases by about 80 per cent; for a 6 per cent increase, damage rises by over 500 per cent. This scenario holds vulnerability constant at the 1990 level; in fact, vulnerability is likely to increase. Figure 10.10 depicts the corresponding damage distributions.

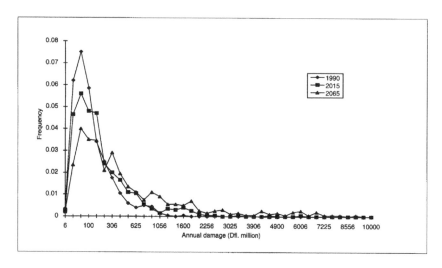

Figure 10.10 Probability density function of the annual storm damage for the 1990 situation and increases in modal wind speed of 2 per cent (2015) and 6 per cent (2065). The *x*-axis is on a quadratic scale

Figure 10.11 displays the average storm damage over a period of 101 years. Modal wind speed rises 6 per cent from the 1990 value over 75 years. For comparison, the damage under constant risk and that under rising storm frequency (with the same assumptions as for intensity) are also displayed. An increase of storm frequency hardly affects average damages, as damages are assumed to be linear relative to the number of storms. The averages in Figure 10.11 are based on only 1,000 replications, illustrating the tremendous variability of storminess over the Netherlands. Figure 10.12 displays the increase in damage as a function of the increase in modal wind speed. The smooth line is the best-fit quadratic function: $\Delta \text{TD} = 42.5 \Delta \hat{V}_{max}^{2}$. Assuming that modal wind speed increases by 2.4 ms^{-1} (6 per cent) for a global mean temperature (GMT) rise of 2.5°C, the damage cost function would be $\Delta \text{TD} = 38.1 \Delta \text{GMT}^{2}$. The upper curve reflects an increase in modal wind speed of 6 per cent from 1990 to 2065. The two lower curves are for a scenario in which storm frequency increases (6 per cent of the 1990 value over 75 years) and there is no change in risk. Average annual damages under constant storm risk and under increasing storm frequency are hardly distinguishable. Averages are for 1,000 replications.

Average storm damage in western Europe

The Netherlands is but one country, and certainly not the largest in the world. It would be interesting to know the impact of climate change on larger regions, such as western Europe. Again, time and money constraints prevented us from

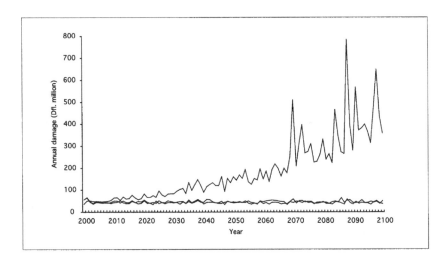

Figure 10.11 Average annual damages for three scenarios: modal wind speed increases 6 per cent from the 1990 value over 75 years (upper curve); constant risk and increasing storm frequency (bottom curve)

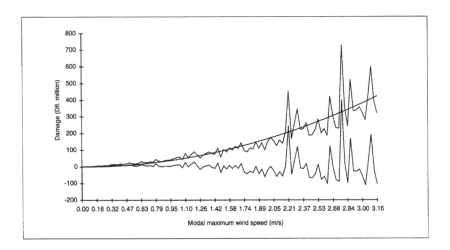

Figure 10.12 Increase in average annual damage as a function of increase in modal wind speed. The bumpy line displays the increase in average damage (cf. Figure 10.10), the smooth line displays the best-fitting quadratic function. The difference between the two is also displayed

271

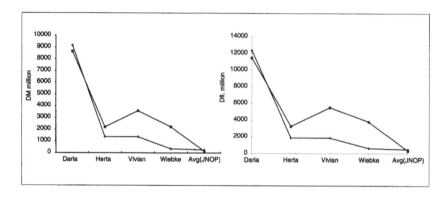

Figure 10.13 Observed (filled squares) and predicted (pluses) insured (left) and total economic (right) losses in millions of Deutschmarks for eight 1990 storms. For the minor storms Judith, Nana, Otille, and Polly only the average (avgJNOP) is known

doing serious research, but some crude estimates can be made. Munich Re (1993) contains some data on storm losses in western Europe (Austria, Belgium, Denmark, France, Germany, Great Britain, Luxembourg, Scandinavia, Switzerland and the Netherlands). The data comprise the eight storms in 1990. Regressing European losses on Dutch losses leads to a regression coefficient of 6.8 (1.1) for insured losses, with an R^2 of 77 per cent, and a regression coefficient of 6.2 (1.2) for total economic losses, with an R^2 of 62 per cent. Figure 10.13 displays the observed and predicted losses. Thus, it is roughly estimated that the impact in absolute terms of changes in storm intensity will be six to seven times as large in western Europe as a whole as in the Netherlands.

10.5 Policy options

A distinction can be made between three types of disaster management policy: preparedness, protection and prevention. The following sections discuss these options for storms in the Netherlands. An extensive literature search was carried out to assess what the policy responses to the Daria event were, if any. It appears that analysis or action was undertaken in only a few sectors. The main conclusions from the analysis and additional observations are given below.

10.5.1 Preparedness

Preparedness includes insurance, entitlements to government relief, early warning systems, etc; that is, the socio-economic coping mechanisms to absorb and nullify the consequences of a disaster (Tol and Olsthoorn, 1994).

The Nederlandse Spoorwegen (Dutch Railways), which was criticised for providing inadequate information to stranded travellers, produced a report with recommendations about how to manage such an event in future. A number of the reported recommendations have already been implemented. All other recommendations should be adopted as soon as possible. The Bosschap, an association of forest owners, had already produced a disaster plan for managing storm risks in forestry before the 1990 storm events. Further analysis is required to determine if this plan can be implemented in the forestry sector as a whole. Local, provincial and national government should be better prepared for large-scale disasters. This could be achieved by learning from publications of the COT, which is an official team set up to evaluate large-scale disasters. The role of the press, the radio and TV in disaster situations should be nationally regulated. People also have to be made aware that warnings and advice given should be taken seriously. It is questionable, however, whether Dutch society can prepare itself better and decrease insured storm damages. From the Munich Re (1993) study it could be concluded that the reinsurance company is seeking ways to withdraw from broad coverage of sustained damage.

10.5.2 *Protection*

Protection is the structural part of coping. Protection strengthens or shields the stock at risk so that less damage is sustained.

In the Netherlands, setting building standards reduces vulnerability of buildings to storms. The building codes were updated after the 1990 January storm events in the Netherlands. Under the new regulations (NEN 6702, 1993), buildings should be able to withstand higher wind speeds. The national government and not the insurance companies triggered adjustments to these standards after the severe storms in 1990. The building industry embraced this adjustment in order to improve its image.

The Hagelunie monopolises the insurance market for insuring greenhouses in the Netherlands. If extensively analysed the causes of the extensive damages in this sector and demanded technological adjustments for decreasing storm damage after the 1990 events (P.C. van Staalduinen, personal communication, 1995).

However, the effect of these and other protection policies on storm damage in the Netherlands is not known. So far almost no research has been done to evaluate these types of impact management.

10.5.3 *Prevention*

Prevention of storms would mean changing the weather or climate. Weather modification is not considered a serious option at the current level of meteorological knowledge. Potential increases in storm damage are but one argument for limiting anthropogenic climate change by reducing greenhouse gas emissions.

10.6 Conclusions

In early 1990 Europe was hit by a series of severe storms with high socio-economic impact. The event of 25 January 1990 event, Daria, was the worst storm in the Netherlands since 1944. Daria caused a natural disaster in large parts of Europe and killed twenty-one people in the Netherlands alone. The damages in Europe due to the events of early 1990 are provisionally estimated at Dfl. 30 billion (1990 values). The damages in the Netherlands alone are estimated at Dfl. 2.6 billion, of which Dfl. 1.5 billion was covered by insurance. This estimate is about 20 per cent higher than the estimate by insurance institutes because losses due to delays in public transport, loss of life, and a number of small losses, are included here. The actual damages could be even higher because damages such as income losses, interruptions to business and damages to government-owned buildings and most of the infrastructure are not included. From an extensive literature search it has been found that the storms prompted policy response in only a few sectors.

In order to estimate damage from future storms in the Netherlands, which under climate change might be more severe, two storm-damage models have been developed by statistically analysing the meteorological characteristics and insurance claims of five storms in the period 1987–92. The first model is geographically explicit, linking meteorological data and damage data on a two-digit postcode level for the Netherlands. This model indicates that future storm risks (to private houses and business) are very sensitive to maximum wind speed, much more than to socio-economic development. The second model is a stochastic weather generator coupled to aggregated damage data for the whole of the Netherlands.

The extent of climate changes and their influence on maximum wind gusts, which drive the damages in these models, is still highly uncertain. However, a 2 per cent increase in the maximum wind gust in 25 years is conceivable. Scenario analyses performed with these models are, however, very speculative for three reasons. First, climate change uncertainties are large. Second, the models draw on limited databases of storm characteristics and storm damage. Third, the behaviour of insurance companies, policy-holders and governments are hard to predict and therefore not included in the models.

Scenario analysis with the geographically explicit storm model shows that if the Daria event were repeated in the year 2015, with a 2 per cent increase in the intensity of wind gusts due to climate change, the damages would be around 40 per cent higher than they were in 1990. Scenario analysis with the average-damage model indicates that the annual mean insured damages could, owing to climate change, increase by 80 per cent in 25 years (the year 2015). The annual mean damages in western Europe as a whole will probably be around six to seven times as large as the damages in the Netherlands.

Little potential seems to exist for reducing vulnerability to storms in the Netherlands. Preparedness and protection appear to be used to almost their full extent. However, more attention should be given to 'social relief' strategies.

Acknowledgements

Critical comments by Jules Beersma, Dr G. Davidson, Andrew Dlugolecki, Tom Downing, Marjan van Kraaij, Angela Liberatore and Dr R. Muirwood on an earlier version of this chapter (Dorland et al., 1996) are gratefully acknowledged. The authors are of course responsible for any remaining error. Marjan van Kraaij of the Centre for Insurance Statistics and Jürg Trüg of Swiss Re kindly made data available.

Notes

1 The hourly mean potential wind speed data ($\bar{V}_p$) are derived from the measured maximum hourly mean wind speed data by taking into account, among other factors, wind direction, terrain roughness and heights of the measurement points.
2 The exchange rate at 1990 values was Dfl. 1.00 = US$0.49 = ECU 0.43.

References

Agee, E.M. (1991) Trends in cyclones and anticyclone frequency and comparison with periods of warming and cooling over the Northern Hemisphere. *Journal of Climate* 4: 263–267.

Balling, R.C., Cerveny, R.S., Miller, T.A., and Idso, S.B. (1991) Greenhouse warming may moderate British storminess. *Meteorology and Atmospheric Physics* 46: 181–184.

Berz, G. and Conrad, K. (1993) Winds of change. *The Review*, June: 32–35.

Bosschap (1993) *Report on How to Handle Calamities in the Dutch Forests*, The Hague: Bosschap (in Dutch).

Bussink, R.L. (1990) Cleaning and selling storm wood. *Extra Bulletin Bosbouwvoorlichting*, February (in Dutch).

COT (Crisis Onderzoeksteam; Crisis Research Team) (1990) *Alert* 3 (March).

CPB (Netherlands Bureau for Economic Policy Analysis) (1992) *Scanning the Future: A Long Term Scenario Study of the World Economy 1990–2015*, The Hague: Sdu Publishers.

CVS (Centre for Insurance Statistics) (1994) *Data on the January 25/26 1990 Storm Losses*, Zoetermeer: Centre for Insurance Statistics.

Dlugolecki, A.F. (1992) Insurance implications of climate change. *Geneva Papers on Risk and Insurance* 17, 64: 393–405.

Dorland, C. and Tol, R.S.J. (1995) *Storm Damage in the Netherlands: Modelling and Scenario Analysis*, W95/20, Amsterdam: Institute for Environmental Studies, Vrije Universiteit.

Dorland, C., Olsthoorn, A.A., and Tol, R.S.J. (1994) *The 1990 Winter Storms in the Netherlands*, W94/21, Amsterdam: Institute for Environmental Studies, Vrije Universiteit.

Dorland, C., Tol, R.S.J., Olsthoorn, A.A., and Palutikof, J.P. (1996) 'An analysis of storm impacts in the Netherlands', in T.E. Downing, A.A. Olsthoorn and R.S.J. Tol, eds, *Climate Change and Extreme Events: Altered Risk, Socio-economic Impacts and Policy Responses*, Oxford and Amsterdam: Environmental Change Unit, University of Oxford, Vrije Universiteit, Amsterdam, 157–184.

Dutch Red Cross (DRC) (1990) *Evaluation Report for the Payment of the EEG-Funds for the Victims of the January 1990 Storm Events*, The Hague: Dutch Red Cross.

Ekkelboom, J. (1990) Bad cooperation within wood sector due to storm. *Nederlands Bosbouw Tijdschrift*: 253–256 (in Dutch).

Gates, W.L., Mitchell, J.F.B., Boer, G.J., Cubasch, U., and Meleschko, V.P. (1992) 'Climate modelling, climate prediction and model validation', in J.T. Houghton, B.A. Callander and S.K. Varney, eds, *Climate Change 1992: The Supplementary Report to the IPCC Scientific Assessment*, Cambridge: Cambridge University Press, 97–134.

Gooi en Eemlander (G&E) (1990a) Army helps cleaning up glass damage. *Gooi en Eemlander*, 30 January.

Gooi en Eemlander (G&E) (1990b) Salvation Army offers comfort in Jaarbeurs. *Gooi en Eemlander*, 26 January.

Gooi en Eemlander (G&E) (1990c) Storm damage KLM amounts 5 million. *Gooi en Eemlander*, 26 January.

Hadley Centre (1992) *The Hadley Centre Transient Climate Change Experiment*, Bracknell: Hadley Centre, Meteorological Office.

Hall, N.M.J., Hoskins, B.J., Valdes, P., and Senior, C.A. (1994) Storm tracks in a high resolution GCM with doubling CO_2. *Quarterly Journal of the Royal Meteorological Society* 120: 1179–1207.

Hammond, J.M. (1990) The strong winds experienced during the late winter of 1989/1990 over the United Kingdom: Historical perspectives. *Meteorological Magazine* 119: 211–219.

Haverkamp, M., and Schaap, J. (1990) The January 25 storm in Het Gooi. *Weerspiegel* 17, 4: 299–305 (in Dutch).

Isemer, H.J. (1992) Comparison of estimated and measured marine surface wind speeds. Paper in at *Proceedings of International COADS Workshop*, 13–15 January 1992, Boulder, CO, 143–158.

Jones, H.A. (1991) Temperatures and windiness over the UK during the winters of 1988/1989 and 1989/1990 compared to previous years. *Weather* 46: 126–135.

Lamb, H.H. (1991) *Historic Storms of the North Sea, British Isles and Northwest Europe*, Cambridge: Cambridge University Press.

Lambert, S.J. (1995) The effects of the enhanced greenhouse warming on winter cyclone frequencies and strengths. *Journal of Climate* 8: 1447–1452.

Lankeit, F., Ponater, M., Sausen, R., Sogalla, M., Ulbrich, U., and Winkelband, M. (1996) Cyclone activity in a warmer climate. *Contribution to the Physics of the Atmosphere* 69, 3: 393–407.

McCallum, E. (1990) The Burns' Day storm, 25 January 1990. *Weather* 45: 166–173.

McCallum, E. and Norris, W.T.J. (1990) The storms of January and February 1990. *Meteorological Magazine* 119: 1419.

Mitchell, J., Devine, N., and Jagger, K. (1989) A contextual model of natural hazard. *Geographical Review* 17, 4: 391–409.

Mitchell, J.F.B., Manabe, S., Meleschko, V., and Tokioka, T. (1990) 'Equilibrium climate change and its implications for the future', in J.T. Houghton, G.J. Jenkins and J.J. Ephraums, eds, *Climate Change: The IPCC Scientific Assessment*, Cambridge: Cambridge University Press, 131–172.

Munich Re (1990) *Windstorm: New Loss Dimensions of Natural Hazard*, Munich: Münchener Rückversicherungs-Gesellschaft.

Munich Re (1993) *Winter Storms in Europe: Analysis of 1990 Losses and Future Loss Potential*, 2042-E-e, Munich: Münchener Rückversicherungs-Gesellschaft.

Murphy, J.M. (1992) *A Prediction of the Transient Response of Climate*, Climate Research Technical Note 32, Bracknell: Hadley Centre, Meteorological Office.

NEN (Netherlands Standardisation Institute) 6702 (1993) *Loading and Deformation: Technical Basis for Buildings 1990*, Delft: Nederlands Normalisatie Instituut (in Dutch).

NS (Nederlandse Spoorwegen; Netherlands Railways) (1990) *Evaluation of the January Storm 1990*, Utrecht: Nederlandse Spoorwegen, Dienst van Exploitatie (in Dutch).

Organization for Economic Co-operation and Development (OECD) (1992) *Natural Disaster Reduction: Guidelines for Donors*, 000550, Paris: Development Co-operation Directorate, Development Assistance Committee, Working Party on Development Assistance and Environment.

Palutikof, J.P., and Downing, T.E. (1994) 'European wind storms', in T.E. Downing, D.T. Favis-Mortlock and M.J. Gawith, eds, *Climate Change and Extreme Events: Scenarios of Altered Hazards for Future Research*, Oxford: Environmental Change Unit, University of Oxford, 61–78.

Palutikof, J.P., Guo, X., and Halliday, J.A. (1992) Climate variability and the UK wind resource. *Journal of Wind Engineering and Industrial Aerodynamics* 39: 243–249.

Palutikof, J.P., Halliday, J.A., Guo, X., Barthelmie, R.J., and Hitch, T.J. (1993) *The Impact of Climate Variability on the UK Wind Resource*, WN 6029-P1, Harwell: ETSU.

Rijkwaterstaat (RWS) (1990) *Report on the Storm Flood of January 25 and 26 1990*, SR 62, The Hague: Rijkswaterstaat, Dienst Getijdewateren, Stormvloedwaarschuwings-dienst (in Dutch).

Schinke, H. (1993) On the occurence of deep cyclones over Europe and North Atlantic in the period 1930–1991. *Beiträge zur Physik der Atmosphäre* 66: 223–237.

Schmidt, H., and von Storch, H. (1993) German bright storms analysed. *Nature* 365: 791.

Schraft, A., Durand, E., and Hausmann, P. (1993) *Storms over Europe: Losses and Scenarios*, Zurich: Swiss Reinsurance Company.

Swiss Re (1994) *Sigma 2*, Zurich: Swiss Reinsurance Company.

Tol, R.S.J. and Olsthoorn, A.A. (1994) *Climate Change, Wind Storms and the Insurance Industry: Some Notions from a Weather Insurance Simulation Model, Part 1*, W94/15, Amsterdam: Institute for Environmental Studies, Vrije Universiteit.

van Mourik, B. (1990) The rarity of the heavy January 25 1990 storm. *Weerspiegel* 17, 3: 207–211.

Viner, D. and Hulme, M. (1994) *The Climate Impacts LINK Project: Proving Climate Change Scenarios for Impacts Assessment in the UK*, Norwich: Climatic Research Unit.

Volkskrant (1990a) [Dfl] 30 million for restoration coast line thought insufficient. *Volkskrant*, 2 March (in Dutch).

Volkskrant (1990b) Information during January storm was insufficient. *Volkskrant*, 1 March (in Dutch).

Volkskrant (1990c) Severe storm stops public life. *Volkskrant*, 26 January (in Dutch).

Volkskrant (1990d) Storm damage to forests is worse than expected. *Volkskrant*, 1 February (in Dutch).

Volkskrant (1990e) Thousands of Dutch are stuck on the way home. *Volkskrant*, 26 January (in Dutch).

von Storch, H., Guddal, J., Iden, K.A., and Jonsson, T. (1993) *Changing Statistics of Storms in the North Atlantic?*, no. 116, Hamburg: Max-Planck-Institut für Meteorologie.

Ward, M.N. (1992) Provisionally corrected surface wind data, worldwide ocean atmosphere surface fields, and Sahelian rainfall variability. *Journal of Climate* 5: 454–475.

Wieringa, J. (1990) Severe storms on January 25. Paper in *Conference Report of the 5th National Wind Energy Conference*, at CEA and ECN, Lunteren.

11

HEATWAVES IN A CHANGING CLIMATE

Megan J. Gawith, Thomas E. Downing and Theodore S. Karacostas

11.1 Introduction

Periods of extreme temperature and heatwaves can have severe implications for human activities and health. Human morbidity and mortality rates rise sharply during periods of extremely hot weather, particularly in regions where hot weather is uncommon, when the heatwave occurs early in the summer season, or when there is little night-time cooling.

There is every probability that heatwaves will become more frequent in a warmer world (McMichael *et al.*, 1996). If average temperatures increase, more days will occur every year on which tolerable temperature thresholds are exceeded (Martens, 1997). Such conditions are likely to result in increased morbidity and mortality, especially in areas currently unaccustomed to higher temperature regimes (Smith and Tirpak, 1990).

Considerable research has been conducted to investigate the potential implications of climate change for human health. Each major publication on climate change impacts in recent years has contained a chapter, or at least a section, addressing the issue (e.g. Hashimoto and Nishioka, 1991; DoE, 1996; McMichael, 1996; McMichael *et al.*, 1996; Bentham, 1997), and several investigations have been conducted to identify potential impacts of climate change on human morbidity and mortality (e.g. Balling and Idso, 1990; Bentham, 1992; Hashimoto and Nishioka, 1991; Kalkstein *et al.*, 1996; Martens 1996, 1997; Ramlow and Kuller, 1990). Most of these studies have focused on the influence of milder winters, rather than extreme heat, on human health, and draw the general conclusion that milder winters will reduce cold-induced deaths, having a net reduction on human mortality (DoE, 1996). It has been estimated, for example, that 50 fewer deaths will occur per 100,000 population in cold-temperate cities under a warmer climate (Martens, 1997), and that 17,000 fewer deaths will occur in England and Wales if global temperatures increase by 3°C (Bentham, 1997).

Comparatively little attention has been paid to the health impact of heatwaves. This is an important omission. Anomalously warm years in the UK (1976 and 1990) resulted in above-average summer death rates in England and Wales (Bentham, 1997). Increased temperatures in future are likely to increase summer mortality. Potential changes in the frequency and severity of heatwaves and their consequence for human health, must be considered. Appropriate response strategies need to be identified. Assessments are also required for warm, humid regions where the intensity of heatwaves may become more severe. Better understanding of potential future impacts of heatwaves may allow risks of increased morbidity and mortality to be minimised.

This study is concerned with providing such an assessment. The health impacts of heatwaves are first considered. Difficulties associated with quantifying the impact of heat on human health are discussed, as these have important implications for our ability to make accurate assessments for future climatic conditions. Experience from past heatwaves is then reviewed to provide an understanding of conditions leading to heatwaves. Lessons which may help inform responses to future events are also identified. The following section considers potential impacts of future heatwaves. A comparative study of two case-study sites with differing climates and experience of heatwaves is conducted. Possible changes in the frequency of heatwaves under future climate scenarios are investigated. Results from the literature review are combined with experience from the case studies to identify adaptive strategies which may be employed to help reduce impacts of heatwaves on human health in a future climate.

11.2 Quantifying health impacts of heatwaves

Heat-regulatory systems in healthy individuals enable people to cope effectively with a range of thermal conditions. People are most comfortable in average temperatures of 16–30°C (Martens, 1997; Schneider, 1996). If ambient temperatures exceed the comfort range for a given location, individuals experience thermal stress which may result in discomfort, morbidity or substantially increased risk of mortality. A typically V-shaped relationship thus exists between mean temperature and human comfort (Figure 11.1). Residents of cold climates (solid line) are better able to tolerate cold extremes than those of warm climates, while residents of warm, humid regions (dashed line) are better able to cope with hot than cold extremes. Individuals subject to extreme conditions can take action to minimise thermal stress and reduce the risk of morbidity or mortality (dotted lines) by visiting, for instance, air-conditioned or heated public centres to gain physiological relief from extremely hot or cold conditions.

Quantifying the impact of heatwaves on human morbidity and mortality is fraught with difficulties. Firstly, heatwaves are difficult to define as no meteorological definition of a heatwave exists (Giles and Balafoutis, 1990). Rather, periods of excessive heat are considered in terms of discomfort indices which relate temperature and humidity to physiological and sensory impacts. More than

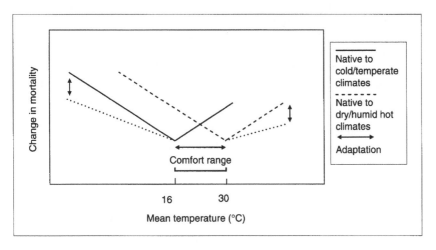

Figure 11.1 Typical V-shaped relationship between temperature and thermal stress
Source: After Martens (1996)

three hundred indices have been devised (Wenzel, 1984), as individual responses vary considerably between individuals in both space and time (Karacostas and Downing, 1996). Definitions of heatwaves are thus relative, based on prevailing climatic conditions in a given location. Conditions which cause residents of Oxford, England, considerable discomfort and heat stress may be regarded as normal by residents of Thessaloniki, Greece.

The second problem is that of defining a heat-related death. Heat-related deaths are often classified as the number of 'extra deaths' occurring above the mean for a particular population during a heatwave (Changnon *et al.*, 1996). Exposure to extreme heat may be either a primary cause of death or a secondary cause, the death in fact being caused, for example, by associated smog or poor air quality. The 1987 heatwave in Athens is reported to have claimed 1,115 lives, but it is possible that many of those deaths were caused by the high levels of smog rather than by heat (Giles and Balafoutis, 1990). Few studies have managed to separate and quantify weather- and pollution-induced deaths, and the synergistic impacts of weather and pollution are as yet poorly understood (McMichael *et al.*, 1996). A study in London, for example, showed that weather was a less important cause of death than pollution (Ito *et al.*, 1993); however, similar work in the relatively less polluted Netherlands (Kunst *et al.*, 1993) revealed that temperature extremes in summer were the primary determinants of mortality risk, which was affected little by variations in pollutant levels.

Uncertainty as to whether heat is the primary or secondary cause of death means that statistics of heat-related deaths are often unreliable and may underestimate the number of deaths which have occurred (Changnon *et al.*, 1996). Quantitative interpretation of the health impacts of extreme temperature events is hampered further by what is termed 'mortality displacement': many deaths during such periods occur among vulnerable people, particularly those older than 65 (Kalkstein, 1991; McMichael, 1996), who would have been likely to die from other causes during subsequent weeks.

Finally, it is not possible to identify a single threshold temperature above which mortality can be expected to increase. The exact temperature at which mortality occurs varies from place to place, reflecting a degree of physiological adaptation to local conditions (Martens, 1997). Different temperature thresholds also exist for different causes of death (Martens, 1996), further complicating analyses. Threshold temperatures extracted from the literature are listed in the Appendix to this chapter, highlighting the range of temperatures which invoke increased mortality.

11.3 Impacts of, and lessons from, past heatwaves

The severe social and economic costs generated by past heatwaves demand that every effort is made to better understand these events and develop appropriate response strategies. Heatwaves have the potential to cause severe loss of human life: 1,280 deaths during the Greek heatwave of 1987, and 830 in the mid-July 1995 heatwave in the American Midwest were attributed to extreme temperatures (Giles and Balafoutis, 1990). Examination of historical records reveals that heatwaves have claimed approximately 1,000 deaths annually in the USA since 1900 (Changnon *et al.*, 1996). Whilst many deaths are caused by cardiovascular failure, some result from secondary effects such as increased drownings (Brugge, 1991) and road accidents (Stern and Zehavi, 1990).

Economic and financial impacts are numerous. The 1990 heatwave in the UK, for example, led to road closures owing to melting tar; Heathrow Airport closed one runway as newly laid tar failed to set; trains were forced to travel more slowly to avoid damage to rails; and the entire stock at a Liverpool chocolate factory melted as refrigeration systems failed (Brugge, 1991). As with all extreme weather events, there are winners and losers. Sales of air conditioners in the US Midwest rose by 52 per cent during 1995 in comparison to the previous year; £15 million worth of ice cream was sold on 3 August 1990 in the UK (Young, 1991).

What have we learned about synoptic conditions leading to heatwaves?

While heatwaves are an extreme event not commonly associated with the UK, that country has nevertheless experienced a number of such events since the middle of the nineteenth century. Widespread events occurred in 1868, 1881, 1906, 1911, 1923, 1976 and 1990, with regional heatwaves occurring in some parts of the country in 1932, 1952, 1957, 1983 and 1995 (Bentham, 1997; Wheeler, 1984; Brugge, 1991; Young, 1991; Burt, 1992). The similarity in meteorological conditions during a number of these events suggests that certain conditions are a prerequisite for the occurrence of very high temperatures in the UK. According to Shaw (1977) and Brugge (1991), such events are most likely to occur:

- in late July/early August when land and sea surface temperatures are near their maxima, and if temperatures in northwestern Europe are high;
- when a high-pressure system is positioned over northern Britain/southern Scandinavia, to establish an easterly airflow to draw hot air off the Continent;
- in cloud-free conditions associated with the anticyclone which permit maximum levels of incoming solar radiation to reach the ground; and
- when dry conditions prevail, preventing loss of heat through surface evaporation.

The Greek heatwaves of 1987 and 1988 were caused by the location of a deep ridge of high pressure extending from Libya across Greece to the Black Sea, with a closed high developing in the lower atmosphere over Sicily (Giles and Balafoutis, 1990; Prezerakos, 1989). Anticyclonic airflow around this ridge drew hot, dry North African air across Italy and the Mediterranean into Greece, producing very hot conditions.

The heatwave in the American MidWest in 1995 began with the development of an upper-level ridge over the Great Plains. The ridge gradually moved northwards and eastwards before forming a closed circulation system centred over Illinois (Kunkel *et al*, 1996). Air temperatures increased to the high thirties. The system gradually moved eastwards, and by 14 July excessive heat had been transported to the northeastern USA. This sequence of events was closely monitored and a heatwave predicted days in advance, as an article in the *Chicago Sun Times* on 8 July 1995 illustrates:

> A heatwave is brewing for Chicago – a suffocating mass of air, the hottest for the season – a massive ridge of high pressure is stalling over the Midwest, a superheated dome of dry, sinking air with the mercury to go above 100°F by [11 or 12 July].
>
> (quoted in Changnon *et al.*, 1996: 1502)

Synoptic understanding of past events suggests that heatwaves are predictable. It follows that timely responses could be implemented to minimise the impacts of such events.

Who was vulnerable?

Despite the different cultural characteristics of each of the three countries investigated here, those most vulnerable to extreme heat exposure displayed a similar profile. In each case, the most vulnerable groups were:

- the elderly, particularly those older than 65 (according to Changnon *et al.* (1996), more than 70 per cent of heat-related deaths in the USA have been among this age group);
- the very young;
- those with impaired mobility; and
- low-income socio-economic groups residing in poorly ventilated, high-density housing in the inner city, which experience additional heating owing to the urban heat island.

What problems were experienced?

Impacts of heatwaves are most clearly documented through the recent experience in Chicago. Demands for power to drive refrigeration and air-conditioning systems led to power failures; inadequate hospital and ambulance services delayed people's access to medical attention and lives which might have been saved were lost. Many elderly people without air-conditioning and unable to afford fans were too afraid to ventilate their homes for fear of crime, thus increasing their risk of mortality (Changnon *et al.*, 1996). In Greece, high level of pollutants trapped below the high-pressure system dramatically increased death rates.

What coping mechanisms were employed?

Individuals may take many actions to bring relief from excessively hot weather. These include swimming or bathing in cool water; avoiding strenuous physical activity or outdoor labour; drinking plenty of fluids; visiting air-conditioned shopping centres; ventilating their homes; or using fans or air-conditioning. The latter is not necessarily a response which can be widely adopted. In 1971, for example, only 1 per cent of homes in the UK had air-conditioning; by 1983 only 3 per cent of homes were air-conditioned (Euromonitor, 1985). It is unlikely that systems could be installed rapidly enough to meet immediate demands. Even were this possible, additional problems might arise owing to excessive power demands.

More organised responses to past heatwaves have included closing polluting factories and introducing work shifts in Athens in 1987, when the biggest 100 factories were ordered to cut output by a third to help reduce smog. Public staff in offices without air-conditioning were also put on to part-time work to reduce exposure to the extreme heat (Giles and Balafoutis, 1990). In Chicago, a state of emergency was declared, enabling increased resources to be allocated towards addressing the crisis situation, and impacts of the event to be reduced. Had this step been taken sooner, many lives might have been saved (Changnon *et al.*, 1996).

The effectiveness of responses to heatwaves depends on the time available to implement them, the resources required to do so, and the cultural and socio-economic environment in which they are adopted. It is necessary to identify changes in the frequency, intensity and distribution of heatwaves in a future climate if appropriate response strategies are to be developed.

11.4 Heatwaves in Oxford and Thessaloniki in a future climate

In this section, potential changes in the frequency, intensity and distribution of heatwaves in a future climate at Thessaloniki, Greece, and Oxford, England, are investigated. Heatwaves during the 30–year period from 1961 to 1990 at the two case-study sites are identified; the persistence and severity of those events at each site are related to morbidity and mortality; and changes in the frequency and intensity of heatwaves in a future climate are considered. The methodology employed to conduct the study is first outlined. The distribution of heatwaves for the climate baseline and a scenario of climate change are compared, to enable possible impacts of climate change on future heatwaves to be considered. Finally, implications for impacts and policy responses are discussed.

11.4.1 Methodology

An extreme-values model was used to identify the probability of occurrence of heatwaves. The model accurately identified past events known to have occurred at both study sites between 1961 and 1990 (for further details refer to Karacostas and Downing, 1996).

The impact of high temperature and heatwave events on individuals can be depicted through the evaluation of thermal discomfort. Thermal discomfort is caused by the combined effect of air temperature, humidity, wind speed and radiation. Thom's (1959) temperature–humidity index (THI) is the most widely used and accepted index of heat discomfort. The index is defined as the temperature of saturated and almost still air which causes a thermal sensation in affected individuals.

The THI may be derived from the formula:

$$THI = 0.4(T + T_w) + 15 \tag{1}$$

where T is the dry-bulb temperature and T_w is the wet-bulb temperature, both in °F. Although the equation may be resolved using numerous methods, assessment of each revealed little difference between them. The appropriate formula was thus selected according to data requirements and availability.

The most suitable data for the calculation of THI are hourly values of air temperature and relative humidity (RH). Hourly values of THI were calculated using the following equation:

$$THI = T - (0.55 - 0.55RH)(T - 58) \tag{2}$$

where T is in °F and RH is expressed in decimal format (i.e. 60% is 0.60). Various statistical analyses were then performed. Raffner and Bair (1977) relate the degrees of sheltered air temperature and relative humidity to equivalent THI values. For categorising health effects, general bands of the THI may be delineated, on the basis of discomfort of a clothed person engaged in sedentary activity in nearly calm air (Table 11.1). It should be emphasised, however, that the variation between individuals can be significant, and those from different regions may have different thresholds owing to their body types, behaviour and environment.

Hourly measurements of air temperature and relative humidity were not available for Oxford. It was therefore necessary to develop a methodology based on daily data for this site. Hourly data for Thessaloniki extend from 1960 to 1992, for the months of May, June, July, August and September. The data for 24-hourly values of temperature and humidity for each were processed by software developed to calculate hourly THI values.

The use of Thessaloniki's hourly THI data was achieved in the following way:

1   The time series was separated into three decades: (i) 1960–9, (ii) 1970–9, (iii) 1980–9.
2   Each month was separated into three 10-day intervals of almost equal size (May: 1–10, 11–20, 21–31; June: 1–10, 11–20, 21–30; July: 1–10, 11–20, 21–31; August: 1–10, 11–20, 21–31; September: 1–10, 11–20, 21–30).
3   A linear regression model was calculated for each decade and for each 10-day interval, relating each of the 24-hour values of THI (used as dependent variables) with the independent variables: maximum daily temperature (T_{max}), minimum daily temperature (T_{min}), average daily temperature (T_{mean}) and average daily relative humidity (RH). The form of the regression models was:

$$\text{THI} = a_0 + a_1 * T_{max} + a_2 * T_{min} + a_3 * T_{mean} + a_4 * RH \tag{3}$$

T_{max}, T_{min}, T_{mean} and RH were available from datasets for each site and were thus used as the independent variables for the regression models. The regression models were applied to both sites. Thessaloniki's data for the

Table 11.1 The temperature–humidity index

THI	Impact
70–74	Few people feel uncomfortable
75–79	About one-half of all people feel uncomfortable
80–84	Nearly everyone feels uncomfortable
85–89	Rapidly decreasing work efficiency
>90	Extreme danger

years of 1990, 1991 and 1992 were used to assess the goodness of fit and efficiency of the models implemented with the hourly regressions. Figure 11.2 shows that the results provide a very good approximation of the THI.

11.4.2 Heatwaves in Thessaloniki and Oxford from 1961 to 1990

The THI was calculated for the climatological station of Thessaloniki, Greece, for 1960–79. The relative frequency distribution at Thessaloniki, for the months of May, June, July, August and September, is depicted in Figure 11.3. July and August are the two hottest months, while May has fewer and milder events. For the severe and extreme classes in Table 11.1, the THI values show considerable variation between years (Figure 11.4). Numerous severe heatwaves have occurred, with notable seasons in 1964, 1969, 1987 and 1988.

As for Thessaloniki, July and August are the warmest months in Oxford, and most prone to heatwaves (Figure 11.5). In contrast, however, very few heat events occur in May, June and September. This is consistent with Brugge's (1991) analysis of heatwave conditions in the UK. For Oxford, a lower threshold of THI (moderate or higher) is appropriate since fewer extreme heatwaves occur. Figure 11.6 shows that 1975, 1976, 1983, and 1989 and 1990 all recorded significant events. These also tended to be drought years, although 1984 was worse than 1989.

Although individual responses to heat vary considerably from person to person and time to time, THI values can be used to express approximately a person's reaction to heat and humidity. An attempt was made to collect data on heat-related illnesses and correlate them to the THI in order to quantify the impact of heatwaves on human morbidity and mortality at the study sites.

Data were collected from three hospitals in Thessaloniki for 1988–92 for admissions due to cardiac illnesses (coronary heart disease, heart failure, cardiac arrhythmia, hypertension and other) and pneumonological diseases (airway inflammation, chronic obstructive pulmonary disease, lung infections, pleural effusion and other). No clear relationship was found between these hospital admissions and the THI. Although the August of 1988 was warmer than average, it was not associated with a significant increase in admissions for illnesses that are related to heat stress. The other heat events in 1988–92 were relatively minor.

No hospital admissions were recorded for heat stress or heat exhaustion in Oxford. This is not surprising, since Oxford experiences relatively few serious heatwaves. It is quite likely that people at risk are treated by paramedics and general practitioners without referral to a hospital. It was not possible to define a relationship between temperature and humidity conditions and human health at either study site due to the lack of data. Any assessment of potential impacts of future heatwaves thus has to remain qualitative.

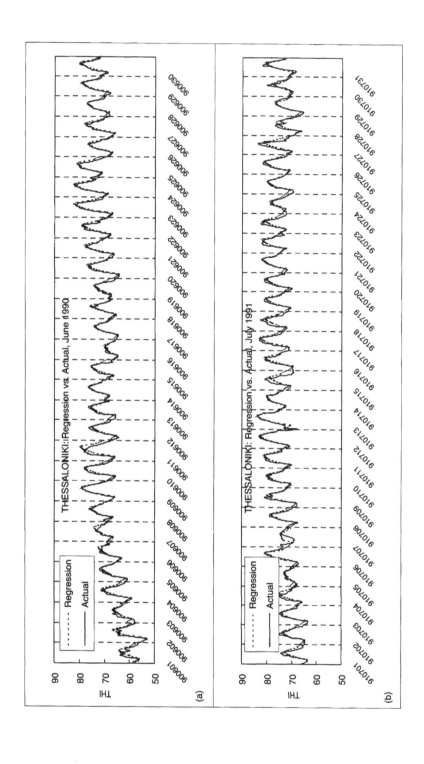

(a) THESSALONIKI: Regression vs. Actual, June 1990

(b) THESSALONIKI: Regression vs. Actual, July 1991

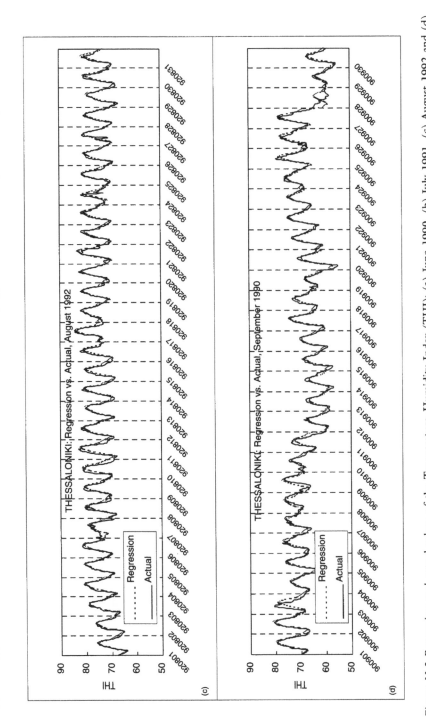

Figure 11.2 Regression and actual values of the Temperature–Humidity Index (THI): (a) June 1990, (b) July 1991, (c) August 1992 and (d) September 1990

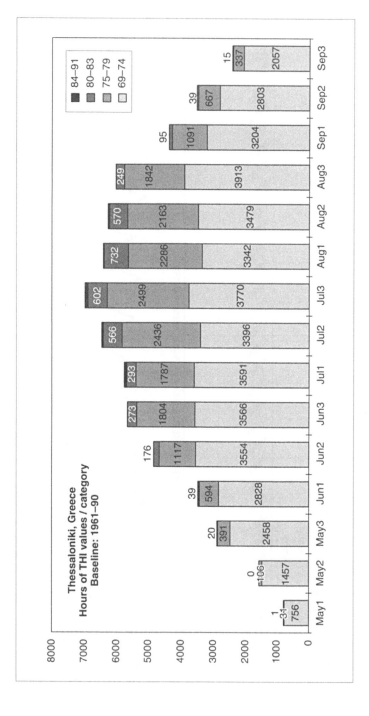

Figure 11.3 Distribution of the Temperature–Humidity Index at Thessaloniki: May to September, 1961–90

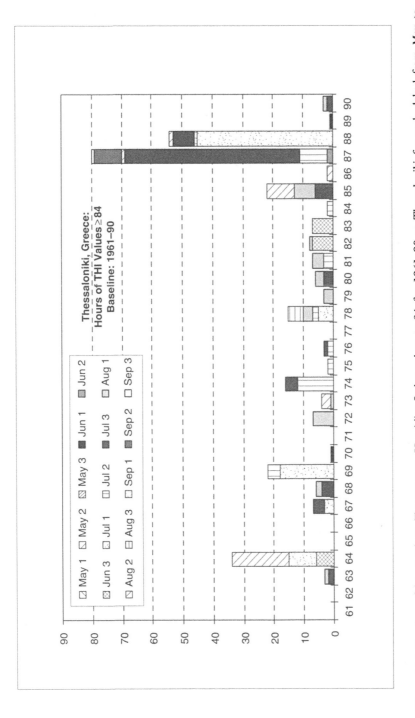

Figure 11.4 Number of hours where Temperature–Humidity Index values ≥ 84 for 1961–90 at Thessaloniki for each dekad from May to September

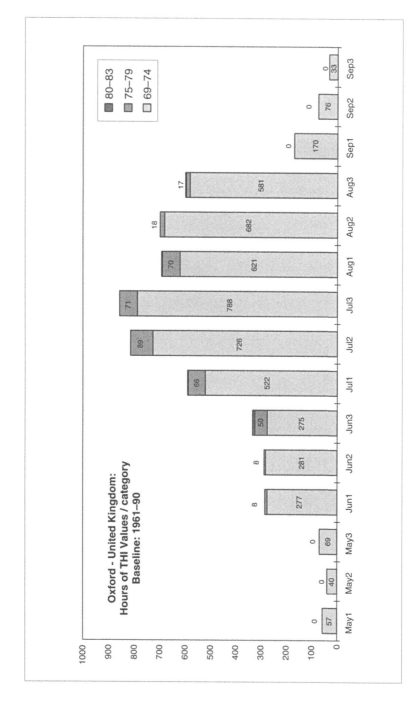

Figure 11.5 Distribution of the Temperature–Humidity Index at Oxford: May to September 1961–90

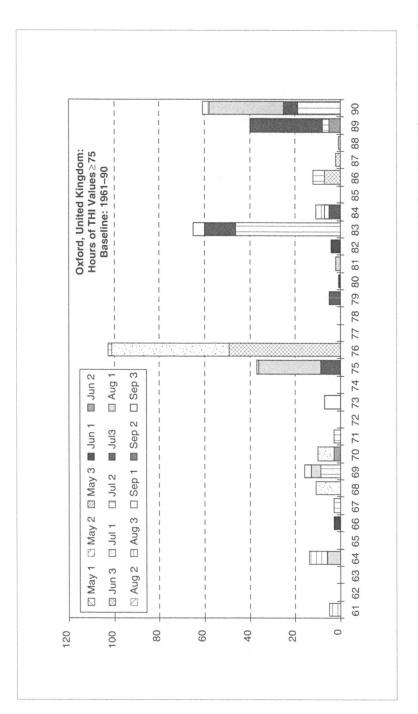

Figure 11.6 Number of hours where Temperature Humidity Index values ≥ 75 for 1961–90 at Oxford for each dekad from May to September

11.4.3 Scenarios of heatwaves in a future climate

Whilst considerable uncertainty is associated with scenarios of climate change, there is broad agreement that temperature is the one element of future climates that may be most accurately 'predicted'. It is almost certain that temperatures will increase in a warmer world. The frequency and intensity of heatwaves will thus almost inevitably increase. This certainty enables greater confidence to be placed in projections of heatwaves under altered climates than for other extreme events.

Spatial scenarios of maximum temperature

Scenarios of the spatial distribution of changes in maximum temperature, derived from the United Kingdom Meteorological Office high resolution (UKHI) equilibrium global circulation model (Mitchell et al., 1990) are shown in Figure 11.7b, against the current baseline of T_{max} in Figure 11.7a. As is common for climate change impacts assessments (see Carter et al., 1994), output from this GCM was scaled to estimates of global average temperature change resulting from a 'non-interventionist' scenario (the IS92a) of greenhouse gas emissions (Houghton et al., 1992) produced by the Model for the Assessment of Greenhouse Gas-Induced Climate Change (MAGICC) (Wigley, 1991; Wigley and Raper, 1987, 1992). The 2025 scenario represents a 'minimum shift' using MAGICC's low estimate for 2025 (+0.34°C global average temperature change). The higher estimate of climate sensitivity, scaled for 2050 (+1.37°C), represents an upper bound of warming expected in the next few decades (the 'maximum shift' scenario), at least for changes in average climatic conditions.

Thessaloniki and Oxford both experience increased summer temperatures under each scenario, with Thessaloniki experiencing a greater temperature increase than Oxford. The greatest temperature increase in Thessaloniki occurs in July, while in Oxford this occurs in August. This pattern suggests an intensification of the temperature of the hottest month in each site. Temperatures at both sites are also likely to increase throughout the summer, indicating a lengthening of this season.

Site scenarios of changes in THI

The extreme-event model was used with a scenario of future climate change to investigate the extent to which the frequency and intensity of heatwaves might change in a future climate. The extreme-event model requires daily temperature and relative humidity data. Climate-change scenarios are only able to produce changes in maximum daily temperatures. Relative humidity was therefore assumed to be constant. Daily temperature data were available for downscaled scenarios based on the United Kingdom Meteorological Office's transient GCM experiment (UKTR) (Barrow et al., 1995). The model decade centred on 2070 was used.

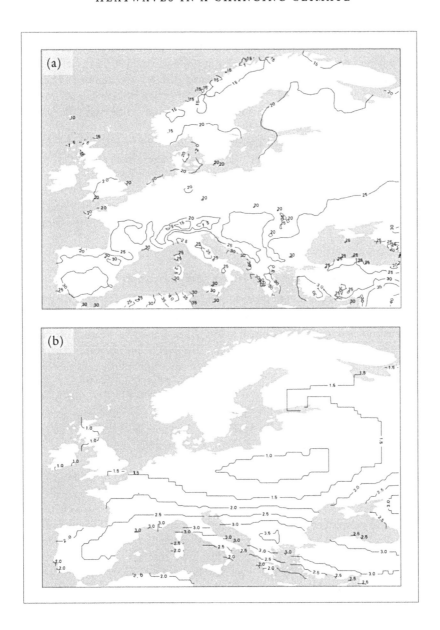

Figure 11.7a Current distribution of T_{max} in Europe

Figure 11.7b Scenario of climate change for T_{max}, based on the UKHI scenario, scaled to the IS92a global warming for 2050 and high estimate of climate sensitivity

It is clear, comparing figures 11.4 and 11.8 that the frequency of THI values above 84 in Thessaloniki will increase markedly by the middle of the twenty-first century whereas few years exceed 50 hours at present, over 100 years will be the norm by 2050. In the peak period (third dekad of July), the number of hours above 69 (the lowest threshold for the index) increases by almost 500. More significant changes, however, are evident in the increases in the shoulder warm periods, suggesting a lengthening of the summer season.

Figure 11.9 shows the effect of expected warming in the middle of the twenty-first century for Oxford. As in Greece, the annual average escalates, from less than 20 at present to over 100 by 2050. Heat events shift from a peak in July to a peak in August. The number of hours over 69 increases from less than 1,000 in any 10-day period to over 2,000 in each period in July and August. This order-of-magnitude shift in heat events would represent a significant change in the climate of southern England.

11.4.4 Potential impacts of future heatwaves at the study sites

It has been shown that the duration of hot weather has a greater impact on the general population than does its intensity, except for people older than 65 (Kalkstein and Davis, 1990; Kalkstein, 1991). The impact of weather on mortality is greatest in cooler climates where extreme heat is uncommon and people are less acclimatised to heat. Increases in Oxford temperatures may therefore have greater health implications than increases in Thessaloniki, which is more accustomed to high temperatures. However, if heatwaves in Oxford occur later in the season, as suggested by both scenarios, people may become better acclimatised and heatwaves may have less severe impacts. The sheer scale of potential temperature increases in Thessaloniki (up to 5°C) suggests, though, that climate change may have a significant impact on human mortality in this city.

In addition to direct health effects, secondary impacts may result from heat events. If direct effects are significant, pressure on medical and health care facilities would increase. A reduction in economic activity may occur as workers are unable to perform at optimum levels, or fail to attend work. The implications for energy supply and consumption are serious, particularly if air-conditioning is employed for cooling. This is clearly illustrated through the experience of Chicago, where excessive energy demands for cooling systems led to power failures. The lack of cooling which resulted was widely regarded as contributing to the large number of fatalities. Similar demands may be exerted on water supplies as consumption rates increase through people's efforts to remain cool.

11.5 Adaptation and responses to future heatwaves

A number of responses are possible to reduce the impact of heatwave events, each functioning over a different time scale, requiring differing levels of expenditure and resulting in differing levels of disruption to normal activity (see Table 11.2).

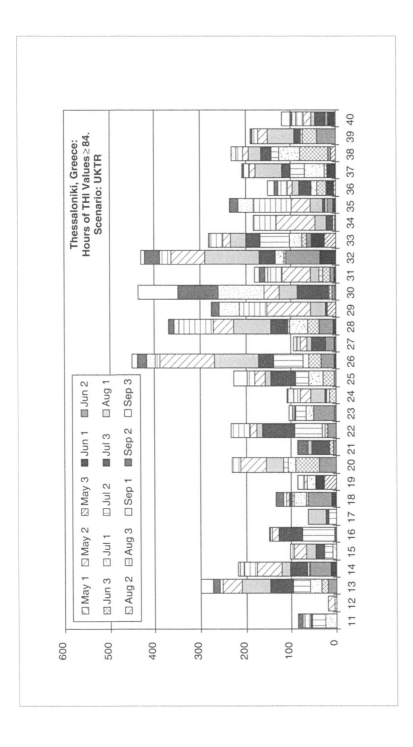

Figure 11.8 Change in the frequency of temperature–humidity index values > 84 in Thessaloniki using the UKTR scenario. Model years 11–40 are shown

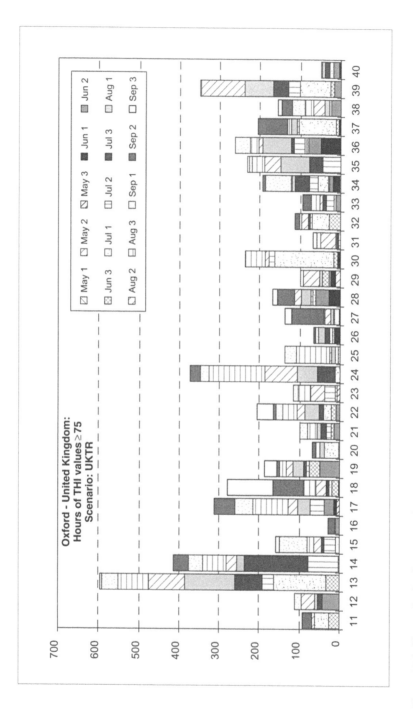

Figure 11.9 Change in the frequency of temperature–humidity index values > 75 in Oxford using the UKTR scenario. Model years 11–40 are shown

Table 11.2 Responses to reduce the impact of heatwaves

Adaptation/response	Response time	Resource input	Disruption
Bathe in cool water/swim	Short	Medium	High
Use air-conditioning	Short	High energy needs	None
Use cooling fans	Short	Moderate energy needs	None
Shift work patterns	Short to medium	Low	Medium
Heatwave disaster plan	Medium	Low	Varies
City-specific heatwave indices	Medium	Low	None
Design energy-efficient buildings	Long	High	None

Low-cost, short-term responses may include submerging oneself in cool water (swimming or bathing), installing fans and remaining in cool shady places. These are not, however, very practical responses, and mean halting other (productive) activities. High cost, medium-term but convenient responses may include installing air-conditioning in one's home and workplace. Research in the USA, for example, estimated that 21 per cent of heat-related deaths in New York between 1964 and 1988 could have been prevented by air-conditioning (Kalkstein, 1993). This response has the important advantage of enabling individuals to continue with their normal indoor activities. However, the purchase and use of air-conditioning requires capital expenditure and may not be viable for the poor. Widespread use of air-conditioning would also increase energy consumption and heat generation in cities. While convenient, it may therefore not prove to be a long-term alternative.

Effective long-term responses require the design of better-insulated and energy-efficient buildings that would reduce the need for active cooling systems or their operating costs. It will take both time and significant demand for such measures to be adopted, and for existing building stock to be renovated to function more efficiently.

A useful lesson to emerge from the Chicago experience is that heat disaster plans should be developed to respond to heatwaves in high-risk areas. Extreme heat events can be predicted, suggesting that appropriate response strategies can be set in place before the event occurs. Such plans might include putting hospitals on alert, ensuring that additional staff and resources can be obtained to address emergency situation, and rationalising water and energy use so as to maximise their efficiency. Certain public centres close to low-income residential areas could be identified as 'cooling centres' which are kept open to enable those in poorer households and other vulnerable groups to escape the worst of the heat.

An important component of a heatwave response plan should be to develop specific heat indices for cities at risk of heatwaves. Given that human tolerance to heat varies considerably between geographical locations, it is important that fairly complex indices be developed which accurately reflect the degree of human discomfort that might be experienced under certain meteorological conditions. Such response strategies could play a major role in reducing negative impacts.

11.6 Conclusion

Empirical analyses of climatic time series often suffer from the limited length of available records. This is particularly problematic if the focus of interest is on values in the tails of the distributions. The present study is no exception. The analytical procedure described is highly data intensive, particularly since it requires hourly values of ambient air temperature and relative humidity. Additional climatological parameters are also desirable. Analysis is consequently limited to those stations which have sufficiently long climatic series for the identification and analysis of the heatwaves. Unfortunately, few European data sets with sufficiently long records are available, thereby limiting analysis of the spatial distribution of heatwaves over Europe. With further data, it would be possible to couple the site-probability model with a spatial assessment of heatwaves in Europe.

Site-specific analyses of these extreme events are, however, possible. Comparisons between these analyses may yield further insights into appropriate responses. Previous studies have suggested that the impact of weather on human mortality is relative rather than absolute, and depends on seasonal and regional variations of weather (Kalkstein, 1991). The direct health impact of additional heat stress related to climate change is likely to be greater in northern Europe than in southern Europe. In southern Europe, an extension of the hot months for another month and additional heatwaves are within the current expectations. For northern Europe, the means to cope with heatwaves are limited and an increase in frequency without adequate education and preparation may lead to increased morbidity and mortality. Such site-specific studies are important if specific heat thresholds are to be identified for different geographic locations.

There is every probability that heatwaves will become both more widespread and more frequent as a result of climate change. Heat waves are predictable extreme events. Considerable experience has been gained in responding to past events and the technology exists to develop appropriate response strategies. It is thus essential that every effort is made to further our understanding of the nature and distribution of these events and to employ past experience to ensure that the impact of heat events on future societies is minimised.

Acknowledgments

The authors would like to thank Chris Mitas, Christina Rizou, Constantina Spyrou and Michael Papadopoulos from the Aristotle University of Thessaloniki for their contribution to the analysis of 1961–90 climatic data. We are grateful to John Orr of the Environmental Change Unit, University of Oxford, and Elaine Barrow of the Climatic Research Unit, University of East Anglia, for climatic data and preparation of European scenarios. This chapter draws upon Karacostas and Downing (1996).

Appendix

Selected studies for the estimation of the effect of temperature changes on morbidity

Health endpoints	Study area and design	Change in health endpoint per 1°C increase in average temperature, %		Temperature range	References
		Age	Change (95% CI; σ)		
Cardiovascular diseases	Combined	<65	−1.6 (−2.8 to −1.4; 0.6) 0.7 (−0.1 to 1.5; 0.4)	cold (a) warm	
		>65	−4.1 (−5.5 to −2.7; 0.7) 1.6 (−0.2 to 3.4; 0.9)	cold warm	
Coronary heart disease	England, Wales; 1968–88 monthly time series		(b) −2.4 (−0.4 to −0.8; 0.8)	−2 to 20°C	Langford and Bentham (1995)
Cerebrovascular disease			−2.8 (−4.6 to −1.0; 0.9)	−2 to 20°C	
Coronary heart disease	Taiwan; 1981–91 daily time series	45–64 >64	(c) −3.0 (−6.5 to 0.0; 1.6) −2.8 (−5.1 to −0.1; 1.3)	9 to 26°C	Pan et al. (1995)
		45–64 >64	5.8 (0.0 to 13.8; 3.5) 3.7 (0.0 to 8.5; 2.2)	26 to 32°C	
Cerebral haemorrhage		45–64 >64	−4.3 (−7.0 to −2.1; 1.3) −3.5 (−5.9 to −1.5; 1.1)	9 to 28°C	
		45–64 >64	−8.5 (−12.8 to −3.0; 2.5) −6.8 (−11.0 to −1.3; 2.5)	28 to 32°C	

Table 11.3 Continued

Health endpoints	Study area and design	Change in health endpoint per 1°C increase in average temperature, %			References
		Age	Change (95% CI; σ)	Temperature range	
Cerebral infarction		45–64	−1.8 (−6.6 to 1.6; 2.1)	9 to 29°C	
		>64	−3.8 (−6.5 to −1.3; 1.3)		
		>64	22.0 (11.0 to 36.0; 6.4)	29 to 32°C	
Coronary heart disease	Israel (1976–85); monthly time series	(d, e)			Green et al. (1994)
		45–64	−9.3 (−10.8 to −7.1; 0.9)	5 to 21°C	
		>65	−10.6 (−11.8 to −9.8; 0.5)		
Cerebrovascular disease		45–64	−10.0 (−12.3 to −7.6; 1.1)		
		>65	−9.7 (−11.5 to −8.0; 0.9)		
Cardiovascular diseases	The Netherlands; 1979–87 daily time series	(f)			Kunst et al. (1993)
			−1.2 (−2.4 to 0.0; 0.6)	<16.5°C	
			1.1 (−0.1 to 2.3; 0.6)	>16.5°C	
Coronary artery disease	England, Wales: 1969–71 monthly time series	(f)			West and Lowe (1976)
Total mortality	Combined		−2.5 (−5.4 to 0.4)	<17.9°C	
			−1.0 (−2.0 to 0.0; 0.5)	cold (a)	
			1.4 (−0.8 to 3.6; 1.1)	warm	
Total mortality	England, Wales; 1968–88 monthly time series	(b)			Langford and Bentham (1995)
			−2.1 (−3.5 to 0.7; 0.7)	−2 to 20°C	
Total mortality	The Netherlands; 1979–87 daily time series	(f)			Kunst et al. (1993)
			−0.9 (−2.0 to 0.2; 0.5)	<16.5°C	
			1.9 (−0.4 to 4.2; 1.0)	>16.5°C	

Cause	Study	Age	Estimate	cold (a) / warm	Reference
Respiratory diseases	Combined		−3.8 (−6.9 to −0.7; 1.6) 10.4 (0.0 to 20.8; 5.3)		
Chronic bronchitis Pneumonia	England, Wales; 1968–88 monthly time series		(b) −10.5 (−17.2 to −3.8; 3.4) −3.4 (−5.6 to −1.2; 1.1)	−2 to 20°C −2 to 20°C	Langford and Bentham (1995)
Respiratory diseases	The Netherlands; 1979–87		(f) −2.7 (−5.4 to 0.0; 1.4) 10.4 (0.0 to 20.8; 5.3)	<16.5°C >16.5°C	Kunst *et al.* (1993)
Pneumonia	England, Wales; 1963–6 daily time series	<60 >60	(f) −5.4 (−10.9 to 0.1; 2.8) −4.7 (−9.4 to 0.0; 2.4)	<20°C	Bull and Morton (1975, 1978)
Coronary heart disease (myocardial infarction)	England, Wales; 1963–6		(d, f) England and Wales		Bull and Morton (1975, 1978)
		<60 >60	−1.2 (−2.4 to 0.0; 0.6) −1.9 (−3.9 to 0.1; 1.0)	<20°C	
	New York; 1965–8 daily time series		New York		
		<60 >60	−0.9 (−1.9 to 0.1; 0.5) −1.0 (−2.0 to 0.0; 0.5)	<20°C	
		<60 >60	0.5 (−0.1 to 1.1; 0.3) 8.7 (0.1 to 17.3; 4.4)	>20°C	
Cerebrovascular disease			England and Wales		
		<60 >60	−1.3 (−2.7 to 0.1; 0.7) −2.0 (−4.0 to 0.0; 1.0)	<20°C	
			New York		
		<60 >60	−0.6 (−1.2 to 0.0; 0.3) −1.7 (−3.5 to 0.1; 0.9)	<20°C	
		<60 >60	2.8 (0.1 to 5.5; 1.4) 15.1 (0.0 to 30.1; 7.7)	>20°C	

Table 11.3 Continued

Health endpoints	Study area and design	Change in health endpoint per 1°C increase in average temperature, %		Temperature range	References
		Age	Change (95% CI; σ)		
Cerebrovascular disease	New York; 1959–63 Tokyo; 1960–4 London; 1960–4 seasonal time series		(f, g) New York: −0.8 (−1.8 to 0.2; 0.4) Tokyo: −2.4 (−5.3 to 0.5; 1.2) London: −3.1 (−6.8 to 0.6; 1.6)	4 to 22°C 4 to 22°C 4 to 16°C	Sakamoto-Momiyama and Katayama (1971)

Source: Martens (1997)

Notes:

a Cold: average temperature below the comfort temperature; warm: average temperature above the comfort temperature as defined in text.

b Calculated as percentage difference between mean mortality at yearly average temperature between 1968 and 1988 (=6°C) and 1° increase. No influenza included and in the regression equation 'year' was set to 1988. Standard error was estimated as in (f) with a *p*-value of 0.001.

c Calculated between the difference of the Odds Ratio at comfort temperatures and 10°C below this temperature divided by 10 and the difference between the Odds Ratio at comfort temperature and the maximum temperature used in the study (32°C) divided by the difference in temperature. Effect of cerebral haemorrhage not included in the combined estimate.

d Study results averaged.

e Calculated as mean ratio of winter to summer mortality divided by the difference in winter and summer minimum temperature (=15°C).

f The standard error was estimated on the basis of *p*-values and assuming a normal distribution, with a *p*-value of 0.05. The standard error was calculated on the basis of the upper limit of the cutoff probability: $z = 1.65$ for a *p*-value of 0.05; standard normal distribution: $z = (x - \mu)/\sigma$; H_0: $\mu = 0$; 95% CI $= \mu \pm 1.96 * \sigma$

g Average of spring and autumn values.

h Calculated as percentage difference between mortality at most comfortable temperatures and mortality at lowest and highest temperatures, divided by the temperature difference between comfort, lowest and highest temperatures (using the weighted least-squares model and fixed CO_2 concentrations).

References

Balling, R.C., Jr., and Idso, S.B. (1990) Effects of greenhouse warming on maximum summer temperatures. *Agricultural and Forest Meteorology* 53, 1–2: 143–147.

Barrow, E., Hulme, M., and Semenov, M.A. (1995) 'Scenarios of climate change', in P.A. Harrison, R.E. Butterfield and T.E. Downing, eds, *Climate Change and Agriculture in Europe: Assessment of Impacts and Adaptations*, Oxford: Environmental Change Unit, University of Oxford.

Bentham, G. (1992) Global climate change and human health. *GeoJournal* 26, 1: 7–12.

Bentham, G. (1997) 'Health', in J. Palutikof, S. Subak and M.D. Agnew, eds, *Economic Impacts of the Hot Summer and Unusually Warm Year of 1995*, Norwich: University of East Anglia.

Brugge, R. (1991) The record-breaking heat wave of 1–4 August 1990 over England and Wales. *Weather* 46, 1: 2–10.

Bull, G.M. and Morton, J. (1975) Relationships of temperature with death rates from all causes and from certain respiratory and arteriosclerotic diseases in different age groups. *Age and Ageing* 4: 232–246.

Bull, G.M. and Morton, J. (1978) Environment, temperture and death rates. *Age and Ageing* 7: 210–224.

Burt, S. (1992) The exceptional hot spell of early August 1990 in the United Kingdom. *International Journal of Climatology* 12, 6: 547–567.

Carter, T.R., Parry, M.L., Harasawa, H., and Nishioka, S. (1994) *IPCC Technical Guidelines for Assessing Climate Change Impacts and Adaptations*, London: UCL.

Changnon, S.A., Kunkel, K.E., and Reinke, B.C. (1996) Impacts and responses to the 1995 heat wave: A call to action. *Bulletin of the American Meteorological Society* 77, 7: 1497–1506.

Department of the Environment (DoE) (1996) *Review of the Potential Effects of Climate Change in the United Kingdom*, London: HMSO.

Downing, T.E. (1995) Damn this heat! *New Statesman and Society*, 11 August, 18–19.

Euromonitor (1985) *Data and Statistics*, London: Euromonitor.

Giles, B.D. and Balafoutis, C.J. (1990) The Greek heat waves of 1987 and 1988. *International Journal of Climatology* 10, 5: 505–517.

Green, M.S., Harari, G. and Kristal-Boneh, E. (1994) Excess winter mortality from ischaemic heart disease and stroke during colder and warmer years in Israel: An evalution and review of the role of environmental temperature. *European Journal of Public Health* 4: 3–11.

Hashimoto, M. and Nishioka, S. (1991) 'Potential impacts of climate change on human settlements; the energy, transport and industrial sectors; human health and air quality', in J.A. Jäger and H.L. Ferguson, eds, *Climate Change: Science, Impacts and Policy*, Proceedings of the Second World Climate Conference, Cambridge: Cambridge University Press, 109–122.

Houghton, J.T., Callander, B.A., and Varney, S.K., eds (1992) *The Supplementary Report to the Intergovernmental Panel on Climate Change Scientific Assessment*, Cambridge: Cambridge University Press.

Ito, K., Thurston, G.D., Hayes, C. and Lippmann, M. (1993) Associations of London, England, daily mortality with particulate matter, sulphur dioxide, and acidic aerosol pollution. *Archives of Environmental Health* 48, 4: 213–220.

Kalkstein, K.S. (1993) Direct impacts in cities. *The Lancet* 342, 4: 1397–1399.

Kalkstein, L.S. (1991) 'Potential impact of global warming: climate change and human mortality', in R.L. Wyman, ed., *Global Climate Change and Life on Earth*, New York: Routledge.

Kalkstein, L.S., and Davis, R.E. (1990) Weather and human mortality: An evaluation of demographic and international responses in the United States. *Annals of the Association of American Geographers* 79, 1: 44–64.

Kalkstein, L.S., Jamason, P.F., Greene, J.S., Libby, J., and Robinson, L. (1996) The Philadelphia hot weather–health watch/warning system: Development and application, summer 1995. *Bulletin of the American Meteorological Society* 77, 7: 1519–1528.

Karacostas, T.S., and Downing, T.E. (1996) 'Heat wave events in a changing climate', in T.E. Downing, A.A. Olsthoorn and R.S.J. Tol, eds, *Climate Change and Extreme Events: Altered Risk, Socio-economic Impacts and Policy Responses*, Oxford and Amsterdam: Environmental Change Unit, University of Oxford and Vrije Universiteit, Amsterdam.

Kunkel, K.E., Changnon, S.A., Reinke, B.C., and Arrit, R.W. (1996) The July 1995 heat waves in the Midwest: A climatic perspective and critical weather factors. *Bulletin of the American Meteorological Society* 77, 7: 1507–1518.

Kunst, A.E., Looman, C.W.N., and Mackenbach, J.P. (1993) Outdoor air temperature and mortality in the Netherlands: A time-series analysis. *American Journal of Epidemiology* 137: 331–341.

Langford, I.H. and Bentham, G. (1995) The potential effects of climate change on winter mortality in England and Wales. *International Journal of Biometeorology*, 38: 141–147.

Martens, W.J.M., ed. (1996) *Vulnerability of Human Population Health to Climate Change: State-of-Knowledge and Future Research Directions*, Dutch National Research Programme on Global Air Pollution and Climate Change, Maastricht: University of Limburg.

Martens, W.J.M. (1997) *Health Impacts of Climate Change and Ozone Depletion: An Eco-epidemiological Modelling Approach*, Maastricht: Maastricht University.

McMichael, A.J. (1996) 'Human population health', in R.T. Watson, M.C. Zinyowera, R.H. Moss and D. Dokken, eds, *Climate Change 1995. Impacts, Adaptations and Mitigation of Climate Change: Scientific-Technical Analyses*, Contribution of Working Group II to the Second Assessment Report of the Intergovernmental Panel on Climate Change, Cambridge: Cambridge University Press.

McMichael, A.J., Haines, A., Slooff, R., and Kovats, S. (1996) *Climate Change and Human Health: An Assessment Prepared by a Task Group on Behalf of the World Health Organization, the Meteorological Organization and the United Nations Environment Programme*, Geneva: World Health Organization.

Mitchell, J.F.B., Manabe, S., Meleshko, V., and Tokioka, T. (1990) 'Equilibrium climate change and its implications for the future', in J.T. Houghton, G.J. Jenkins and J.J. Ephraums, eds, *Climate Change: The IPCC Scientific Assessment*, Cambridge: Cambridge University Press, 131–172.

Pan, W.H., Li, L.A. and Tsai, M.J. (1995) Temperature extremes and mortality from coronary heart disease and cerebral infarction in elderly Chinese. *Lancet* 345: 353–355.

Prezerakos, N.G. (1989) A contribution to the study of the extreme heat wave over the south Balkans in July 1987. *Meteorology and Atmospheric Physics* 41, 4: 261–271.

Raffner, J.A., and Blair, F.E. (1977) *The Weather Almanac*, 2nd edition, Detroit: Gale.

Ramlow, J.M., and Kuller, L.H. (1990) Effects of the summer heat wave of 1988 on daily mortality in Allegheny County, PA. *Public Health Reports* 105, 3: 283–289.

Sakamoto-Momiyama, M. and Katayama, K. (1971) Statistical analysis of seasonal varia-tion in mortality. *Journal of the Meteorological Society of Japan* 49 (6): 494–508.

Schneider, S.H. (1996) 'The future of climate: Potential for interaction and surprises', in T.E. Downing, ed., *Climate Change and World Food Security*, Heidelberg: Springer-Verlag.

Shaw, M.S. (1977) The exceptional heat-wave of 23 June to 8 July 1976. *Meteorological Magazine* 106, 1264: 329–346.

Shumway, Azari, A.S. and Paewitau, Y. (1988) Modelling mortality fluctuations in Los Angeles (California, USA) as functions of pollution and weather effects. *Environmental Research* 45(2): 224–241.

Smith, J.B., and Tirpak, D.A. (1990) *The Potential Effects of Global Climate Change on the United States*, London: Hemisphere.

Stern, E. and Zehavi, Y. (1990) Road safety and hot weather: A study in applied transport geography. *Transactions of the Institute of British Geographers* NS 15, 1: 102–111.

Thom, E.C. (1959) The discomfort index. *Weatherwise* 12: 57–60.

Wenzel, H.G. (1984) Beurteilung von Hitzebelastungen des Menschen durch Klimasum-menmabe. *Moderne Unfallverhütung* 28: 47–54.

West, R.R. and Lowe, C.R. (1976) Mortality from ischaemic heart disease: Inter-town variation and its association with climate in England and Wales. *International Journal of Epidemiology* 5 (2): 195–201.

Wheeler, D.A. (1984) The July 1983 'heat wave' in north-east England. *Weather* 39: 178–181.

Wigley, T.M.L. (1991) A simple inverse carbon cycle model. *Global Biogeochemical Cycles* 5: 373–382.

Wigley, T.M.L. and Raper, S.C.B. (1987) Thermal expansion of sea water associated with global warming. *Nature* 330: 127–131.

Wigley, T.M.L. and Raper, S.C.B. (1992) Implications for climate and sea level of revised IPCC emissions scenarios. *Nature* 357: 293–300.

Young, M.V. (1991) Hottest ever – worse to come! Exceptional heat waves – how newspapers report the events. *Weather* 46: 137–141.

12

ECONOMIC ANALYSIS OF NATURAL DISASTERS

Richard S.J. Tol and Frank P.M. Leek

12.1 Introduction

Other chapters in this volume examine in detail the impact of (changes in) specific extreme weather events for a number of case studies. Tol (Chapter 13, this volume) discusses the implications of changes in extreme weather for the insurance industry. This chapter has a broader goal: the impact of weather disasters and disaster management on the economy as a whole, in both the short and the long term. The literature pays little attention to this complicated issue, so most of the analysis below is original. It should be noted that the larger part of the discussion in this chapter is strictly economic. Attention is restricted to the impact of (changes in) extreme weather on the economy through the loss and reallocation of economic assets (see also Zeckhauser, 1996).

After a short treatise on the significance of the direct impact of extreme weather events, the argument of the chapter is along the following lines. First of all, economists traditionally think in terms of activity and flows of goods and services. Weather events affect stocks rather than flows; indeed, in the aftermath of a disaster, economic activity may well increase. Economic indicators do not measure the damage to the stock at risk, or the misery of the disaster, but do measure the restoration effort. Hence economic indicators are bound to give a misleading picture of the impact of weather disasters. Second, the economy is a complex, dynamic, adaptive system, constantly faced with shocks and changes. A weather disaster is one such shock, which the system will absorb as quickly as possible. Indeed, economic agents will do their best to nullify the negative consequences of the disaster, and to take advantage of it whenever possible, although success is never guaranteed. This implies that the impacts of most weather disasters rapidly get lost in the noise of the system, and that for the few cases where impacts are more apparent, the longer-term impact can be disadvantageous as well as advantageous. Third, economic agents dislike uncontrollable shocks. They will try their best to smooth the impact of shocks. These efforts have real and lasting economic effects in that defensive expenditure

308

on protection, mitigation and insurance competes with other investments in limited budgets.

For the larger part, the chapter contains conceptual and basic mathematical reasoning. The arguments are supported by some descriptive statistics, and by examples drawn from the other chapters of this book. Economic indicators and dynamic adaptation are treated in Section 12.2. Section 12.3 is on disaster management and its economic consequences. Section 12.4 provides a more dynamic analysis of the interplay between natural disaster, disaster management, human expansion and climate change. Section 12.5 employs the findings of the other sections to assess the broad economic consequences of the results of the case studies in this volume. Section 12.6 presents some conclusions.

Sections 12.3 and 12.4 rely on mathematical models of the economy, whereas in the other sections the argumentation is more verbal. The reason is this. Although mathematics is preferred because of its rigour, mathematical models are not to be used for things they are not meant to be used for. Because of the time and financial constraints, only simple economic models could be analysed. Simple economic models are a sort of metaphor to describe broad tendencies, or attractors towards which reality evolves if left undisturbed for quite a while. Events have little meaning in this idealised world, but the yearly recurrent possibility of an event has. Hence the impact of events is treated verbally whereas the impact of risk is treated mathematically.

12.2 The economics of natural disasters

Natural disasters are by definition costly. Table 12.1 illustrates the differences in vulnerability by type of event and by continent. The table records losses of life, the only impact for which reliable databases exist. Many problems exist with sampling data of sufficient quality with regard to the impact of natural disasters, particularly regarding economic data (Otero and Marti, 1995). Disasters are irregular, hit different places, and are never the same. Hence it is hard to compare the impact of one event with that of another; the particular problems this poses in comparisons over time are discussed in Section 12.4. In addition, common economic indicators give a very distorted picture of what actually is going on after a disaster; more of this below. Further, direct damages are hard to measure, not only because most people have other things on their mind after an event, but also because many disasters affect the poor, who are without insurance, book-keeping and formal markets, and not infrequently are killed in the event. Finally, the human impact of disasters commands more attention than the economic impact, for everyone but economists, who, as pointed out above, have shown little interest in natural disasters.

When studying higher-order economic impacts, translation of direct damages into economically meaningful dimensions is the first problem. This is obviously hard, if not impossible, for the elements that measure human hardship. However, this is not of utmost importance for the analysis of the medium- and long-term

Table 12.1 Loss of life by disaster type and by continent, 1947–80

Event	Number of events	Asia	Oceania	Africa	Europe	South America	Central America	North America
Earthquake	180	354,521	18	18,232	7,750	38,837	30,613	77
Tsunami	7	4,459	—	—	—	—	—	—
Volcanic eruption	18	2,805	4,000	—	2,000	440	151	34
Flood	333	170,664	77	3,891	11,199	4,396	2,575	1,633
Hurricane	210	478,574	290	864	250	—	16,541	1,997
Tornado	119	4,308	—	548	39	—	26	2,727
Severe storm	73	22,008	—	5	146	205	310	303
Fog	3	—	—	—	3,550	—	—	—
Heatwave	25	4,705	100	—	340	135	—	2,190
Avalanche	12	335	—	—	340	4,350	—	—
Snowfall and extreme cold	46	7,690	17	—	2,780	—	200	2,510
Landslide	33	4,021	—	—	300	912	260	—
Total	1,059	1,054,090	4,502	23,540	28,694	49,275	50,676	11,471

Source: After Alexander (1993)

economic effects. It is argued below that the 'only' thing that counts here is how these matters affect the propensity to save and (re)invest in the affected region. Less obvious, at least to non-economists, is the problem with the translation of the economic damages into economic indicators. The problem here is that the damages due to natural disasters are damages to stocks, whereas economists tend to think in terms of flows.

The few studies which systematically assess the higher-order economic impacts perform their analyses directly using common economic indicators (Albala-Bertrand, 1993; Otero and Marti, 1995; Dacy and Kunreuther, 1969). Table 12.2 (after Albala-Bertrand, 1993; Otero and Marti, 1995, report similar findings) is the result of twenty-eight case studies of natural disasters between 1960 and 1979 with respect to the impact on key economic indicators in the year of the disaster and the two subsequent years. Surprisingly, the table indicates that the 'received view' (according to Albala-Bertrand) of the economic impact of disaster is contradictory to the empirical findings.

Gross domestic product (GDP) is found to improve. This is easily explained. The larger part of the damage consists of loss of capital and durable consumption goods. Stocks are not measured in GDP, while replacing them is. GDP is an indicator of economic activity, not of economic wealth.[1]

Unfortunately, Albala-Bertrand was not able to find data on unemployment. The received view is that it increases, but this is in contrast with the reported GDP growth. Gross fixed capital formation is found to increase. This is no surprise, as this indicator measures reconstruction activities but not capital losses.

No significant rise in inflation is reported, although it does seem to accelerate. The explanation is that the overall supply of goods falls because of the damaged

Table 12.2 Received view of economic impacts of disasters versus analysis results

Variable	Received view	Analysis results
Gross domestic product	Fall, stagnate	Improvement
Inflation	Rise, accelerate	No significant change
Unemployment	Increase	Not available
Gross fixed capital formation	Decrease, depress	Improvement
Manufacture	Depress	No discernible change
Agriculture	Fall	Improvement
Construction	Increase, boom	Improvement
Public deficit	Increase	Small increase
Trade deficit	Increase	Sharp increase
Reserves	Deteriorate, loss	Sustained increase
Capital flow	No comments	Sharp increase
Unrequited transfers	No comments	Sharp increase
Exchange rate	Devaluation	No discernible change

Source: Albala-Bertrand (1993: 87)

Note: Shaded rows indicate contradictions. Analysis for agriculture applies particularly to earthquakes infrastructure and production, while the overall demand rises, because of reconstruction activities. Otero and Marti (1995) also point out speculation. Part of the demand for reconstruction is paid from reduced consumption of other goods, however, and part can be met by increased imports. Additional factors pushing up inflation are decreased savings, insurance payments and influx of foreign capital.

Albala-Bertrand does not report empirical material on interest rates. Inflation pushes interest rates up. The demand for capital for reconstruction is an additional pressure upwards. On the other hand, decreased savings, insurance payments and influx of foreign capital expand the supply of capital.

The construction sector shows an expected boom. Manufacturing appears to suffer in the year of the extreme weather event, but this is generally compensated for in the year after. A tentative explanation is that natural disasters *postpone* supply of and demand for manufactured goods rather than *reduce* it. Manufacturing is of course disrupted by the disaster, and part of the capital stock may be damaged. The impact on the national economy depends on the share of the capital stock affected, and on the degree of occupation before the event. Severe shock can of course lead to prolonged loss of output, and associated economic suffering.

Albala-Bertrand (1993) reports a generally positive impact on agriculture. However, agriculture is very diverse. Timing is important with regard to the growing season. For the longer-term impact, it is also important whether annual or perennial crops (such as forest plantations) are affected, and whether the soil is damaged (wind erosion, saltwater inundation) or fertilised (cf. Olsthoorn *et al.*, Chapter 9, this volume, and Brignall *et al.*, Chapter 4, this volume).

Public-sector spending grows to meet the needs of the emergency and rehabilitation phases, while tax revenues may shrink because of diminished exports. Combined, they may create or increase fiscal budget deficits. Counteracting this is the increased GDP growth. Tax collection may increase revenues if part of the disaster response is monetary aid (such as cash for work) that is recorded in official forms, replacing informal activities that are not readily taxed. On balance, however, Albala-Bertrand (1993) reports a small increase in the public deficit for his twenty-eight case studies.

The trade deficit sharply increases. This is due to falling exports, resulting from diminished output, and increasing import requirements to face unmet internal demands and requirements for rehabilitation and reconstruction. Reconstruction efforts may involve acquiring or increasing foreign or local indebtedness. Moreover, depending on the economic position of the country prior to the disaster, it is possible that the country's international reserves and its ability to meet external commitments are jeopardised (Otero and Marti, 1995). Albala-Bertrand (1993) indeed reports an increase in capital inflow, unrequited transfers and foreign reserves. Capital inflow reflects new loans and foreign investments (probably in reconstruction); unrequited investments reflect foreign aid (most of the countries in the study are so-called developing ones); reserves reflect extended credit facilities. The exchange rate does not change discernibly.

Keeping these empirical findings in the back of our minds, let us try to connect the direct damages to their economic consequences. As noted above, stocks are not taken up in economic indicators. Hence the loss of a durable consumption good will not show up in conventional indicators, such as GDP. The flow of services the durable consumption good provided is also not measured, and its loss will not be indicated either. These losses are also not reflected in economic accounts, but are felt by those who suffer the loss. What will be measured is the replacement of the lost good.

This replacement can be paid for from three sources: insurance, savings and assistance. If the good is fully replaced through insurance, be it designated reserves, mutual insurance or commercial insurance, no 'economic' loss is ascribed to this specific disaster. The economic loss is the annual insurance premium, discussed in the next section. In the case that the good is fully replaced through savings, the victim does suffer a loss, as the goal for which he or she saved will now be forgone. The economy as a whole experiences this as a redistribution from the original goal to reconstruction, but this is not measurable, as the situation without the disaster is never realised. In the case that the good is replaced through domestic compensation, through either charity or some government action, the situation changes for the victim; for the economy, it depends on whether the losses are paid from designated or general means. In the case that the good is replaced through foreign assistance, the economy as a whole grows through the additional influx of capital, but simultaneously causes an international redistribution of endowments.

Financial markets are affected by such redistributions. The supply of assets increases as individuals, businesses and insurance companies may sell property to meet their needs; at the same time, the demand of the affected individuals and companies falls. In addition, the profitability of (insurance) companies may be negatively affected by the disaster. If the disaster is not too large, the decrease in prices due to excess supply will be redistributed through the global financial system. The possibility of a crash following a large natural disaster, however, cannot be excluded.

The situation is more complicated if production is affected. The replacement mechanisms for capital goods are the same as for durable consumption assets. The lost good is not measured, the replacement is. If designated reserves are available, no loss can be ascribed to the event; otherwise, opportunity costs and redistribution result. But there is more. Assume an economy with one capital good. Each year, part of the total capital stock is replaced by new machines. Assuming technical progress, the new machines are in some way better than the old ones. If an extreme weather event hits this hypothetical economy and destroys part of the capital stock, this will be replaced by new, better machines. Thus the productivity of the capital stock is improved, although in the short term output falls. In addition, the wave of investment right after an extreme weather event leads to an 'aftershock' at the end of the economic lifetime of the capital. See Box 12.1 for a stylised numerical illustration of this.

On the other hand, the buffer capacity – that is, the sum of designated and undesignated reserves and assistance – might be insufficient. This can be the case if funds are simply not there, or the holder of the funds is not willing to invest them again in the disaster area. In that case, both consumption and production are reduced for a longer period of time.

Provided that buffer capacity is sufficient, the damage recovery activities constitute an impetus in the (local) economy. Savings, designated or not, government reserves and foreign money flow into the economy to spur reconstruction. This can be viewed as an old-fashioned Keynesian impulse. Depending on the economic circumstances and the actual impulse, the result can be negative, positive or negligible. Negative consequences result in the case that demand

---

Box 12.1 Extreme weather and capital stocks

Assume an economy with 2,000 machines, each producing exactly the same product. Assume that the lifetime of machinery is 20 years. Assume further that each machine, once depreciated, is replaced by a new one. Assume that these replacement investments are institutionalised as a fraction (0.056) of the total output. Assume that a machine built in year t is 1 per cent more productive than a machine built in year $t-1$. Hence, each new machine is 1.01^{20} times as productive as the machine it replaces. Then, the output grows with productivity; that is, 1 per cent per year. Demand is assumed to grow at the same pace. Figure 12.1 displays the development of total output over 30 years.

Suppose now that in year 5, a storm hits the capital stock and destroys 5 per cent of all machinery, evenly spread over all vintages. Assume that the owners of the capital are not insured and do not have reserves. Assume further that investment rules do not change. Then, output falls between years 4 and 5 by 4.05 per cent, but grows afterwards again at 1 per cent per year. Total output is permanently lowered (pluses in Figure 12.1).

Suppose now that the owners of the capital stock are insured. Assume that in the year after the disaster the value of all destroyed machines is invested in new ones. Assume further that investment rules do not change. Then, output falls between years 4 and 5, but rises sharply between years 5 and 6. In addition, since the new machinery is more productive, the growth rate rises above the 1 per cent in the scenario without a disaster. In year 26, the post-disaster vintage is depreciated, and the output shows an aftershock. After this, the growth rate converges (oscillatory) to 1 per cent (asterisks).

Figure 12.1 also shows the consequences of reconstruction which replaces half of the destroyed machinery in year 6, a quarter in year 7, and one-eighth in years 8 and 9. The initial depression in output is longer in this case, the post-disaster growth rate is slightly lower, and the aftershock is smoother.

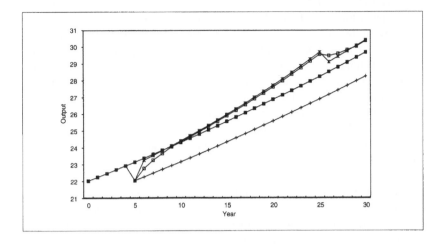

Figure 12.1 Total output of a hypothetical capital stock without disaster (filled squares), with disaster but without insurance (pluses), with disaster and immediate replacement by insurance (asterisks), and with disaster and partly delayed replacement by insurance (empty squares)

already exceeds supply; the additional impulse is likely to spur inflation, without having real effects. Positive consequences result if the economy is either short in demand or short in capital. Negligible consequences result if the disaster or the impulse it creates are relatively small or if the economy is not sensitive to real impulses. Note that the reconstruction impulse is not confined to the reconstruction period; multiplier effects make it last longer.

Another point is that disasters may induce structural changes. An important impetus to structural change is the enormous, but short-lived, exposure of the disaster area and its problems to interventions, as in rehabilitation (e.g., Cuny, 1983). Again, structural changes might be for the better or the worse.

12.3 The economics of disaster management

The previous section *inter alia* concludes that natural disasters, although implying an initial loss of wealth, do not necessarily have negative economic consequences in the longer term, the crucial determinant being appropriate disaster management. Disaster management is the broad term which describes actions in anticipation of and in reaction to natural disasters. Disaster management spans prevention, protection and preparedness, and involves a great number of actors at various levels, with different means and ends.

The benefits of disaster management are the reduced and avoided damages, but management is obviously not without costs. Investments need to be made in protective measures, such as dikes. Sunk capital entails opportunity costs, and of course maintenance and depreciation. Investments also need to be made in human capital; for example, in the training of a corps of engineers. Obviously, investments in disaster management have positive side-effects, as does any other investment, such as employment and technological progress. However, the motivation behind investments in disaster management is to avoid something bad; investments in disaster management tend to be defensive. In more abstract terms, considering economics as an exercise in optimisation, the presence of a natural hazard adds a constraint to the optimisation, and if the hazard is sufficiently serious, the constraint is binding. Hence the optimum will be lowered, unless this constraint lifts other constraints (in the longer term).

The other part of disaster management is creating buffer capacity and flexibility. Buffer capacity implies building up reserves and insurance, but also social networks and even nations. (Some historical evidence suggests that the earliest states in China, Egypt and Mesopotamia evolved from the need for centralised flood hazard management; flood control certainly played a substantial role in the building of the state of the Netherlands.) Flexibility implies keeping options to secure life and livelihood in case of disaster. Again, buffer capacity and flexibility have positive side-effects, but the reason is to avoid the negative; such expenditures are of the defensive kind.

Defensive investment, and so disaster management, implies sacrificing opportunities, using scarce resources not to enhance but to protect welfare. However, comprehensive databases of such costs are absent, at least to the knowledge of the authors. Some indications can be found in the literature. Alexander (1993) reports a world-wide expenditure of US$20 billion per year on prediction, prevention and mitigation of all natural disasters. Dlugolecki *et al.* (1996) estimate that, world-wide, US$11 billion per year is spent on windstorm and tropical cyclone insurance. The total premiums of non-life insurance are typically in the range of 5 to 6 per cent in industrialised countries (OECD), and less than 1 per cent in developing countries like Bolivia (0.45 per cent in 1983), India (0.49 per cent in 1986), Indonesia (0.49 per cent in 1986) and the Philippines (0.58 per cent in 1986); Fiji, however, stands out with 1.5 per cent in 1986 (Outreville, 1992).

The economic implications of investments in disaster management are readily demonstrated in the following example. Commonly, the core of a neo-classical growth model looks like this.[2] Let Y_t denote total output at time t, L_t the labour force and K_t the capital stock. Then, assuming a Cobb–Douglas production function for convenience,

$$Y_t = \alpha_t L_t^\gamma K_t^{1-\gamma} \tag{1}$$

where α is a cofficient. Output Y_t is divided over consumption C_t and investment I_t

$$Y_t = C_t + I_t \tag{2}$$

and capital K_t develops as

$$K_t = (1 - \delta)\, K_{t-1} + I_{t-1} \tag{3}$$

with $0 < \delta < 1$ as the rate of depreciation.

If natural disasters and their management are introduced, (3) becomes

$$K_t = (1 - \delta - \beta_{t-1})K_{t-1} + I_{t-1} \tag{3'}$$

where $0 < \beta_t < 1$ denotes the damage, which is a stochastic function, declining in relation to the stock of defensive capital D_t:

$$\beta_t = f(D_t, u_t); \; \frac{\partial f}{\partial D} < 0 \tag{4}$$

with u_t the stochastic element. Defensive capital needs to be maintained as well:

$$D_t = (1 - \delta^d)D_{t-1} + I_{t-1}^d \tag{5}$$

so (2) is replaced by

$$Y_t - I_t^d = C_t + I_t \tag{2'}$$

Hence defensive investments lower consumption and investment, thereby lowering current and future welfare, while damages affect the capital stock. Moreover, the composition of effective demand is altered by the introduction of defensive investments.

In the case that the damage due to natural disaster is covered through insurance (or designated savings and reserves), (3') becomes

$$K_t = (1 - \delta - \beta_{t-1})K_{t-1} + I_{t-1} + \varepsilon_{t-1}\beta_{t-1}K_{t-1} \tag{3''}$$

where $0 < \varepsilon_t < 1$ denotes the level of insurance cover (note that for convenience immediate replacement is assumed), and of course (2) changes to

$$Y_t - P_t = C_t + I_t \tag{2''}$$

where P_t is the insurance premium paid. Here again, insurance lowers consumption and investment. On the other hand, insurance premiums are temporarily invested, and the reserves of the insurance companies more permanently. So, in non-disaster periods, saving and investment are higher with than without the risk, but the dissavings after a disaster more than compensate for this in the long run (provided, of course, that the insurance premiums are more or less actuarially

317

fair). However, insurance premiums and reserves are not likely to be invested in vulnerable regions or sectors.

To summarise, some of the limited resources of society are invested in disaster management, and some of the resources are lost because of disasters. Hence, *ceteris paribus*, in an expanding society, sectors or regions with above-average natural hazards are below-averagely capable of generating and utilising new capital, and/or have below-average standards of living; in a steady-state society, sectors or regions with above-average natural hazards have below-average standards of living.

However, in the previous section we argued that natural disasters could have positive economic effects in the longer term. The mechanism through which this happens is through the 'technology' coefficient α_t in equation (1). Basically, α determines how much output can be produced, given a mix of labour and capital. Hence, α is a constellation of the state of technology, the institutional settings, the political situation, etc. The argument that a natural disaster could be an opportunity to update the capital stock implies in this model indeed a change in α. Disaster management can also change α, though the mechanism is different. A natural disaster could be the shock needed to remove barriers to efficiency. Disaster management is not likely to break down barriers. Instead, it smoothes consumption and production by taking away financial risks; as such, it enhances efficiency and reduces the manifestations of risk aversion.[3]

Risk aversion is one of the main driving forces for disaster mitigation, and one of the main hindrances for investment. In economic theory, a decision-maker is risk averse if, of two projects with equal expected outcomes, he or she prefers the one with the smaller variance. In project evaluation based on cost–benefit analysis (cf. Tol *et al.*, 1995), risk aversion is often translated into a risk premium, which, under suitable conditions, is approximately proportional to the standard deviation (Arrow, 1970; Pratt, 1964). Risk aversion manifests itself in two ways. First, it inflates optimal disaster mitigation. A risk-neutral decision-maker would choose that mix of disaster mitigation for which the marginal annual expected costs equal the marginal annual expected benefits (avoided damage); a risk-averse decision-maker would choose the mix such that the marginal annual expected costs equal the marginal annual expected benefits plus the annual risk premium. Mathematically, the optimal decision follows from something like

$$\mathrm{E}\ MC = \mathrm{E}\ MB + \gamma\ \mathrm{Var}\ MB \tag{6}$$

where E denotes the expectation operator, Var the variance operator, MC the marginal costs, MB the marginal benefits, and γ the risk aversion parameter; clearly, $\gamma < 0/\gamma = 0/\gamma > 0$ corresponds to risk seeking/neutral/averse. In the common case of risk aversion, the optimal level of disaster mitigation is higher than in the case of risk neutrality or certainty.

The second manifestation of risk aversion is that rule (6) applies not only to investments in disaster mitigation, but to all investments. Disaster management reduces uncertainty, thereby lowering the risk premium, and hence the requirements a potential project should meet. This has two consequences. First, areas or sectors with a higher level of disaster mitigation attract, *ceteris paribus*, more investments. Second, a generally higher level of disaster mitigation raises the savings rate, as saving becomes more worthwhile relative to consumption. Again, the safer an area, the more capital it attracts and generates.

Capital is the engine of economic activity and growth. The previous paragraphs conclude that disaster mitigation requires additional capital, although it may also generate additional savings. The implications of this depend on investment decisions and financial markets, which consider a much broader range of issues. Empirical analyses indicate that financial development, of which commercial insurance is a substantial part, is crucial for successful economic growth in its earlier stages because capital scarcity is then the major barrier to growth (Jung, 1986; Patrick, 1966). Increased exposure to natural disasters in certain regions or sectors increases capital scarcity. Potential additional savings to form disaster reserves are relatively safely invested. The relatively vulnerable region or sector is hence hampered in development and reducing vulnerability.

There are two evaluation approaches in addition to cost–benefit analysis, namely absolute and safe minimum standards. In these approaches, a minimum level of safety is set, on the basis of considerations other than economic ones. In a safe minimum standard approach, the costs of meeting the standard are taken into account; in an absolute standard approach they are not. Setting safe minimum standards is the approach which is probably the most suitable for disaster mitigation. Absolute and safe minimum standards have impacts similar to those of risk aversion and uncertainty in a cost–benefit analysis (conservative neo-classical economists claim that standards are nothing but a short cut to risk aversion). Standards determine the level of disaster mitigation. Standards also determine whether a generic investment is acceptable or not. Disaster mitigation standards imply that fewer projects are viable for investment, or that more needs to be invested in a project to make it meet the standard.

12.4 Disasters and disaster management in a dynamic perspective

The two previous sections provide a more or less static picture of disasters and disaster management. In reality, society and its (decreasingly) natural environment are highly dynamic. This section first discusses the influence of such dynamics on data analysis, then goes into the determinants of changes in socio-economic vulnerability, and finally treats the consequences of change for disaster management.

12.4.1 Changing vulnerability

We are interested in those extreme events which cause damage in excess of a certain threshold. Often, this is translated as a socio-economic threshold, such as

100 people killed or direct economic damage over $1 million or 1 per cent of GDP. Figure 12.2 illustrates what happens to the chance of crossing such a threshold (i.e. the frequency of the event) in a dynamic situation. The growth rate of the threshold-crossing chance can be an order of magnitude different from the growth rate of the economy or population; this difference varies substantially over time and between distributional assumptions, and can indeed be positive as well as negative. For example, in year 0, the mean grows by 1 per cent, while the chance to cross the fixed threshold grows by 11 per cent (normal distribution). In year 100, the mean still grows by 1 per cent, while the chance to cross the threshold grows by 3 per cent. In year 200, the mean grows by 1 per cent, while the threshold-crossing chance grows by 0.5 per cent. So, showing trends in the number of disasters, with the threshold expressed in dollars, number of people or even percentage of GDP (GDP per square kilometre grows) is rather meaningless without additional detail on the shape of the distribution and the actual nature of the threshold.

As shown above, it is hard to interpret trends in damage if the underlying socio-economic conditions are changing. The most important factors are economic and population growth. More and richer people imply increased exposure to severe weather. Growth, and migration, lead to higher concentrations of goods and people, so that the same event is more damaging. Settlement may occur in more vulnerable areas, either through attraction to certain environments (such as the sunbelt in the USA) or because the most suitable areas for human

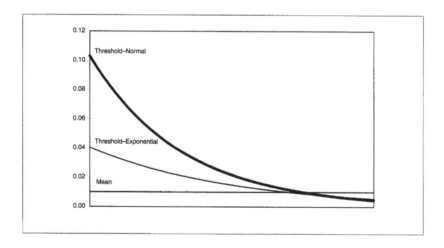

Figure 12.2 Growth rates of the mean of a stochastic process and the probability that it exceeds a fixed threshold. The figure displays the growth rate of the mean, which is set at 1 per cent for 200 years (the horizontal line). The figure also depicts the probability that a normal or exponential process exceeds a fixed threshold. This threshold is set at three times the standard deviation plus the mean in year 0

occupation have already been taken. Also, the susceptibility of economic activities may be increasing (Berz and Conrad, 1993; Anderson, 1994).

Opposite to the pressures pushing vulnerability up is the power of increasing economic and technological potential. The richer a society gets, the more money it can spend on hazard mitigation. In addition, the higher the income level, the more people are willing to spend on mitigation. Provided that this money is spent in a sensible manner, economic growth may reduce vulnerability. Technological potential pushes in the same direction, provided it is properly used, of course. The increasing existence and accessibility of techniques for monitoring, modelling and controlling nature, and the organisation and management of capital, imply a greater variety of disaster reduction options, and hence lower costs and greater benefits.

So, we have factors pushing relative vulnerability up, and factors pushing relative vulnerability down. Which tendency dominates depends, as always. Figure 12.3, adapted from Burton *et al.* (1993), displays income per capita versus per capita deaths due to natural hazards for 1973 and 1986 for some twenty countries. Only Bangladesh, Pakistan and Kenya grew richer and less vulnerable at the same time. A number of countries just grew richer, without obvious changes in vulnerability (e.g., Iran, Germany, Indonesia). Some countries grew more vulnerable and richer (e.g., Romania, the USA, Australia). Peru, Guatemala and Honduras just became more vulnerable.

At first sight, this suggests that there is no relationship between affluence and vulnerability. However, the richer countries are generally less vulnerable than the poor. A number of explanations can be given for this pattern. (1) The (positive) correlation between poverty and vulnerability is a long-term issue. Thirteen years is too short a period to detect the secular relationship between income and vulnerability. The pattern in Figure 12.2 is largely noise. (2) There is no secular correlation, but a rapid phase transition between relatively vulnerable and invulnerable, occurring somewhere in the 'modernisation' process. This is true for river floods in the Netherlands, which were controlled between 1860 and 1890 without causing much trouble afterwards (see Langen and Tol, Chapter 6, this volume). (3) The tendency to grow less vulnerable with income is (partly) reversed by increasing population density, environmental degradation, etc. (4) Societies tend to grow more vulnerable, irrespective of income, because of waning awareness and degrading protection. This process continues until a certain threshold is crossed, and vulnerability is in a relatively short time reduced to an acceptable level.

12.4.2 Changing weather

Let us use a cost–benefit framework, touched upon in Section 12.3, to get a firmer grip on the relationship between disaster management and changes in weather events. Suppose that damage D_t in year t as a result from extreme weather follows the relation

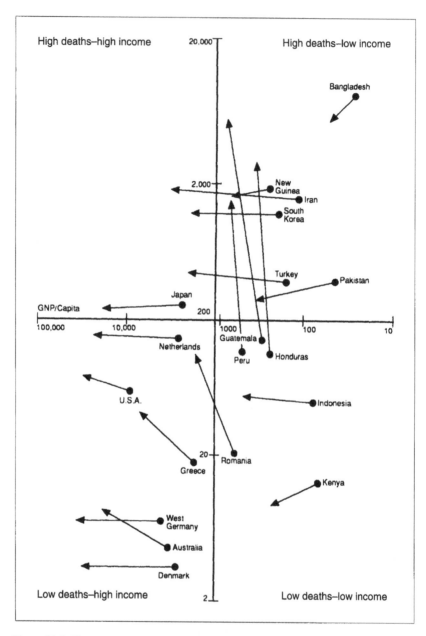

Figure 12.3 Changes in per capita deaths from natural hazards and in per capita income in selected countries. Income is in GNP per capita (US$), normalised to 1981. Population is also normalised to 1981

Source: Burton *et al.* (1993: 14)

$$D_t = W_t P_t^{-\alpha} \tag{7}$$

with $0 < \alpha < 1$, where W_t is some indicator of the weather and P_t is the level of protection (prevention, preparedness). With a slight loss in generality, D_t is assumed to be expressed in money. Protection is assumed to behave like ordinary capital goods: it slowly degrades over time unless reinvestment and maintenance prevent that. So,

$$P_t = (1 - \delta)P_{t-1} + I_{t-1} \tag{8}$$

The total yearly cost is the sum of D_t and I_t. The net present value of the costs, C, follows the relation

$$C = \sum_{t=0}^{\infty} \frac{W_t P_t^{-\alpha} + I_t}{(1 + \rho)^t} \tag{9}$$

Assume now that everything is in a stationary state. Then

$$P_t = P_{t-1}, \ I_t = I_{t-1} \Rightarrow P_t = (1 - \delta) \ P_t + I_t \Rightarrow P_t = \frac{I_t}{\delta} \tag{10}$$

so that

$$C = \sum_{t=0}^{\infty} \frac{W \delta^\alpha I^{-\alpha} + I}{(1 + \rho)^t} = \frac{1 + \rho}{\rho} (W \delta^\alpha I^{-\alpha} + I) \tag{9'}$$

The optimal level of investment I^*, and hence protection, follows from

$$\frac{\partial C}{\partial I} = \frac{1 + \rho}{\rho} (W \delta^\alpha \ \alpha \ I^{-\alpha - 1} + I) = 0 \tag{11a}$$

resulting in

$$I^* = (W \delta^\alpha \ \alpha)^{1/(\alpha + 1)} \tag{11b}$$

The yearly expenditure on weather disasters in a steady state is then

$$W \delta^\alpha \ I^{*-\alpha} + I^* = W \delta^\alpha (W \delta^\alpha \ \alpha)^{-\alpha/(\alpha + 1)} + (W \delta^\alpha \ \alpha)^{1/(\alpha + 1)} \tag{12}$$

Suppose now that the weather indicator is V instead of W, but that the decision-maker does not know this. If $V < W$, then damage D is lower, but the investments and sunk capital in disaster mitigation (still based on W) are higher than under optimal behaviour (based on V). These opportunity costs outweigh the savings in damage, because the real investment deviates from the optimum investment. If $V > W$, damage is higher, and investments and sunk capital are lower, but the damages outweigh the savings in investment. In a cost–benefit framework, any deviation of reality from the situation considered under optimisation incurs costs. Note that the term W stands for damage done by weather events; hence, climatic or vulnerability changes have the same effect.

In a climate change scenario, the cost–benefit analysis of the above becomes more complicated (provided that the decision-maker realises that the climate is changing), but the message does not differ. As a consequence of optimisation, the desired level of protection is a function of the perceived characteristics of the weather hazard. If the perceived characteristics deviate from the real characteristics, the resulting situation is by definition suboptimal. So, unless climate is forecast perfectly, change will induce transition costs. This effect is more pronounced if W is highly non-linear in climate change.

The opportunity costs (12) of a certain state of climate are increasing in W. So, if extreme weather worsens (W increases), climate change induces both opportunity and transition costs. If weather becomes less extreme (W decreases), climate change induces opportunity benefits and transition costs (on the balance of which no definitive statement can be made).

The second consequence of change for optimal disaster management has to do with the increased risks. The term W can be interpreted as the expected damage due to extreme weather if there had not been any protection, or, with risk aversion, the expected damage plus the (scaled) standard deviation of damage. Climate change introduces additional uncertainty, thereby increasing W and so opportunity costs, even if expected damage stays the same.

So much for optimal disaster management. Let us leave the metaphor of 'economic men' (or women) and return to the real world. A weak modification of the assumption of the optimising economic agent is the satisficing agent. This agent does not seek the optimum, but stops searching the moment a satisfactory solution is reached. In optimisation terms, the pay-off function levels off at satisfaction. This has two implications. First, a small change in climate or vulnerability does not necessarily impose a loss. It could, but change may just as well be an improvement, for example by justifying an initial level of protection that was too high (in an optimisation sense). It requires a larger change to push the situation out of the satisfactory range. Because of this, a satisficer will react later to a change than an optimiser. This is advantageous in the case of a false alarm, but disadvantageous in the case of a real alarm.

Other decision criteria, such as risk standards, lead to the same phenomena. Differences between hazards one is adapted to and expecting and actual hazards lead either to overprotection (i.e. overexpenditures of scarce resources) or to underprotection (i.e. too vulnerable, leading to too much damage). This statement is false only for those agents who act completely haphazardly.

12.5 Conclusion

This chapter analyses the broader economic implications of natural hazards. Although the direct and indirect impact of a natural disaster is, by definition, negative, and can be substantial and dramatic, the higher-order impacts of an event are mixed, depending on circumstances. Conventional economic indicators, such as GDP or unemployment rates, usually conceal the real impact. The loss of capital in the longer term can have a positive effect, provided that sufficient reinvestment from designated reserves takes place.

The true economic impact of weather hazards is revealed in the means spent on disaster management. Scarce resources are allocated to things not intrinsically worthwhile, an opportunity cost. Unfortunately, no reliable or complete estimate has been found of the actual expenditure on disaster management.

Should the risk of extreme weather change, for instance because of the enhanced greenhouse effect, then the level of adaptation is bound to lag behind, because prediction of the impact of climate change will never be perfect. In addition to the change in opportunity costs (either for better or for worse), adaptation induces a transition loss which is invariably for the worse (unless decisions are made haphazardly). In addition, the increase in uncertainty could change investment, at the expense of the most vulnerable sectors and regions.

Climatic hazards have distinct characteristics for future risks. The clearest projection of climate change in Europe is for increased temperatures, and these would dramatically increase the frequency and duration of heatwaves. However, the effects of heatwaves are mixed. The direct effects on health are apparent, but not economically large, and are covered, if serious, through medical insurance or health care. Significant costs could accrue through increased cooling costs, particularly as northern Europe passes a threshold for adoption of air-conditioning in stores, workplaces and homes.

The effect of climate change on regional water balances and their extremes – droughts and floods – are uncertain. The scenarios used in this study generally show increased precipitation compensating for increased temperatures in most of Europe. Thus the reported impacts on agricultural droughts and land subsidence are relatively modest. Further research is needed to evaluate a wider range of scenarios; some regional differences are likely to emerge, although the climatic changes predicted for the next few decades are likely to be modest. Floods are sensitive to small variations in precipitation; in two of the three case studies, flood damage is also very sensitive, but in one case study, flood damage hardly increases with increasing flood heights. The direct economic impacts of droughts, subsidence and floods can be large (exceeding \$1 billion for some episodes). Most drought impacts are not insured, while, in the UK, insurance for subsidence has improved over the past few years. Except in the UK, floods are not insured, or are insured with substantial government support. Mitigation can be effective, but costly.

The effect of climate change on windstorms is least understood, especially for tropical cyclones. The scenarios used in this study show a modest increase in wind hazard in the Netherlands. No credible scenario is available as yet for tropical cyclones. In the Netherlands, the economic costs of windstorms are small compared to GDP; insurance cover is widespread. Islands in the southwest Pacific are very vulnerable; the cyclone risk already limits development. Specific disasters consume a significant portion of private and public financial reserves.

Acknowledgements

The authors hereby express their gratitude to Kees Dorland, Tom Downing, David Favis-Mortlock, Angela Liberatore, Xander Olsthoorn and Aart de Vos for discussion and comments. This chapter is a revised version of Tol and Leek (1996).

Notes

1 Gross domestic product is the sum of all measurable monetary flows in the domestic economy. Net domestic product (NDP) does measure depreciation. Note that national accounts that measure economic stocks are starting to be used.
2 Note that this model is not suited for the study of the impact of natural disaster, although damages are included below. The model describes long-term tendencies, whereas the economic impact of a disaster depends to an important extent on the short-term characteristics of the economic situation at the time of the event.
3 Obviously, disaster management is itself a manifestation of risk aversion. What is meant here is that, once a risk is properly managed, risk aversion is less important for other economic decisions.

References

Albala-Bertrand, J.M. (1993) *Political Economy of Large Natural Disasters*, Oxford: Clarendon Press.
Alexander, D. (1993) *Natural Disasters*, London: UCL Press.
Anderson, M.B. (1994) *Disaster Vulnerability and Sustainable Development: A General Framework for Assessing Vulnerability*, Yokohama: IDNDR.
Arrow, K.J. (1970) *Essays in the Theory of Risk-Bearing*, Amsterdam: North-Holland.
Berz, G., and Conrad, K. (1993) Winds of change. *The Review*, June: 32–35.
Burton, K., Kates, R.W., and White, G.F. (1993) *The Environment as Hazard*, 2nd edition, New York: Guilford Press.
Cuny, F.C. (1983) *Disasters and Development*, Oxford: Oxford University Press.
Dacy, D.C., and Kunreuther, H.C. (1969) *The Economics of Natural Disasters*, New York: Free Press.
Dlugolecki, A.F., Clark, K.M., Knecht, F., McCauley, D., Palutikof, J.P., and Yambi, W. (1996) 'Financial services', in R.T. Watson, M.C. Zinyowera and R.H. Moss, eds, *Climate Change 1995: Impacts, Adaptations and Mitigation of Climate Change: Scientific-Technical Analyses – Contribution of Working Group II to the Second Assessment Report of the Intergovernmental Panel on Climate Change*, Cambridge: Cambridge University Press, 539–560.

Jung, W.S. (1986) Financial development and economic growth: international evidence. *Economic Development and Cultural Change* 34: 333–346.

Otero, R.C., and Marti, R.Z. (1995) 'The impacts of natural disasters on national economies and the implications for the international development and disaster community', in M. Munasinghe and C. Clarke, eds, *Disaster Prevention for Sustainable Development: Economic and Policy Issues*, Geneva and New York: International Decade for Natural Disaster Reduction and World Bank, 11–40.

Outreville, F.J. (1992) The relationship between insurance, financial developments and market structure in developing countries: an international cross-section study. *UNCTAD Review* 3, 53–69.

Patrick, H.T. (1966) Financial development and economic growth in underdeveloped countries. *Economic Development and Cultural Change* 12: 174–189.

Pratt, J.W. (1964) Risk aversion in the small and in the large. *Econometrica* 32, 1–2: 122–136.

Tol, R.S.J., and Leek, F.P.M. (1996) 'Economic approaches to natural disasters', in T.E. Downing, A.A. Olsthoorn and R.S.J. Tol, eds, *Climate Change and Extreme Events: Altered Risk, Socio-economic Impacts and Policy Responses*, Amsterdam: Institute for Environmental Studies, Vrije Universiteit.

Tol, R.S.J., Van der Werff, P.E., and Misson, C. (1995) *An Evaluation Tool for Natural Disaster Management*, W95/37, Amsterdam: Institute for Environmental Studies.

Zeckhauser, R. (1996) The economics of catastrophes. *Journal of Risk and Uncertainty* 12: 113–140.

13

AN ANALYTICAL REVIEW OF WEATHER INSURANCE

Richard S.J. Tol

13.1 Introduction

Weather insurance is one of the first commercial sectors to confront changing patterns of natural hazards. In the past few years, Greenpeace (1994; cf. also Leggett, 1993) started alarm bells ringing over the rapid increase in damage and claims due to natural hazards, which would be, if not the first sign of climate change, at least a warning of a warmer future. More careful analyses show that the larger part of this increased damage can be explained as increased socio-economic vulnerability (Berz and Conrad, 1993; Vellinga and Tol, 1993; Changnon and Changnon, 1992; Clark, 1988, 1991), and that the troubles in the insurance sector resulted from improper price setting and damage mitigation (Paish, 1994; Hindle, 1994). But changes in the average weather are bound to include changes in extremes, and these will affect the insurance industry in some way or another. Given the other problems of the sector, climate change places additional stress on the system, but the precise effect will depend on circumstances. This chapter looks at these issues in more detail. The two central questions here are: (1) what are the likely effects of climate change on the commercial weather insurance sector, and (2) to what extent is commercial weather insurance a suitable coping mechanism for climate change?

Section 13.2 discusses the nature of the weather insurance sector, its practices, the impact of climate change, and possible responses. Section 13.3 presents a simulation model of the weather insurance market, which supplements the more qualitative analysis of section 13.2. The major advantage of quantitative analysis is its rigour and the ease with which many cases can be analysed. However, such advantages are obtained only at the expense of abstraction of reality. Section 13.4 continues with the results obtained by this model, and compares its findings to the theoretical notions of Section 13.2. Section 13.5 summarises and concludes. The impossibility of insuring against climate change, the relationship between insurance and greenhouse gas emission reduction policies, and the relationship between insurance and liability are discussed in Boxes 13.1, 13.2 and 13.3, respectively. The Appendix provides a micro-economic treatment of solidarity and mutual insurance, applied to a small island state (cf. Olsthoorn *et al.*, Chapter 9, this volume).

13.2 The weather insurance sector

13.2.1 Insurance in theory

The aim of insurance is to spread the economic consequences of an adverse event from the actually affected over a group of those potentially affected.[1] Classic examples are fire and theft. A large collection of individuals is at risk of fire and theft, but only a few are affected at any time. To the affected individual, the result can be quite dramatic, but for the group the total damage is limited. Hence it is rational to transfer the adverse consequences of an individual risk to a homogeneous group of individuals running the same risk, because the group is more diversified than the individual. This is a formal form of solidarity in times of disaster within and between communities. From this point on, we restrict ourselves to monetary consequences. Because people are in general risk averse, most individuals prefer to pay a small and certain amount per annum, say P, than paying damage D with chance p, with $P = pD$. Indeed, risk aversion leads to the fact that most people are willing to pay an annual amount, the insurance premium, greater than the expected damage, $P > pD$. The difference between the insurance premium and the expected damage (or actuarially fair premium) should cover the transaction costs of transferring the risk from the individual to the collective. These transaction costs include the costs of acquiring information, setting premiums, determining policies, the costs of collecting premiums, and the costs of handling and paying claims. Henceforth such costs are referred to as administration costs. As the collective is represented by an institution, and often a commercial company, the administration costs include the costs of running the company (managers, buildings) and keeping it running (competition, reserves). On top of that, profits are required to compensate for the entrepreneurial efforts of the insurance company and its stockholders.

Three basic types of insurance can be distinguished. First, individuals can agree to share the costs of a disaster befalling one of them. Without effective enforcement, such an agreement can prove worthless following a serious event. Second, individuals can agree on mutual insurance; that is, to pay premiums periodically to a fund under their collective supervision. A major disadvantage is that the diversification of the risk is restricted to this particular collection with a limited number of participants. Third, individuals can hand over the risk to a commercial agency; that is, an insurance firm. A disadvantage is that the profit of the insurance company does not flow back to the insured. The main advantage is the high degree of specialisation and diversification, resulting in much lower transaction costs. The focus is here on commercial insurance; a micro-economic analysis of solidarity and mutual insurance applied to small island states can be found in the Appendix.

Should a risk fall on a group rather than on an individual, for example as in floods and earthquakes, two alternatives are possible. First, the risk can be spread over time. This is not trivial, because of fiscal constraints on capital accumulation and control over funds over a longer period of time. Second, the collective could act as an individual and spread the risk over a group of similar collectives. This is usually done through (local) insurance companies which themselves insure through (regional or global) reinsurance companies. The main problems are that there are many such groups; each group has distinct properties and faces distinct risks; and that natural hazards can be rather devastating. On top of that, extreme weather events are rare by nature while circumstances continually change. Thus, it is extremely hard to quantify the risks.

So, a risk is insurable if it concerns a diversifiable event; that is, an event of which the consequences for the affected few can be spread over a large number of unaffected people. Furthermore, in order to calculate an appropriate premium, the risk should be more or less quantifiable. If a risk is hardly quantifiable, insuring it is problematic. No insurance company would consider (fully) insuring a risk with an unknown probability and unknown consequences. The reason is that under these circumstances premiums cannot be relied upon to cover the claims. The prime goals of any insurance company are to stay in business and be profitable. Insuring unquantifiable risks endangers these goals. Lloyd's of London is less restrictive in this respect, the prime reason being that Lloyd's does not have a limited capital provided by shareholders, but unlimited capital provided by 'names'.[2] Hence, Lloyd's has lower risk aversion than other (re)insurance companies.

On the other hand, precisely quantifying a risk can hamper its insurability. Sound premiums are based on the average claim of the insured. However, each insured individual is unique. Hence, each population of insured can be split into two or more subpopulations with (slightly) different average risk characteristics. At first sight, doing so provides a better basis for premium calculation. However, diversification has the same effect as antiselection: 'bad' risks drive out the 'good' ones; that is, the bad risks drive up the price the good risks have to pay. In the extreme, each insured person pays his or her own risk as a premium, plus the administration costs and profits of the insurance company. In this case, keeping personal reserves ('self-insurance') is cheaper. The tremendous cost of acquiring the information needed prohibits individual price-setting. But the higher the level of diversification, the more erratic the annual stream of claims (per type of policy) to the insurance company becomes, and the less profitable the policy becomes to the insurer. This calls for another level of insurance, and hence more administration costs, risk premiums and profits. On the other hand, an unduly low level of diversification implies a redistribution of risks, a situation to which the less risk prone may object.

Next to quantifiability and diversifiability, the third, readily overlooked, condition for insurance is enforceability. An insurance policy is a contract between the insurer and the insured. It is valuable only if each party can be relatively

Box 13.1 Insuring the consequences of climate change

One of the first published proposals to insure damage costs arising specifically from climate change is by Chichinilsky and Heal (1993). They characterise the risks of climate change as *poorly understood, endogenous, collective* and *irreversible*, each of which would make it hard to insure commercially. Indeed, knowledge regarding climate change and, in particular, its impact is poor, and excessive uncertainty regarding the actual risks makes any form of insurance difficult; it is unquantifiable. For example, it is not easy to say that a specific flood or a probability of a flood is caused by climate change. The other three characteristics are inadequate descriptions of climate change, and not necessarily problems for commercial insurance.

The importance of the concept of endogeneity of risks is unclear to the present author. A risk is described as endogenous if the person at risk has a certain level of control over the actual damage. With this definition, virtually all risks are at least partly endogenous. This does not stop commercial insurance, but it does stop full cover because full cover would be a disincentive for damage mitigation (moral hazard). The optimal way of coping with risk is often a mix of damage mitigation, insurance and acceptance of losses (Pate-Cornell, 1996a, b).

It is also unclear what Chichinilsky and Heal exactly imply with their concept of collectivity. If damage befalls groups of people simultaneously, these groups should be treated as collective individuals and their risks should be pooled with the risks of other groups. If the number of groups is severely limited, spreading risks is impossible. If one interprets climate change as an *event* (which it is not) affecting almost everybody, there is only one group, and so no basis for diversification. Climate change is likely to affect all of humanity in some way of another, but obviously not in the same fashion. Indeed, there may well be winners and losers, and some losers will lose only a bit while others will be affected substantially. So, although the risk of climate change is a collective one (in that it is shared by everyone), it is in no sense homogeneously collective.

Finally, while climate change itself may be irreversible, a large share of its impact is not. Studies on the economic impact of climate change suggest that it is the transition that induces costs rather than the changed climate itself (Tol, 1995b). Adaptation to the changed climatic circumstances lowers the impact dramatically. But successful adaptation takes time, so that, generally, the impact is recurrent but declining. Of course, irreversible damages may also occur, particularly with regard to damage to ecosystems and loss of species. However, commercial insurance of intangible, irreversible losses, such as permanent disability, is available; notice that this form of insurance, and the related life insurance, requires a much higher degree of quantifiability than property insurance. In addition, the loss of a species or a unique ecosystem is a loss to humankind, not to a particular country or region. Hence diversification of such a loss is impossible.

According to Chichinilsky and Heal, the four characteristics discussed above leave climate change uncoverable through commercial insurance. The above analysis supports this conclusion, although the arguments have been refined: diversifiability and quantifiability are both problematic.

The alternative to commercial insurance, proposed by Chichinilsky and Heal, is a combination of Arrow securities for the collective risk (e.g., the actual nature of

Box 13.1 Continued

climate change) and mutual insurance contracts for the individual remainders. As discussed in the previous section, the great advantage of commercial over mutual insurance is the lower transaction costs of the former. Indeed, the higher transition costs of mutual insurance prevent it from diversifying risks to a large extent, making it a less viable option for the insuring of larger risks, such as climate change. Chichinilsky and Heal do not fully develop the idea of using Arrow securities for covering the collective part of the risk. An Arrow security pays one dollar (say) should a certain state of the world occur. Such securities are tradable; the market price reflects the buyers' perception on the actual chance of the associated state of the world occurring. One disadvantage is that it requires discretisation of a high-dimensional continuum – the many possible states of the world. If it is required that the approximation of this discretisation be close, the number of states of the world, and hence the number of different securities, is huge (Morgan and Henrion, 1991). This leads again to high transaction costs.

However, the main problems with insuring, in whatever way, the risks of climate change are its time scale and its lasting nature. As noted in Section 13.2, the usual way of spreading risks is over space rather than over time. The reasons for this have to do with time preference (discounting), uncertainty and taxes. With climate change, these reasons are particularly profound. Buying insurance today for a risk which is not likely to result in a claim for the next forty years or so is extremely problematic in this sense, and it also implies that such insurance contracts should be inheritable.

A more serious problem is the lasting impact of climate change, which in fact has two properties. First, climate change leads to recurrent damages, and hence to recurrent climate change insurance claims. These claims could well last for a long period of time, eroding the funds of the insurance scheme. Thus the premiums and capital accumulation should be rather high, or the paid claims will either be low or last for only a short while. Second, at the onset of climate change, the losers may quickly be identified, making the winners withdraw from the insurance. Binding contracts could in theory prevent this, but these must then be enforced over a long period of time, which is rather problematic, particularly in those regions lacking stable and strong authorities. Agreeing to such binding contracts is also problematic, as current studies already indicate which regions and sectors are most vulnerable to climate change.

Nordhaus (1994: 171–172), in his claim that part of the damage due to climate change can be insured, overlooks exactly the same point. Those vulnerable to the enhanced greenhouse effect are already partly identified, and will quickly become fully identified once damages start to occur. Consider this small example: suppose two people, A and B, are both aware of having a chance of one-half of being hit by climate change in 20 years' time. In principle, it is rational for both to agree on mutual insurance. However, because of time preference (including myopia, economic growth and the taxes which must be paid on the accumulated capital), initial premiums are low. Over time, premiums will grow as climate change damages come closer. However, at the same time knowledge is likely to grow. Suppose that in year

15 it is clear that A has a chance of 75 per cent of being hit, and B only a 30 per cent chance. Then B will be inclined to pay less in premiums, shifting the risk back to A. Suppose now that in year 20 climate change hits A, and not B; that is, A has a chance of 100 per cent, and B of 0 per cent. Then B will stop paying premiums, and A will stop paying as well (as the contract comes to an end) and quickly erode the accumulated capital. Would B have agreed on a binding contract in the first year? Probably not, because B would be obliged to pay premiums starting today, and have only a chance of receiving claims in the future. B would be willing to do so only provided current premiums are relatively low, which requires large diversification (compensating for the increased transaction costs) and quick capital accumulation (at least, quicker than B could achieve). These conditions are unlikely to be met. Even if the best guess of a particular group is that no climate change will occur at all, few would be willing to sell insurance, as the size of the total claim, the vast uncertainties, and their own risk aversion would deter them.

To summarise, climate change induces risks which are not easily diversified and not readily quantified. In addition, the necessary insurance contract cannot be enforced. Hence the proposals of Chichinilsky and Heal (1993) and Nordhaus (1994) to insure part of the damage costs of climate change should be interpreted with even greater caution than they themselves do. The fact that climate change is likely to hit identifiable regions and sectors over and over again makes risk-spreading unlikely and ineffective. Nevertheless, the thought that climate changes are insurable seems to be catching on in the policy and science debates, and with less caution than was attached to the original proposals. This should be avoided, as it may well imply a false sense of security or the impression that ready solutions exist to adapt to all impacts of climate change.

certain that the other party meets its obligations under the contract (note that the insured is generally the weaker party). This requires legislation and regulation. In formal economies, this goes without saying, but in informal economies commercial insurance is much harder to establish.

13.2.2 Insurance in practice

Commercial insurance of natural hazards is largely restricted to tropical cyclones and extratropical storms (note that the degree of cover in the tropics is generally lower than in higher latitudes), while river floods are typically insured, if at all, with some form of government support. Here are some examples of state participation in the cover of natural disasters (Albala-Bertrand, 1993). Baden-Württemberg, Germany, has obligatory inclusion of earthquake and flood cover in the state monopoly insurance for buildings, financed by obligatory claims-based additional premiums with a state guarantee. In the other German states, floods were excluded from insurance cover, but the situation is changing under new European Union regulations and under market and government pressure (in reaction to the recent floods). In the Netherlands, an attempt to set up flood insurance (in lieu of government liability) recently failed. In France, since 1982 property and motor insurance has mandatory natural disaster cover; it is financed by additional premiums of 1 to 5 per cent with state reinsurance. Premiums are differentiated in three zones; prevention is also stimulated via other mechanisms. Japan has private earthquake insurance, which is supplementary to fire insurance, with obligatory state reinsurance, covering about 85 per cent of the overall loss limit. New Zealand used to have state natural disaster insurance under the Earthquake and War Damage Act of 1944 as an automatic supplement to fire cover, financed by an additional premium. However, the country is now rapidly privatising insurance. In Norway there is a state support fund for natural disasters. In Switzerland there is a mix of state funds financed by casino incomes and state natural disaster cover financed by canton-level building-fire insurance monopolies, reinsurance coverage, and a private fire-insurers' natural disaster pool. In Spain, since 1954 there has been state disaster cover (for natural and man-made disasters) with the participation of the insurance industry, paid for by obligatory additional premiums of 1 to 10 per cent. Nation-wide disasters are excluded, but can be covered by decree and state subsidies. In the USA, since 1968 federal flood insurance has been heavily subsidised. However, earthquake insurance is available throughout the country on a private basis. The coverage can be written as an endorsement to a standard homeowner policy. In the UK, disaster insurance is sold in the private market. Here, all natural disasters are included in a standard homeowner's policy. Flood risks are insured in a standard package for small businesses; flood insurance is sold separately to large enterprises (Arnell, 1987).

13.2.3 Recent troubles

Recent, large losses due to hurricanes, floods and storms aroused the attention not only of the climate-change community (e.g., Schuurman et al., 1995; Kattenberg et al., 1996) but also of the (re)insurers (Dlugolecki et al., 1994, 1996). Figure 13.1 (p. 344–345) depicts the upward trend in insured damage due to natural hazards. The upward slope is explained by various factors, all relating to expanding human activities. Obviously, population and economic growth increase the stock at risk and so damages. Population and economic growth also implies stronger concentration of people and goods at risk, and increased settlement in vulnerable areas. Further, humans may change their environment, making them more vulnerable to weather hazards. Consumers become better informed and more willing to make claims. Protection measures are not always optimally implemented. In addition, climate may have changed, but the signs for this are still weak and ambiguous (for a further discussion, see Tol and Leek, Chapter 12, this volume).

The problems in the insurance sector are not caused simply by expansion of the stock at risk. Premiums grow as well. Obviously, the insurance companies get into trouble when the claims rise faster than the premiums. This can be because the insurers miscalculate the risk. Premiums are partly based on claims history. If, as the previous paragraph suggests, the risk per dollar insured increases, a premium based on historical information will be too low.[3] Premiums are also influenced by competition. The increasing information at the disposal of the insured enhances competition, pushing down prices. But even if premiums grow as fast as the claims there would be a problem. The damages due to natural disasters are partly diversified over time, however hard that may be, through the general reserves of (re)insurance companies. Expanding business implies that the reserves lag behind the risk.

Responses are now vividly discussed within the sector and already partly implemented (Dlugolecki *et al.*, 1994, 1996; Lohmeijer-van der Hul, 1994). Three response types can be distinguished. (1) The product can be adjusted (raising premiums, restricting cover). (2) the buffer capacity can be enlarged (more reinsurance, closer co-operation with banks and governments). (3) the vulnerability can be reduced (enforcement of building regulations, land zonation).

Adjusting the product so that premiums match claims shifts the risk back to the insured (see Priest, 1996, for a full discussion). At the end of the day, the insurance sector only spreads risk, never takes it away. Premiums can be raised on an *ad hoc* basis but 'sound technical pricing' is an increasingly popular option. The main advantage of sound technical pricing is that premiums are based on (scientific) calculations of the risk (which is the definition of sound technical pricing; note that this corresponds to the actuarially fair premium above). The main disadvantages of sound technical pricing are that it is expensive to do it right but it is misleading (and hence potentially expensive) if done wrong, and that it affects the basis for diversification if done in too great detail (which is the only way of doing it really well). Trade-offs obviously need to be made here. In a rapidly changing world, it is of crucial importance that insurers get a grip on what drives changes in their exposure.

Restriction of cover can be done by increasing the deductibles, by restricting the insured sum, by placing (individual or collective) caps on claims, or by excluding specific hazards, areas, sectors, or categories of insured from insurance. All these options limit insurance, and none of them is very popular with the insured or their governments. Product adjustments reduce the socio-economic vulnerability of the insurer but enlarge the socio-economic vulnerability of the insured.

Enlarging the buffer capacity reduces the insolvency risk the insurance companies run, and hence the profit they need to make per policy. Extending reinsurance is an obvious option, but restricted by reinsurance companies which are increasingly aware of the risks they themselves run. An alternative is the recently introduced catastrophe futures market in Chicago (Rodetis, 1994; de Roon and Veld, 1994; McCullough, 1995; Anon., 1996; Kunreuther, 1996), on which insurance companies sell futures that are later refunded (plus the necessary rents, of course) in the event that no catastrophe has occurred. Catastrophe futures are generally fairly small, of the order of magnitude of $25,000. Buyers are large, highly diversified investors. A third option to increase the buffer capacity follows from the increasing concentration of the financial markets. Companies grow and merge, and hence have more internal diversification and reserves. The fourth option is extending co-operation with governments, as described above, although governments are, for obvious reasons, not very keen on this. *Ceteris paribus*, this option reduces the socio-economic vulnerability of the insurer and the insured. The exposure of the above-mentioned third parties is increased, but this should not outweigh the initial reduction.

Risk management (e.g., Williams and Heins, 1985) reduces the risk to both the insurer and the insured. Examples of risk management include building codes and land zonation, but also anti-erosion policies and greenhouse gas emission reduction. Risk management gets harder the less direct or immediate the link between action and result. Insurance companies can establish risk management by providing information, excluding unmanaged risks, or reducing premiums for managed ones. The last two options directly provide incentives to the insured to better manage their risk; the first option assists the insured, assuming that the incentive already exists, for example, with a substantial deductible or a large intangible risk. The insurance companies can also lobby the government to stimulate risk management, or use their reserves to invest in risk management. Lobbying is appropriate in cases where the insured cannot individually manage their risk, for instance in the event that large infrastructural projects (e.g., dikes) or adjustments in the legislation (e.g., building regulations) are needed. Using the large resources of the insurance industry directly for risk management is not trivial. Reserves of insurers must be profitable and easily liquidised. Investments in infrastructure hence are excluded. In their investment policy, insurers could consider the consequences for the exposure of their clients. For example, energy efficiency could be a reason to decide in favour of a certain investment.

Reactions to the great hurricanes of the early 1990s suggest that the insurance sector is likely to be able to adjust to climate change without disproportional damage, provided that changes are not too fast or drastic and responses are well judged. All of the three policy options described above were applied. After these events, premiums were rapidly raised (e.g., (re)insurance rates for storms in Europe – Dorland *et al.*, Chapter 10, this volume; tropical cyclone rates in the South Pacific – Olsthoorn *et al.*, Chapter 9, this volume) and cover restricted (e.g., tropical cyclone insurance in the South Pacific; Olsthoorn *et al.*, Chapter 9, this volume; Rokovoda and Vrolijks, 1993). Sound technical pricing (e.g., flood risk in Germany; cf. von Seggern, 1993) and risk management (e.g., in the Netherlands; see Schuurman, 1995) were stimulated, and catastrophe futures introduced. The floods in Europe revived the discussion between governments and insurance companies (e.g., in the Netherlands; see Schuurman, 1995). In addition, reinsurance companies actively lobbied for greenhouse gas emission control at the Conference of the Parties in Berlin, March 1995. This indicates the great flexibility of the insurance sector, at least to a similar sequence of events. The consequence is that risks are shifted from the insurance sector to society at large. More rapid changes, or completely unanticipated ones, could impair the

---

Box 13.2 Insurance and greenhouse gas emission reduction

One of the first and most comprehensive analyses of the economics of greenhouse gas emission control is Manne and Richels's (1992) *Buying Greenhouse Insurance*, based on a series of highly stylised examples, and further elaborated using the model Global 2100. Manne and Richels argue that, given the large uncertainties, it is economically rational to abate more than a best-guess analysis would suggest. The prime reason for this is risk aversion (see Tol, 1995b, for more reasons), and the additional action could be called a risk premium (Arrow, 1970; Pratt, 1964). Risk aversion is also a reason for insurance, but in the common Arrow–Pratt sense of the word, the insurance premium is the amount one is willing to pay on top of the expected value to transfer a risk to another party. So, a risk premium concerns an economic agent's own decision, whereas an insurance premium concerns a transfer between two agents. Obviously, the two premiums are closely connected. Unfortunately, the two premium concepts have been confused in the climate debate, perhaps starting with Manne and Richels's book. This work and others deal with the question of how uncertainties should affect the emission reduction decision of a central planner. Thus Manne and Richels calculated the risk premium, not the insurance premium. Insurance may sound more appealing to politicians than risk, but catchy phrases only cause confusion. Insurance has (misleading) appeal in that it implies that damage will actually be reimbursed, which is definitely not the case in this proposal.

   Nordhaus (1994) mentions three types of 'insurance'. His second type, consumption smoothing over time, and his third type, precautionary investments, are again unnecessarily and misleadingly placed under the heading 'insurance'. Consumption smoothing by an individual over his or her lifetime is a very common phenomenon including instruments such as insurance, hedging and pensions. Consumption smoothing by a society also appears to be desirable, but as insurance is only one instrument, and not the most effective one, the present author fails to see the point in calling 'consumption smoothing over time' 'insurance'. Precautionary investments are measures to avoid losses or to reduce risks whereas insurance accommodates for losses or spreads risks. Precautionary investment may be very wise, but there is little reason to use the same name for two distinct instruments.

sector. But in that case too the real costs are borne by society. Tol and Leek (Chapter 12, this volume) analyse the possible implications of that.

13.3 A simulation model of the weather insurance market

In this section, a simulation model of the weather insurance market is described. The next section discusses its analytical and numerical outcomes. What is presented is not a model in the usual understanding of the word. It does not describe actual behaviour; its outcomes are not predictions. Instead, the model is a heuristic device; its mechanisms are metaphors of real-world processes. Consequently, it yields qualitative insights, not quantitative results.

The model comprises only the major features of the market. The core of the model is formed by the behavioural assumptions of the principal agents (i.e., insured, insurer, reinsurer) from which supply of and demand for insurance is derived. Cover and premium follow from assuming market equilibrium. The simplicity of the model allows the evaluation of many different climate change scenarios, model settings and responses. This approach is chosen to be able to identify the robustness of the outcomes with respect to changes in the assumptions, and thus to yield insight as to what might happen to the real insurance market under certain climate change scenarios, given certain responses.

The simulation model in this section consists of three parts. The first part generates synthetic storms. The second part calculates the damage caused by the storms. The third part assesses the impact of storm damage on the insured and (re)insurers, and the insurance market. The model is highly stylised; only the major processes are modelled. It is called the Weather Insurance Simulation Model, version 1.1. Version 1.0 is described by Tol and Olsthoorn (1994), version 1.1 is further elaborated by Tol (1995a).

13.3.1 Storm generation, damage and scenarios

The stochastic storm generator draws realisations from a compound stochastic process. The compound process consists of a Poisson process to generate the number of storm days in a certain region in a fixed period (a year, say) and a Pareto IV process to generate the intensity of the storms. A Poisson process is a standard model to describe the number of events in a fixed period. A Pareto IV distribution is a very flexible model to describe the behaviour of stochastic variables in excess of a certain threshold.

Thus, the number of storms N_t in year t equals n with probability

$$P\{N_t = n\} = \frac{\kappa_t^n e^{-\kappa_t}}{n!} \tag{1}$$

with κ_t the expected number of storms in year t.

The Pareto IV distribution, $P_{IV}(\mu, \sigma_t, \omega\alpha)$, for storm intensity $I_{h;t}$ in year t (h is the index for the storms within 1 year) is defined by

$$P\{I_{h;t} \le i\} = F(I_{h;t}) = 1 - \left(1 + \left(\frac{I_{h;t} - \mu}{\sigma_t}\right)^{1/\omega}\right)^{-\alpha} \tag{2}$$

for μ, σ_t, ω, $\alpha > 0$ and $I_{h;t} > \mu$, with density

$$f(i_{h;t}) = \frac{\alpha}{\sigma_t \omega} \left(\frac{I_{h;t} - \mu}{\sigma_t}\right)^{1/\omega - 1} \left(1 + \left(\frac{I_{h;t} - \mu}{\sigma_t}\right)^{1/\omega}\right)^{-(\alpha + 1)} \tag{3}$$

The location parameter μ is interpreted as the threshold value above which wind becomes storm. The tail and shape parameters, ω and α, respectively, are held constant whereas the scale parameter σ_t varies over time, according to the scenarios described below.

Drawings are generated with the (discrete) inverse transform method (Law and Kelton, 1991), using Uniform $(0,1)$ realisations from the random-number generator of TurboPascal 7.0.

The damages caused by the storms are assumed to be proportional to wind energy – that is, the third power of wind speed (Dorland and Tol, 1995), with μ the no-damage threshold and υ the total destruction threshold. Thus, the damage share $\delta_{h;t}$ of storm h in year t follows

$$
\delta_{h;t} = \begin{vmatrix} 0 & I_{h:t} \leq \mu \\ \left(\dfrac{I_{h:t} - \mu}{\upsilon - \mu} \right)^3 & \mu < I_{h;t} < \upsilon \\ 1 & I_{h;t} \geq \upsilon \end{vmatrix} \tag{4}
$$

The total damage equals $\delta_{h;t} S_t$, with S_t the stock at risk in year t (expressed in US dollars, say).

Frequency and intensity have separate scenarios in order to distinguish between their impacts. The basic run is $\kappa_t = 10$; that is, the expected number of storms is ten per year. In another scenario, κ is raised linearly from 10 at the start of the run to 30 at the end. The next two storm frequency scenarios are: κ_t follows a cycle with mean 10, amplitude 10 and length 20, and the trend superimposed on the 20-year cycle in κ_t. Trends are of interest to investigate the sector's reaction to gradual changes, as climate change is often perceived to be. Cycles are of interest because (1) they are observed (Changnon and Changnon, 1992) and (2) they can be used to test the market's reaction to temporary changes.

The scale parameter σ_t of the Pareto IV distribution follows a similar scheme; so, Pareto's σ_t equals 10, σ_t follows a linear trend from 10 to 30, σ_t follows a cycle with mean 10, amplitude 10 and length 20 years, and σ_t follows a cycle with a length of 20 years superimposed on a linear trend in the mean.

13.3.2 Market behaviour and equilibrium

The damage $\delta_{h;t} S_t$ of storm h in year t is divided between the insured, the insurer and the reinsurer as follows. All damage below the deductible threshold $\lambda_t S_t$ is paid by the insured. The damages above $\lambda_t S_t$ are partly $(1 - \gamma_t)$ covered by the insured and partly (γ_t) by the insurer and the reinsurer. The insurer and reinsurer share the claim in a similar fashion. Below the damage threshold $\varepsilon_t S_t$, all claims are covered by the insurer. Above that, the insurer bears $(1 - \theta_t)$ of the claims, and the reinsurer θ_t.

Thus, for the insured the damage per storm $D_{h;t}$ equals

$$D_{h:t} = S_t \left|
\begin{array}{ll}
\delta_{h:t} & \delta_{h:t} \le \lambda_t \\
 & \text{for} \\
(1 - \lambda_t)(\delta_{h:t} - \lambda_t) + \lambda_t & \lambda_t < \delta_{h:t}
\end{array}
\right. \tag{5a}$$

For the insurer, the damage per storm equals

$$D_{h:t} = S_t \left|
\begin{array}{ll}
0 & \delta_{h:t} \le \lambda_t \\
\gamma_t (\delta_{h:t} - \lambda_t) & \text{for} \quad \lambda_t < \delta_{h:t} \le \varepsilon_t \\
(1 - \theta_t) \gamma_t (\delta_{h:t} - \varepsilon_t) + \gamma_t (\varepsilon_t - \lambda_t) & \varepsilon_t < \delta_{h:t}
\end{array}
\right. \tag{5b}$$

which leaves

$$D_{h:t} = S_t \left|
\begin{array}{ll}
0 & \delta_{h:t} \le \varepsilon_t \\
 & \text{for} \\
\theta_t \gamma_t (\delta_{h:t} - \varepsilon_t) & \varepsilon_t < \delta_{h:t}
\end{array}
\right. \tag{5c}$$

for the reinsurer.

Insurance demand follows

$$\gamma_t^D = 1 - \lambda_t - \frac{1 - \lambda_t}{\psi_t^*} \pi_t \tag{6}$$

with π_t the insurance premium. For $\lambda_t = 0$ and $\pi_t \ge \psi_t^*$, $\gamma \le_t 0$; ψ_t^* is thus the maximum premium the insured is ready to pay to transfer all his or her risk.

Insurance supply follows

$$\gamma_t^S = \frac{1 - \lambda_t}{(1 - \theta_t)(1 - \varepsilon_t) \beta_t^*} \pi_t \tag{7}$$

where β_t^* is the minimum premium required by the insurer to cover the entire stock-at-risk ($\lambda = \varepsilon = \theta = 0$). This corresponds to textbook prescriptions of insurance behaviour (cf. Dong et al., 1996).

Reinsurance demand is assumed to be

$$\theta_t^D = 1 - \varepsilon_t - \frac{1 - \varepsilon_t}{\gamma_t (1 - \lambda_t) \beta_t^*} \chi_t \tag{8}$$

where χ_t is the reinsurance commission. The minimum premium β^* is the same as in (7), here corresponding to $\lambda = \varepsilon = 0$, $\gamma = 1$.

Reinsurance supply follows

$$\theta_t^S = \frac{1 - \varepsilon_t}{\gamma_t (1 - \lambda_t) \eta_t^*} \chi_t \tag{9}$$

with $\eta_t{}^*$ the minimum commission of the reinsurer for covering all losses $(\lambda = \varepsilon = 0, \gamma = 1)$.

Equilibrium follows from equating supply and demand, which results in

$$\pi_t = \frac{(1 - \varepsilon_t)(1 - \theta_t)\ \beta_t^*\ \psi_t^*}{(1 - \varepsilon_t)(1 - \theta_t)\ \beta_t^*\ \psi + \psi_t^*}$$

$$\gamma_t = \frac{(1 - \lambda_t)\ \psi_t^*}{(1 - \varepsilon_t)(1 - \theta_t)\ \beta_t^*\ \psi + \psi_t^*}$$

$$\chi_t = \frac{(1 - \lambda_t)\ \gamma_t\ \beta_t^*\ \eta_t^*}{\beta_t^* + \eta_t^*}$$

and

$$\theta_t = \frac{(1 - \varepsilon_t)\ \beta_t^*}{\beta_t^* + \eta_t^*}$$

λ_t and ε_t are assumed to be proportional to β_t^* and η_t^*, respectively.

This leaves us to define the minimum premiums. Following Arrow (1970) and Pratt (1964), a 'fair' insurance premium approximately equals the expectation loss plus a risk aversion constant times the variance of the loss. Since the total damage per year consists of the number of storms and the damage per storm, the maximum premium the insured are willing to pay equals

$$\psi_t^* = E\ N_I\ E\ D_I + \psi\ S_t\ (E\ N_I^2\ E\ D_I^2 - E^2\ N_I\ E^2\ D_I) \tag{10a}$$

and the minimum premiums of the insurers and reinsurers are taken as

$$\beta_t^* = E\ N_J\ E\ D_J + \beta\ S_t\ (E\ N_J^2\ E\ D_J^2 - E^2\ N_J\ E^2\ D_J) \tag{10b}$$

and

$$\eta_t^* = E\ N_R\ E\ D_R + \eta\ S_t\ (E\ N_R^2\ E\ D_R^2 - E^2\ N_R\ E^2\ D_R) \tag{10c}$$

respectively. The expectations are indexed so as to indicate that the market is driven by the perceived storm damage.

The expected storm number is estimated by the insured as follows

$$E\ N_{L;t+1} = (1 - \varphi)N_t + (1 - \varphi)\varphi\ N_{t-1} + (1 - \varphi)\varphi^2\ N_{t-2} + \ldots +$$
$$(1 - \varphi)\ \varphi^i\ N_{t-i} + \ldots \tag{11a}$$

so that the recent past counts more heavily than the distant past, and that, if the yearly number of storms remains constant, its expectation tends towards this value; (11a) can be rewritten as

$$E\ N_{L;t+1} = (1 - \varphi)\ N_t + \varphi\ E\ N_{L;t} \tag{11b}$$

at a rate of φ; a high φ implies a slow adjustment of expectations and a long memory of past events. So, the expected value is updated yearly with the most recent observation. For the expected storm number squared the same procedure is used. For the expected damage per storm, (11) is applied to the average damage in year t. Insurer and reinsurer use the same method, with memory parameters ρ and ζ, respectively.

13.4 The model's results

13.4.1 Some analytical properties

In this section we analyse some of the properties of the model. Table 13.1 displays the primary reaction of the six market variables (deductibles, retention, premium, commission, cover and cession) to changes in minimum and maximum risk premiums. These changes may originate from changes in the underlying risk (storm number, storm intensity, damage function), risk aversion or risk memory. The primary reactions of the market variables can unambiguously be assessed in sign. (Re)insurance cover is the only variable to show dichotomous behaviour in the presence of changes in extremum risk premiums: not surprisingly, cover rises when the maximum risk premium of the insured rises, and cover falls when the minimum risk premium of the (re)insurer falls. The other variables move in the same direction no matter whose risk perception changes; all variables rise with increasing extremum risk premiums, except for reinsurance cession, which falls as the minimum risk premium of the reinsurer rises.

Carrying the interactions through leads us to Table 13.2, displaying the final reactions of the market variables. But for the reactions to the insured's maximum

Table 13.1 Primary reactions of market variables to exogenous shocks in risk premiums

Shock	Reaction	Shock	Reaction	Shock	Reaction
$\psi* \uparrow$	λ —	$\beta* \uparrow$	$\lambda \uparrow$	$\eta* \uparrow$	λ —
	ε —		ε —		$\varepsilon \uparrow$
	$\pi \uparrow$		$\pi \uparrow$		π —
	χ —		$\chi \uparrow$		$\chi \uparrow$
	$\gamma \uparrow$		$\gamma \downarrow$		γ —
	θ —		$\theta \uparrow$		$\theta \downarrow$

Table 13.2 Final reactions of market variables to exogenous shocks in risk premiums

Shock	Reaction	Shock	Reaction	Shock	Reaction
$\psi* \uparrow$	$\lambda -$	$\beta* \uparrow$	$\lambda \uparrow$	$\eta* \uparrow$	$\lambda -$
	$\varepsilon -$		$\varepsilon -$		$\varepsilon \uparrow$
	$\pi \uparrow$		π ?		π ?
	$\chi \uparrow$		χ ?		χ ?
	$\gamma \uparrow$		γ ?		γ ?
	$\theta -$		θ ?		θ ?

premium, and the deductibles and retention, no statements can be made on the direction in which a reaction takes place. Hence, numerical analysis is needed. Some scoping exercises (see Tol, 1995a) lead to the conclusion that the market variables are not very sensitive to the assumptions on the underlying parameters, with an exception for the memory parameters. The assumption regarding which market party has the longer memory does affect the direction of the change in cover or cession.

13.4.2 Scenario analysis

The storm scenarios investigated are described above: $\psi = 0.050$, $\beta = 0.025$, $\eta = 0.01$; we use two sets of memory parameters: $\varphi = 0.7$, $\rho = 0.8$ and $\zeta = 0.9$, as well as $\varphi = 0.9$, $\rho = 0.8$ and $\zeta = 0.7$. The experiment consists of 1,001 runs of 126 time units (years, say). The first run is used to initialise the endogenous variables. The remaining 1,000 runs are used to estimate the expectation of the market variables. Within each run, the first 25 years are run at a constant risk, so as to further remove initialisation noise. The last 101 years are reported.

Figure 13.1 shows a number of characteristics of the first experiment. Here, a constant number of storms is compared to cycles and trends. The insured have the shortest memory. The characteristics shown are the total annual damage, the insured's net expenses on storms (premiums plus damages minus claims), the cash flow into the insurance sector (premiums minus claims), the cash flow of the reinsurer (cession minus claims[4]), insurance deductibles, reinsurance retention, insurance premium, reinsurance commission, insurance cover and reinsurance cession.

Total annual damage more or less follows the storm number scenarios but is seen to be highly volatile; note that the presented figures are the average of 1,000 Monte Carlo runs. The total storm damage and insurance expenditures of the insured rise with the number of storms. The former relationship is assumed to be linear. The latter relationship turns out to be close to linear. Both storm frequency and insured's expenditures increase by a factor of three in the trend scenario. However, the expenditure cycles are more damped than the underlying cause-number of cycles. The total cash flow of the insurance sector follows the

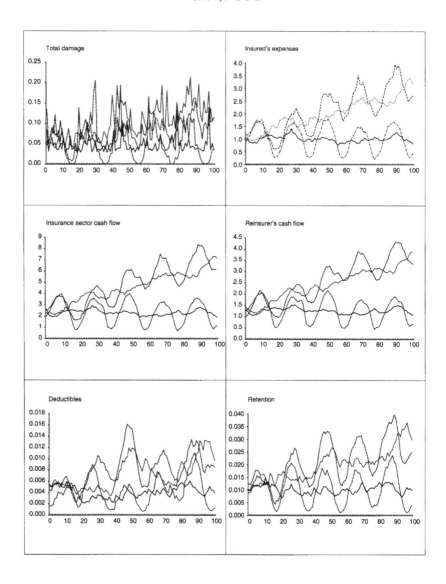

same pattern: a linear reaction to the trend, but a damped reaction to the cycles. The same goes for the cash flow of the reinsurer, and hence for the cash flow of the insurer. Note that the cash flow graphs imply that the insured bear the increased risk; the (re)insurers make extra money because of expanded business. Deductibles resemble more the total damages, being so volatile that the underlying scenarios cannot readily be detected. Retention, premium and commission are smoother, following the driving scenarios. Interestingly, cover and cession are very close for the constant and the trending risk scenarios, but vary widely for the

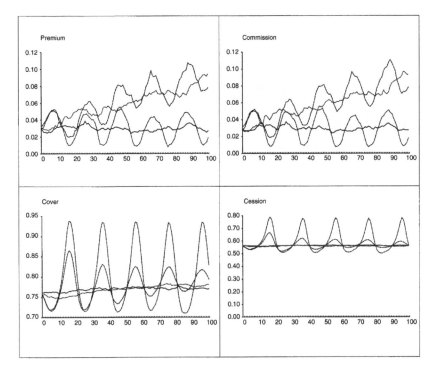

Figure 13.1 The consequences of four storm frequency scenarios for $\psi = 0.7$, $\rho = 0.8$ and $\zeta = 0.9$. Shown are total damage, insured's expenses, insurance sector cash flow, reinsurer's cash flow, deductibles, retention, premium, commission (all as a percentage of the stock at risk), cover and cession (as fractions). Time is on the x-axis. All figures presented are the average of 1,000 random realisations

cycle scenarios. Note that the cycles are markedly asymmetric. The effect of a trend in the cycle plus trend scenario is to dampen the amplitude of the cycle in cover and cession. Constant cover plus increasing premiums and deductibles account for the increasing flow of cash from the insured to the insurer.

Figure 13.2 repeats the experiment with the difference that now the insured have the longest memory. Annual damage has not changed, as mitigation options are excluded. The insured's expenditure looks the same, but the trend is a bit steeper. The shorter memory of the (re)insurer implies a quicker reaction to the changed risk, apparently to the disadvantage of the insured. The cash flow of the (re)insurance sector is a little higher under these memory assumptions. Deductibles are identical as the insurer's memory and the random realisations are identical. Retention is lower for a shorter memory of the reinsurer. This is because the effect of one single disastrous year is lower, owing to the shorter memory. Premium and commission resemble one another under both sets of assumptions concerning the memory of the market parties. Cover and cession

345

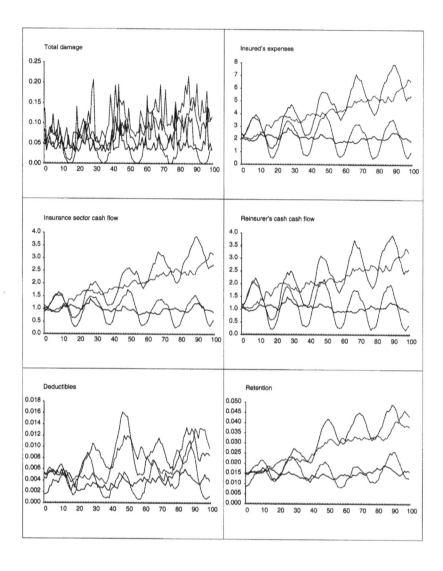

differ markedly. The cycles impact in the opposite direction. The storm frequency trend results first in lower cover, but later in higher cover. Cession
does not react in a systematic or profound way to increasing risk. For this set
of memories, both cover and cession are notably higher.

The next two experiments look at storm intensity rather than storm frequency.
Figure 13.3 displays the results of the scenarios with different features for the case
in which the insured has the shortest memory. The outcomes for total annual
damage are separate, and are substantially higher than for the storm number

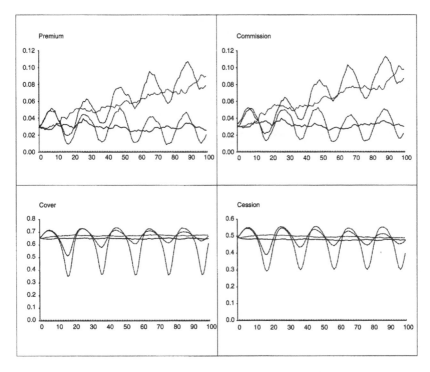

Figure 13.2 The consequences of four storm frequency scenarios for $\psi = 0.9$, $\rho = 0.8$ and $\zeta = 0.7$. Shown are total damage, insured's expenses, insurance sector cash flow, reinsurer's cash flow, deductibles, retention, premium, commission (all as a percentage of the stock at risk), cover and cession (as fractions). Time is on the x-axis. All figures presented are the average of 1,000 random realisations

scenarios. Note that the difference for the market is not just in higher damages; in the presence of deductibles and retention it makes a difference whether the annual damage is caused by a series of small storms or by one big one. The expenditures of the insured increase rapidly with the trend. The cycle is profoundly asymmetric, as a result of the cubic damage function. The cash flows of the insurer and reinsurer also increase rapidly. The (re)insurers are capable of dealing with both increasing storm frequency and increasing intensity. Deductibles, retention, premium and commission straightforwardly follow the trend in the annual damage. The cover graph is very similar to the cover graph for the storm frequency scenario. Increasing risk results in an increased cover, which then falls. This is more profoundly so for cession. Cycles do have a very profound effect, but dampen as the risk increases.

Figure 13.4 evaluates different steepnesses of the intensity trend. Annual damage simply follows the risk. Added are trends of one-half, one and a half

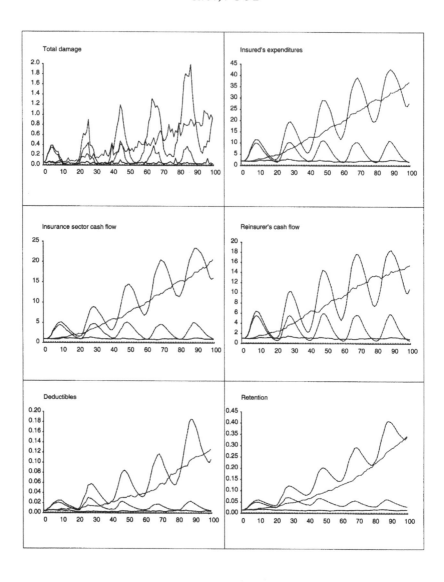

and two times the size of the trend discussed above. Expenditure and cash flows rise steadily in the two lowest cases, stabilise in the long term for the third case, and bend backwards for the steepest case. Deductibles and retention rise monotonously. Premium and commission show the same pattern as the cash flows. The explanation lies in the cover and cession, which increase at first but then start to fall. The insurance sector withdraws from storm risk insurance: this explains the fall in cash flows. The (re)insurers cover less and so receive less premium. The insured pay less cover, and also less of the risk aversion premium.

348

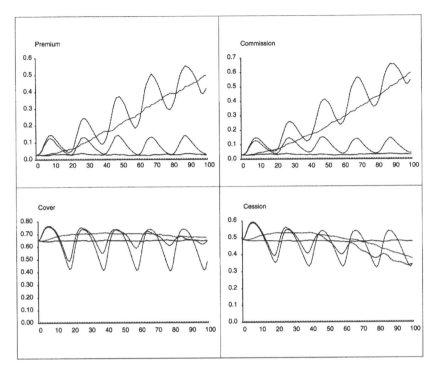

Figure 13.3 The consequences of four storm intensity scenarios for ψ = 0.7, ρ = 0.8 and ζ = 0.9. Shown are total damage, insured's expenses, insurance sector cash flow, reinsurer's cash flow, deductibles, retention, premium, commission (all as a percentage of the stock at risk), cover and cession (as fractions). Time is on the *x*-axis. All figures presented are the average of 1,000 random realisations

In summary, steadily increasing storm risk, due either to increasing frequency or to increasing intensity, leads to increased expenditures for those insured, increased profits in the insurance sector, increased premiums and commissions, and increased deductibles and retention. Cover and cession shift to a different level. This situation is reversed once the risk has increased too drastically, in which case the market and its cash flows start to shrink. The consequences of short-term variations, such as cycles, but also the start of the trend, depend on the speed of the reaction of the market agents. However, the position of the (re)insurers is not affected.

13.5 Windstorms in the Netherlands

Insurance is a major form of risk management for damage due to high winds in the Netherlands (Dorland *et al.*, Chapter 10, this volume). In reaction to the

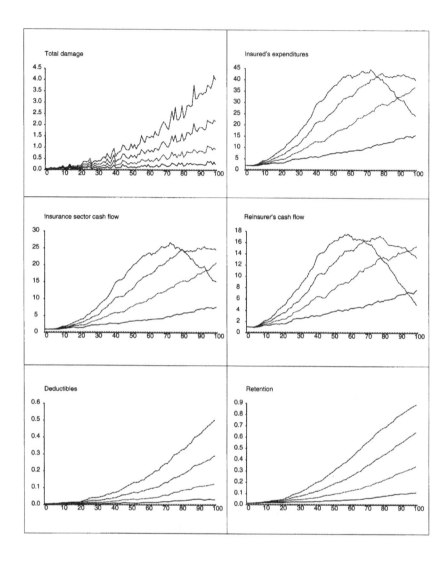

storms of the early 1990s, insurance companies raised deductibles from Dfl. 150–500 to Dfl. 500–1,000 per household policy (Van Kraaij, personal communication, 1995). Dorland *et al.* (Chapter 10, this volume) indicate that storm damage could roughly double for a 6 per cent increase in storm intensity. The additional yearly damage is about Dfl. 40 million, which is relatively small compared to a total of about Dfl. 3,000 million for fire insurance, of which storm insurance is an integral part in the Netherlands (Verbond van Verzekeraars, 1994). What is important is the big losses, such as the Dfl. 1,300 million of Daria. A 6 per cent increase in storm intensity could triple the impact of a large storm such as

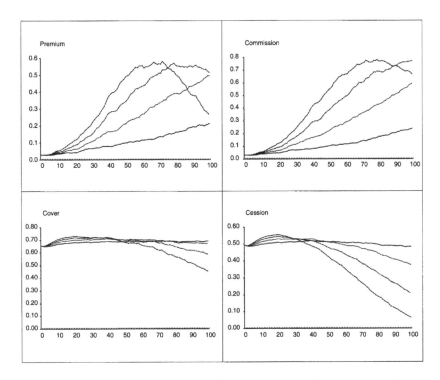

Figure 13.4 The consequences of four trends in storm intensity for $\psi = 0.7$, $\rho = 0.8$ and $\zeta = 0.9$. Shown are total damage, insured's expenses, insurance sector cash flow, reinsurer's cash flow, deductibles, retention, premium, commission (all as a percentage of the stock at risk), cover and cession (as fractions). Time is on the x-axis. All figures presented are the average of 1,000 random realisations

Daria that affected a large area. Reinsurance currently covers such shocks, and is as such a condition for storm insurance. Cession has not (yet?) been restricted, but reinsurance commissions have been raised, apparently partly in anticipation of the enhanced greenhouse effect (cf. Dorland *et al.*, Chapter 10, this volume; Lohmeijer-van der Hul, 1994).

The impact of climate change on wind insurance is also analysed with WISM. The storm generator is the same as used above, with Poisson parameter $\kappa_t = 0.36$ and Pareto parameters $\mu = 32.86$, $\sigma_t = 19.74$, $\omega = 0.33$ and $\alpha = 19.01$. The parameters are fitted to the Dutch storm history of 1911–91; however, the confidence in these estimates is low (see Dorland and Tol, 1995a). The damage function is the sum of the damage functions for houses and businesses in Dorland *et al.* (Chapter 10, this volume), appropriately scaled to serve as a fraction of the stock at risk. Besides the base case, two scenarios are investigated: (1) storm frequency rises to above the 1990 value in 75 years; and (2) modal wind speed rises to above the 1990 value in 75 years. Figure 13.5 depicts the insured's

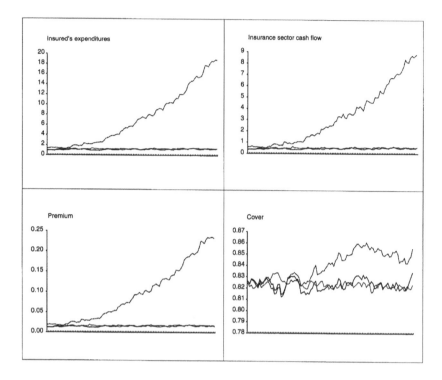

Figure 13.5 The consequences of trends in storm frequency and intensity for $\psi = 0.7$, $\rho = 0.8$ and $\zeta = 0.9$, for WISM adjusted to windstorms in the Netherlands. Shown are insured's expenses, insurance sector cash flow, premium (all as a percentage of the stock at risk) and cover (as fractions). Time is on the x-axis. All figures presented are the average of 1,000 random realisations

expenditures, insurance sector cash flow, premium and cover (cf. Tol, 1995a). Damages are presented in Dorland *et al.* (Chapter 10, this volume).

13.6 Conclusion

This chapter analyses the impact of climate change on the insurance sector, given the sector's reaction to the changed situation. Insurance of weather hazards is problematic because the risk is not readily quantified and diversified. In most cases, weather hazards are not insured, or only partly insured, so often the government is called upon for disaster support. The impact of hurricanes and to a lesser extent floods and winter storms on the insurance sector can be dramatic, however, particularly owing to competitive underpricing in insurance policies and faulty risk analysis.

The insurance sector has a suite of instruments to protect its profitability. Measures regarding price and cover can be applied rapidly, often to the dis-

---

Box 13.3 Insurance and liability: the AOSIS Insurance Scheme

The proposal by the Alliance of Small Island States (AOSIS) has been incorrectly labelled an insurance scheme. Being highly vulnerable to sea-level rise primarily caused by others, AOSIS drafted a treaty to be appended to the Framework Convention on Climate Change. The draft treaty is modelled after international treaties on oil spills and nuclear accidents. It states that all industrialised countries should annually commit funds to a pool, proportional to their greenhouse gas emissions. Should sea-level indeed rise (to be judged by specified measurement techniques, dates and reference points), then the additional damages due to sea surges would be paid from the pool (Wilford, 1993). From the point of view of the claimant directly after the sea surges, this looks indeed like an insurance scheme, the title under which the draft treaty is presented. However, the fact that others pay the premiums makes it a liability rather than an insurance issue.[5] As for the Chichinilsky and Heal case (see Box 13.1), the author does not oppose the proposal, but rather its heading. Again, the term 'insurance' is unnecessarily expanded to something which does not spread risk but, in this case, acknowledges responsibility. Note that responsibility is only partially acknowledged (non-AOSIS members are excluded; the current proposal only looks at sea-level rise) and that responsibility is not openly acknowledged. Note also that the probability exists that 'the North' will be held liable for the impact of climate change on 'the South', possibly leading to damage compensation. However, many barriers exist to formal recognition of liability and global financial mechanisms to compensate for specific climate disasters.

---

advantage of the insured. Measures regarding buffer capacity and risk management require more time and imply structural change, but are to the advantage of both insurer and insured. Empirical evidence from the first half of the 1990s suggests that all instruments are applied; price and cover measures are particularly successful.

The chapter presents a stylised model of the insurance market to obtain a quantitative handle on the impact of climate change. The model, which comprises the interactions of representative, rational insured, insurer and reinsurer, suggests that increases in risk are largely borne by the insured, through price increases and cover restrictions. Short-term reactions of market variables depend on the speed with which actors acknowledge changes.

Although clouded with uncertainty, climate change is not likely to have a large impact on the profitability of the insurance sector. This is because many hazards are not covered by insurance, but also, although the relative change in risk may be substantial, the absolute change in risk is limited, and the insurance sector can adapt effectively and in a timely way.

13.7 Appendix: solidarity and mutual insurance for small island states

13.7.1 Introduction

This appendix presents a micro-economic analysis of solidarity and mutual insurance, tailored to small island states in the southwest Pacific. The analysis follows a formal micro-economic, game-theoretic approach. The mathematics need not deter the reader, as the level is basic. The abstract nature of the arguments need not worry the reader either. Although, strictly speaking, the reasoning applies only to a highly idealised world, it may be argued that, if a particular policy improves welfare in the ideal world, there is something to be gained in the real world as well. In addition, the crucial elements of the pros and cons in the ideal world are important elements in the real world as well. The main merit of reasoning in an ideal, mathematical world is the rigour and the clarity of the findings.

Consider two small islands somewhere in a tropical ocean, Alu and Bawe. Every once in a while both get hit by a tropical cyclone, say with annual chances of p_A and p_B. When hit, the intensity I of the cyclone is distributed as $F_A(I) = P\{\text{intensity} \le I\}$ on Alu and $F_B(I)$ at Bawe, with corresponding probability density functions f_A and f_B. Damage suffered follows the dose–response relationships $D_A(I)$ and $D_B(I)$. Let us assume for convenience that these are deterministic, one-dimensional functions. So, if intensity I is known, damage $D(I)$ is known and can be expressed by a single figure. The welfare loss due to the damage is expressed by the disutility functions $U_A(D)$ and $U_B(D)$, that is, if damage D occurs, the people of Alu feel $U_A(D)$ worse.

The risk – that is, the annual mean welfare loss, EU_A – of Alu is then

$$EU_A = p_A \int_0^\infty f_A(I)\, U_A(D_A(I))\, dI \tag{12}$$

and the annual mean loss of Bawe of course looks exactly the same. The enlightened decision-makers of the island are not interested only in the present risk; future risks count as well. The net present value of the annual mean loss NPVEU^6 is

$$\text{NPVE}\,U_A = \sum_{t=0}^{\infty} EU_A\,(1 + \rho_A)^{-t} = EU_A\,\frac{1 + \rho_A}{\rho_A} \tag{13}$$

where ρ_A is the pure rate of time preference[7] of the Aluan people. The net present risk is the yearly risk over all time. In the chosen specification, the transformation from yearly to net present risk boils down to multiplying the former by a constant. This simplifies the exposition below considerably.

13.7.2 Solidarity

Suppose now that the presidents of Alu and Bawe meet in a bar in Geneva and discuss the problems of their respective islands. Because of the scare-mongering of an environmental pressure group about the relationship between tropical cyclone activity and global warming, it is impossible for the small islands to buy insurance at a reasonable price. The president of Alu, the smarter of the two, proposes that the islands sign an agreement on mutual solidarity in times of disaster.

Let us say that Alu agrees to cover the fraction $0 \le \alpha \le 1$ of Bawe's damage, and Bawe $0 \le \beta \le 1$ of Alu's. The yearly risk, the yearly mean welfare loss $E U^S$, of Alu is then

$$E U_A^S = p_A \int_0^\infty f_A (I) \, U_A ((1 - \beta)D_A (I))dI + p_B \int_0^\infty f_B (I) U_A(\alpha \, D_B (I))dI \qquad (14)$$

The risk of Bawe, and the net present values follow directly. Figure 13.6 depicts an example of the result: Both islands pay more frequently but less each time.

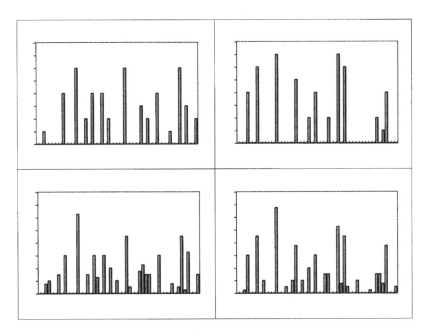

Figure 13.6 Hypothetical sequence of storm damages on 'Alu' (right) and 'Bawe' (left) without (top) and with (bottom) a solidarity contract. In this example, the total damage over 50 periods is 50 for both islands; solidarity covers $\alpha = \beta = 0.25$

The question is of course whether the annual mean welfare loss decreases or increases as a result of the solidarity contract. To that end, we look at the ratio

$$R_A = \frac{\text{NPVE}\,U_A^S}{\text{NPVE}\,U_A} = \frac{E\,U_A^S}{E\,U_A} = \frac{p_A \int_0^\infty f_A\,(I)U_A\,((1-\beta)D_A(I))\mathrm{d}I + p_B \int_0^\infty f_A\,(I)U_A\,(\alpha D_B(I))\mathrm{d}I}{p_A \int_0^\infty f_A\,(I)U_A\,(D_A(I))\mathrm{d}I} \quad (15)$$

Additional assumptions need to be made for it to be possible say to say whether this ratio is smaller than 1 (i.e., that it is in Alu's interest to close the agreement). The disutility functions are generally increasing and convex in damage, that is, twice as much damage is more than twice as bad. Assume a power function for convenience, $U_A = D^a$, $a > 1$. Then

$$R_A = \frac{p_A\,(1-\beta)^a \int_0^\infty f_A\,(I)D_A\,(I)^a \mathrm{d}I + p_B\alpha^a \int_0^\infty f_B\,(I)D_B\,(I)^a \mathrm{d}I}{p_A \int_0^\infty f_A\,(I)D_A\,(I)^a \mathrm{d}I}$$
$$= (1-\beta)^a + \alpha^a\,\frac{p_B \int_0^\infty f_B\,(I)D_B\,(I)^a \mathrm{d}I}{p_A \int_0^\infty f_A\,(I)D_A\,(I)^a \mathrm{d}I} \quad (16)$$

Equation (16) states that the acceptability of the contract to Alu depends on the risk taken away (β), the risk obtained (α), the steepness of the disutility function (a), and the relative exposure of Bawe. R_A is decreasing in β, a and Alu's exposure, and increasing in α and Bawe's exposure. Bawe's ratio R_B behaves exactly in the opposite way (cf. Figure 13.7). The core C of the negotiations – that is, the set of values of α and β which are acceptable to both Awe and Balu, is the space in which both ratios are smaller than unity, $C = \{\alpha, \beta \in [0,1] \times [0,1] \mid R_A \le 1, R_B \le 1\}$. Where exactly in the core the negotiations end up depends on the bargaining power of Alu and Bawe. Figure 13.8 depicts the core.

A real-world example of the above can be found in the European Union. In the Treaty of Maastricht, the member nations intentionally agree on mutual support in case of natural disasters. Less formal examples can be found in 'spontaneous' solidarity, rooting in goodwill and moral standards; for instance, countries such as Bangladesh and China, not the wealthiest nations, supported Japan, very active in emergencies, after the Kobe earthquake.

13.7.3 Non-compliance

The reasoning above is in terms of mean welfare loss or expected disutilities. Just as important as the question as to the conditions under which it is profitable for Alu and Bawe to close an agreement is the question of when one of the two decides to break the contract, or the chance of non-compliance. Alu is tempted not to pay its duties in the event of a disaster at Bawe in one particular year if these exceed Alu's total value of the contract; that is, if

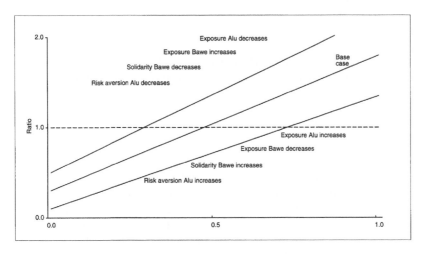

Figure 13.7 Alu's pay-off ratio R_A as a hypothetical function of solidarity parameter (α), plus an indication of the direction in which other factors would push this function. If $R_A < 1$, the contract is acceptable to Alu

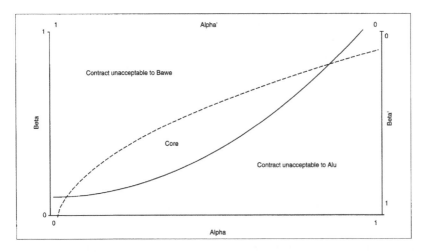

Figure 13.8 Hypothetical Edgeworth box of the solidarity negotiations between Alu and Bawe. Outcomes on α and β below the dotted line are unacceptable to Alu. Outcomes above the solid line are unacceptable to Bawe (in fact, Bawe looks at the figure as if the upper right corner were the origin, and considers α' and β'). The core consists of the outcomes acceptable to both

$$U_A (\alpha D_B) \geq \frac{E U_A^S - E U_A}{\rho_A} \qquad (17)$$

This assumes that if the contract is broken once it will never enter into force again (in game jargon, a tit-for-tat strategy). The change on non-compliance by Alu, q_A, is then

$$q_A = 1 - F_A \left| \frac{E U_A^S - E U_A}{\rho_A} \right| \qquad (18)$$

where F_A is the distribution corresponding to the density f_A.

The issue of non-compliance alters the earlier discussion on the solidarity contract in four possible ways. It was tacitly assumed that the contract is binding and that non-compliance is not possible. Nevertheless, one of the parties may be tempted. The first solution is to make the contract binding; that is, let an authority above Alu and Bawe force both parties to comply, for instance by issuing heavy sanctions. This implies that a part of sovereignty is given up. Sovereignty is highly regarded, so reducing it would be a loss of welfare. However, the benefits of solidarity do not necessarily outweigh this loss of sovereignty. The second solution is to replace the solidarity contract with a mutual insurance pool. This case is extensively analysed below. The third solution is to restrict the payment to the value of the contract. This obviously changes the value of the contract: on the one hand, less aid is given; on the other hand, less aid is received. The expected welfare loss can be expressed mathematically, but the equation is quite complicated and yields little insight. The fourth solution is to accept the risk of non-compliance, and adjust the value of the contract accordingly. The net present expected welfare loss is in this case

$$\text{NPVE}\, U_A^S = \sum_{t=0}^{\infty} (1 - q_A - q_B)^t E U_A^S (1 + p_A)^{-t} +$$

$$+ \sum_{t=0}^{\infty} \left((1 - q_A - q_B)^t \right) E U_A (1 + p_A)^{-t} = (1 + p_A) \left| \frac{E U_A^S - E U_A}{p_A + q_A + q_B} + E U_A \right| (19)$$

which of course coincides with the above for $q_A = q_B = 0$. Note that the chance of non-compliance depends on the value of the contract, and the value of the contract on the chance of non-compliance.

13.7.4 Further parties

Suppose now that Alu and Bawe have finished the negotiations and implemented the solidarity agreement. After a decade or so, the agreement has shown its success in practice. The neighbouring island state Cito watches with envy the more regular but smaller disaster losses of its neighbours. Cito announces that it wants to join the agreement. The evaluation of a three-party solidarity contract follows exactly the same lines as the evaluation of a two-party contract, as above. More of Alu's and Bawe's risk is taken away in return for a part of Cito's risk. This results in smaller annual damage, but more frequent losses. The islands are assumed to prefer this, so Cito's entrance into the contract enlarges the core and enhances the welfare of all three parties. In addition, the risk of non-compliance falls as the duties of individual parties fall and the stakes of remaining involved rise. Generally, the advantages of mutual solidarity rise as the number of parties rises.

13.7.5 Mutual insurance

We have shown in the above that mutual solidarity is in the interest of all parties. However, one of the disadvantages of natural disasters, the irregular pattern of losses, is only partially addressed. A mutual insurance contract would replace the irregular damages with an annual premium. The actuarially fair premium is the expected annual damage:

$$\mathrm{E}D_\mathrm{A} = p_\mathrm{A} \int_0^\infty f_\mathrm{A}(I)D_\mathrm{A}(I)\mathrm{d}I \tag{20}$$

Mutual insurance is clearly preferred over no insurance and no solidarity, as the assumption that the disutility function is convex (see above) guarantees that

$$\mathrm{E}U_\mathrm{A}^I = U_\mathrm{A}(\mathrm{E}\ D_\mathrm{A}) < \mathrm{E}U_\mathrm{A} \tag{21}$$

$\mathrm{E}U^I$ denotes the expected disutility under insurance.[8]

More relevant is the relationship between the expected utility under solidarity $\mathrm{E}U^S$ and under insurance $\mathrm{E}U^I$:

$$p_\mathrm{A}^a \left(\int_0^\infty f_\mathrm{A}(I)D_\mathrm{A}(I)\mathrm{d}I \right)^a \lessgtr p_\mathrm{A}(1-\beta)^a \int_0^\infty f_\mathrm{A}(I)D_\mathrm{A}(I)^a\mathrm{d}I + p_\mathrm{B}\alpha^a \int_0^\infty f_\mathrm{B}(I)D_\mathrm{B}(I)^a\mathrm{d}I \tag{22}$$

The question mark denotes that general statements cannot be made as to whether insurance or solidarity is better. The crucial determinants are the shares that Alu and Bawe cover from each other's damage, or, in fact, the negotiation process of the solidarity contract.

However, (21) guarantees that for Alu and Bawe together (i.e., under full co-operation), insurance is better than solidarity. Hence, a scheme of side payments is conceivable in which Alu compensates Bawe, or the other way around, for the loss incurred by switching from solidarity to insurance. Alternatively, the losing party could negotiate a lower insurance premium.

13.7.6 Insurance failure

The equivalent of the chance on non-compliance with solidarity is the chance of a failure of the mutual insurance pool to pay for the damages. Let F_t be the size of the reserves at time t. These reserves evolve according to

$$F_t = F_{t-1} + ED_A + ED_B - D_{A;t-1} - D_{B;t-1} \tag{23}$$

Taking expectations, this leads to

$$EF_t = EF_{t-1} + EED_A + EED_B - ED_{A:t-1} - ED_{B:t-1} = F_{t-1} + \\ ED_A + ED_B - ED_A - ED_B = F_{t-1} \tag{24}$$

So, the insurance reserves follow a random walk. The conditional variance is

$$\text{Var } F_t \mid F_{t-1} = \text{Var } D_A + \text{Var } D_B; \quad \text{Var } F_t \mid F_0 = t \, (\text{Var } D_A + \text{Var } D_B) \tag{25}$$

The unconditional variance is unbounded.

Using a Normal approximation, this implies that to guarantee a 99 per cent certainty that insurance will not fail in the next year, the reserves have to be as large as 2.6 times the square root of Var D_A+Var D_B. The required reserves grow with the square root of the distance in time. For instance, to guarantee non-failure in 2 years, the reserves have to be as large as 1.4 times 2.6 times the square root; in 25 years, this would be 5 times 2.6 times the square root. The discount rate assures that the future risk remains bounded.

Taking capital markets into account, (24) becomes

$$EF_t = (1 + r) \, F_{t-1}; \quad EF_t = (1 + r)^t F_0 \tag{24'}$$

where r is the interest rate. For convenience, the interest rate is taken to be perfectly known, equal over time, and the same for borrowing and lending. Reserves are expected to increase for ever if the initial reserves are positive; reserves are expected to decrease for ever if the initial reserves are negative. The conditional variance is

$$\text{Var } F_t \mid F_{t-1} = (1 + r)^2 \, (\text{Var } D_A + \text{Var } D_B)$$

$$\text{Var } F_t \mid F_0 = (\text{Var } D_A + \text{Var } D_B) \sum_{i=0}^{t} (1 + r)^{2i} \tag{25'}$$

which is larger than in (25). The unconditional variance is again unbounded.

In the event that the annual premium is not actuarially fair, for instance for coverage of the administration costs or for increasing the reserves, $(24')$ becomes

$$EF_t = (1 + r)\ F_{t-1} + (1 + \pi)\ ED; \quad EF_t = (1 + r)^t\ F_0 + t\ (1 + \pi)\ ED \quad (24'')$$

where π is the wedge between the actual and actuarially fair premiums. The conditional and unconditional variances do not change, but the chance of insurance failure decreases if $\pi > 0$, and increases otherwise.

So, the reserves of the insurance pool need to be large, and cannot guarantee success. The main issue is of course where the initial funds can be found. In a commercial insurance company, the shareholders provide the funds and bear the risks, in return for their part of the π in $(24'')$. However, the problem started with the unavailability of commercial insurance. Alu and Bawe might provide the initial funds, but this would lay a heavy burden on their economies. Alternatively, the World Bank, the UNDP or a foreign government could provide the funds, or back up the pool when needed.

Here again, the larger the number of participants, the smaller the risk of insurance failure, because reserves accumulate faster and the disaster claim of one participant is smaller relative to the pool. Provided that the risks of the participants are uncorrelated, the peak claim thus decreases relative to the reserves.

A real-world example of mutual insurance among states was established under the Lomé Convention. Member states include Fiji, Kiribati, Papua New Guinea, the Solomon Islands, Tonga, Tuvalu and Western Samoa. The convention established a fund of ECU 60 million, from which emergency aid may be granted. Each year, the fund is replenished as appropriate by the European Union.

13.7.7 Conclusion

The analyses of this Appendix show that, under weak assumptions, mutual solidarity and mutual insurance is in the interest of most parties. The more parties, the higher the gain. An important condition for solidarity or insurance to be beneficial for all parties is knowledge of the actual risk. The main hindrance is the possibility of a freak event, a real disaster. In the wake of such an event, non-affected parties may decide to withdraw from mutual solidarity or insurance.

Acknowledgements

The author hereby expresses his gratitude to Use van der Hul and Arnaut Schuurman for invaluable research assistance, to Kees Dorland, David Favis-Mortlock, Frank Leek, Xander Olsthoorn, Angela Liberatore, John Maunder, Bill Nordhaus, Pier Vellinga and Hans van der Zwol for discussion and comments.

Notes

1 This section draws upon Lohmeijer-van der Hul (1994); see also Vaughan (1989). See Arrow (1996) and Kunreuther (1996) for recent discussions of insurance and catastrophes. See Priest (1996) for an alternative view on insurance.
2 Names are the owners of Lloyd's. In contrast to shareholders, whose financial liability is limited to the value of the stocks they own, the financial liability of names is unlimited. In some recent years, their losses have been heavy, and Lloyd's has sought to restructure itself and find other sources of finance.
3 Unless of course historical records report this tendency as well, and this is detected and acknowledged. The latter provision is the crucial one: insurance compnaies tend to use only recent records, likely to be too short to discover trends in their vunerability. Another implication of this is that insurers run the risk of mistaking long cycles for trends.
4 The cash flow of the insurer follows from subtracting the reinsurer's cash flow from the insurance sector's total cash flow.
5 Of course, one can insure oneself for being liable. However, the insurance is in principle against random events, not structural tendencies.
6 As NPV is a linear operator, the expected net present value is identical to the net present expected value.
7 The pure rate of time preference determines by how much welfare now is preferred over welfare later. It is a measure of myopia and impatience. Sometimes it is used to adjust for uncertainty about the future and fear of death.
8 Which equals the disutility of the expected damages because the expected damages are assured to be known with certainty.

References

Albala-Bertrand, J.M. (1993) *Political Economy of Large Natural Disasters*, Oxford: Clarendon Press.
Anon. (1996) Disastrous bonds. *The Economist* 31 August: 68–69.
Arnell, N. (1987) 'Flood insurance and floodplain management.', in J.W. Handmer, ed., *Flood Hazard Management: British and International Perspectives*, Norwich: Geobooks, 117–133.
Arrow, K.J. (1970) *Essays in the Theory of Risk-Bearing*, Amsterdam: North-Holland.
Arrow, K.J. (1996) The theory of risk-bearing: small and great risks. *Journal of Risk and Uncertainty* 12: 103–111.
Berz, G., and Conrad, K. (1993) Winds of change. *The Review* June: 32–35.
Changnon, S.A., and Changnon, J.M. (1992) Temporal fluctuations in weather disasters: 1950–1989. *Climatic Change* 22: 191–208.
Chichinilsky, G., and Heal, G. (1994) Global environmental risks. *Journal of Economic Perspectives* 7, 4: 64–85.
Clark, K.M. (1988) Predicting disasters getting easier. *National Underwriter*, 29 August.
Clark, K.M. (1991) Catastrophe insurers need more, better data. *National Underwriter* 30 September.
de Roon, F., and Veld, C. (1994) Insurance against natural disasters with futures. *Economische Statistische Berichten*, 805–807 (in Dutch).
Dlugolecki, A.F., Clement, D., Elvy, C., Kirby, G., Palutikof, J., Salthous, R., Toomer, C., Turner, S., and Witt, D. (1994) *The Impact of Changing Weather Patterns on Property Insurance*, London: Chartered Insurance Institute.

Dlugolecki, A.F., Clark, K.M., Knecht, F., McCauley, D., Palutikof, J.P., and Yambi, W. (1996) 'Financial services', in R.T. Watson, M.C. Zinyowera and R.H. Moss, eds, *Climate Change 1995: Impacts, Adaptations and Mitigation of Climate Change: Scientific-Technical Analyses*, Contribution of Working Group II to the Second Assessment Report of the Intergovernmental Panel on Climate Change, Cambridge: Cambridge University Press, 539–560.

Dong, W., Shah, H., and Wong, F. (1996) A rational approach to pricing of catastrophe insurance. *Journal of Risk and Uncertainty* 12: 201–218.

Dorland, C., and Tol, R.S.J. (1995) *Storm Damage in the Netherlands: Modelling and Scenario Analysis*, W95/20, Amsterdam: Institute for Environmental Studies, Vrije Universiteit.

Greenpeace (1994) *The Climate Timebomb: Signs of Climate Change from the Greenpeace Database*, Amsterdam: Greenpeace International.

Hindle, J. (1994) The reinsurer's view: Catastrophes – are you being served? Paper read at Greenpeace Business Conference on Climate Change and the Insurance Industry, London, 24 May.

Kattenberg, A., Giorgi, F., Grassi, H., Meehl, G.A., Mitchell, J.F.B., Stouffer, R.J., Tokioka, T., Weaver, A.J., and Wigley, T.M.L. (1996) 'Climate models: Projection of future climate', in J.T. Houghton, L.G. Meira Filho, B.A. Callander, N. Harris, A. Kattenberg, and K. Maskell, eds, *Climate Change 1995: The Science of Climate Change*, Cambridge: Cambridge University Press, 285–358.

Kunreuther, H. (1996) Mitigating disaster losses through insurance. *Journal of Risk and Uncertainty* 12: 171–187.

Law, A.M., and Kelton, W.D. (1991) *Simulation Modelling and Analysis*, 2nd edition, New York: McGraw-Hill.

Leggett, J. (1993) *Climate Change and the Insurance Industry: Solidarity among the Risk Community*, London: Greenpeace.

Lohmeijer-van der Hul, E. (1994) *Climatic Change and the Insurance Sector*, W94/24, Amsterdam: Institute for Environmental Studies.

McCullough, K. (1995) Catastrophe insurance futures. *Risk Management* 42, 8: 31–41.

Manne, A.S., and Richels, R.G. (1992) *Buying Greenhouse Insurance: The Economic Costs of CO_2 Emission Limits*, Cambridge, MA: MIT Press.

Morgan, M.G., and Henrion, M. (1990) *Uncertainty: A Guide to Dealing with Uncertainty in Quantative Risk and Policy Analysis*, Cambridge: Cambridge University Press.

Nordhaus, W.D. (1994) *Managing the Global Commons: The Economics of the Greenhouse Effect*, Cambridge, MA: MIT Press.

Paish, T. (1994) The insurer's response to climate change: An overview. Paper read at Greenpeace Business Conference on Climate Change and the Insurance Industry, London, 24 May.

Pate-Cornell, E. (1996a) Uncertainties in global climate change estimates. *Climatic Change* 33: 145–149.

Pate-Cornell, M.E. (1996b) Global risk management. *Journal of Risk and Uncertainty* 12: 239–255.

Pratt, J.W. (1964) Risk aversion in the small and in the large. *Econometrica* 32, 1–2: 122–136.

Priest, G.L. (1996) The government, the market, and the problem of catastrophic loss. *Journal of Risk and Uncertainty* 12: 219–237.

Rodetis, S.E. (1994) US insurers: Back to the future(s)? *Global Reinsurance*, February: 103–107.

Rokovoda, J., and Vrolijks, L. (1993) *Case study Fiji: Disaster and Development Linkages*, South Pacific Workshop UN Disaster Management Training Programme, 29 November–4 December 1993, Apia.

Schuurman, A. (1995) *An Insurance for the Meuse*, W95/19, Amsterdam: Vrije Universiteit.

Tol, R.S.J. (1995a) *Climate Change, Wind Storms and the Insurance Industry: Some Notions from a Weather Insurance Simulation Model, Part II*, W95/40, Amsterdam: Institute for Environmental Studies, Vrije Universiteit.

Tol, R.S.J. (1995b) The damage costs of climate change: Towards more comprehensive calculations. *Environmental and Resource Economics* 5: 353–374.

Tol, R.S.J., and Olsthoorn, A.A. (1994) *Climate Change, Wind Storms and the Insurance Industry: Some Notions from a Weather Insurance Simulation Model, Part I*, W94/15, Amsterdam: Institute for Environmental Studies, Vrije Universiteit.

Vaughan, E.J. (1989) *Funamentals of Risk and Insurance*, 5th edition, New York: Wiley.

Vellinga, P., and Tol, R.S.J. (1993) Climate change, extreme events and society's response. *Journal of Reinsurance* 1, 2: 59–72.

Verbond van Verzekeraars (1994) *Dutch Insurance Industry in Figures*, The Hague: Verbond van Verzekeraars.

von Seggern, J. (1993) 'Flooding risk potential: A mapping approach', in R.S.J. Tol, ed., *Socio-economic and Policy Aspects of Changes in the Incidence and Intensity of Extreme Weather Events Presentations*, W93/15, Amsterdam: Institute for Environmental Studies, Vrije Universiteit.

Wilford, M. (1993) 'Insuring against sea-level rise', in P. Hayes and K. Smith, eds, *The Global Greenhouse Regime: Who Pays?*, London: UNU Press/EarthScan, 169–187.

Williams, C.A., and Heins, R.M. (1985) *Risk Management and Insurance*, 5th edition, Singapore: McGraw-Hill.

14

IDENTIFYING BARRIERS AND OPPORTUNITIES FOR POLICY RESPONSES TO CHANGING CLIMATIC RISKS

Matthijs Hisschemöller and Alexander A. Olsthoorn

14.1 Introduction

This chapter explores the barriers and opportunities to develop and implement options for policy responses to a possible change in the incidence of extreme weather due to the process of climate change. In this chapter, the term 'policy' primarily refers to government decisions. Since government policies are seldom made by governments alone, the positions and potential behaviour of other social and political actors is important as well. The chapter does not identify local or sector-specific solutions or policies. It focuses on the gap between scientific knowledge and policy practice. A policy sciences approach is taken. The chapter aims at offering a strategy to policy-makers at the European, national and regional levels for developing prevention and adaptation options in close collaboration with scientists and representatives of the sectors and regions at risk.

The term *policy sciences* is used as defined by Lasswell (1951), who in his famous article called upon political scientists, sociologists, economists, psychologists, lawyers and scholars in other sciences to invest in a joint effort to establish a *policy sciences of democracy*. This undertaking aimed at producing a broad body of knowledge in support of improving the quality of policy decisions with regard to major social problems. Hence a policy sciences perspective is interdisciplinary, rather than being confined to just one particular (social) science. It focuses on decision strategies to provide solutions for problems that are considered technically adequate as well as legitimate from a social and political point of view.

Opportunities for furthering the development and implementation of response options are analysed from three perspectives:

- the utilisation of scientific knowledge;
- agenda-setting and implementation; and
- the structure of policy problems.

Section 14.2 examines insights from policy-science literature on the use of knowledge. In section 14.3, the focus is on agenda-setting and implementation. Section 14.4 deals with the way social and political institutions may react to a possible change in extreme weather in the context of climate change. Section 14.5 presents a summary of the chapter's main findings and policy recommendations for strategies with regard to land subsidence, agricultural drought, river floods, heatwaves, windstorms and tropical cyclones.

14.2 From science to policy

Scientists often complain that the knowledge they produce is not used adequately, not used at all or even misused in public policy. This is the case with scientists working in the field of climate change. They may feel that the warnings and solutions that come out of their research work are not sufficiently reflected in policies. Policy-makers, on the other hand, may state that scientists do not always produce usable knowledge. Science outputs may raise questions such as 'So what?' or 'What are we supposed to do about it?' Scientists may not understand that their science problem is not (yet) a problem for policy. Policy-makers are expected to follow directly the guidance of scientists who are, because of their expertise and experience, qualified to recognise and define the problems policy-makers need to address. Such an attitude is understandable, particularly in cases of global environmental problems which were originally recognised by natural scientists, such as ozone depletion and climate change. However, policies never are, or could be, merely science-driven. Notions on how scientific knowledge can be used and explanations for inadequate use or non-use of scientific knowledge by public policy have been developed in many disciplines (Dunn and Holzner, 1988). Without doubt, this multi-disciplinary involvement has contributed to quite a fragmentation of the field. Theory building on knowledge use is considered to be not very well developed (Rich, 1991: 322). Even the concept itself may lead to confusion. Still, different notions on knowledge use have some elements in common: *knowledge use* is usually meant to refer to scientific knowledge having an observable impact on policy.[1] Some of the insights concerning barriers to knowledge use, and the solutions they suggest, are of particular interest for the argument in this chapter.

14.2.1 The two-cultures theory

Probably the most popular explanation for the (non)use of scientific knowledge is the *two-cultures* or *two-communities* theory. According to this view (Caplan, 1979), science and policy-making constitute two different cultures with different rules and values. The scientist's culture values theory construction and the testing of hypotheses. Scientists are reluctant in formulating clear-cut conclusions and policy recommendations. They find it hard to accommodate their research schedules to policy-makers' needs. Last but not least, they do not know what the information needs of policy-makers really are, because they do not think, act or feel like policy-makers. In this view, scientists are primarily interested in analysis rather than in recommending solutions for policy. For that reason, their advice almost always includes the recommendation that more research is needed, even if such a recommendation is not based on an identification of policy-makers' information needs. Policy-makers, on the other hand, first of all need flexibility to use their power most effectively. They are less interested in reliability and validation of knowledge than in having the right solution at the right moment. For policy-makers, the right decision is not necessarily the same as the scientific truth.

A closer look at the two-cultures theory reveals that it actually covers a broad range of explanations for the non-use of scientific knowledge. Rich (1991: 323–324) points to the following factors:

- There is great distrust and even antagonism between the two cultures.
- Different and even competing reward-systems are characteristic for the two cultures.
- The two cultures use different language.
- Researchers and bureaucrats operate under substantially different conceptions of time.
- What is relevant for one community is not necessarily relevant for the other.

Two-cultures theory thus in fact consists of many explanations, which are partly complementary and partly competing. This observation brought Dunn (1980) to conclude that the idea about two cultures is an appealing metaphor rather than a theory.

Given the manifold explanations it embodies, the two-cultures theory yields a variety of recommendations for increasing the utilisation of research findings. All of them relate, in one way or another, to improving the process of communication between researchers and policy-makers. However, improving communication between communities that are defined by entirely different cultures is not an easy thing to do. The cultural gap may be too wide to bridge (Snow, 1960: 67).

In order to overcome problems related to mistrust, different reward systems and languages, it has been recommended that a science–policy interface should be established, explicitly dedicated to communicating policy-makers' research needs to scientists and research findings to policy-makers. Science mediators are expected to smooth the process of knowledge utilisation (Caplan *et al.*, 1975). Recommendations further include fine-tuning the time frames of scientists and policy-makers (van de Vall and Bolas, 1982).

In the area of climate-change policy-making and research, the policy–science interface is highly institutionalised in bodies like the Intergovernmental Panel on Climate Change (IPCC) and research programmes like the International Geosphere Biosphere Programme (IGBP) and the EC Environment and Climate Programme. IPCC communicates research findings to the parties in the Framework Convention on Climate Change. The research programmes mentioned specifically address policy-relevant research topics. The institutionalisation of the science–policy interface in the area of climate change points at a well-established *knowledge system* in this field. A knowledge system consists of interdependent knowledge functions, along with the social structures by which these functions are performed (Dunn and Holzner, 1988). There is evidence suggesting that a knowledge system which combines the fields of climate change and disaster research is, if not completely absent, in a very embryonic stage (Olsthoorn *et al.*, 1994).

The utilisation of scientific knowledge in climate policies may increase once a knowledge system has matured. In both the Netherlands and the European Union this implies, among other things, the linking of atmospheric science to solution-oriented research in the fields of energy and technology (Dinkelman, 1995; Liberatore, 1994). However, whether the mere existence of a knowledge system is a sufficient condition to bridge the gap between science and policy is doubtful. Recent findings in the field of climate change suggest that, in different countries, policy-makers as well as scientists stress the need for improving policy–science communication, especially in 'translating' policy-makers' information needs into research questions and in fine-tuning the time frames of research and policy (Bernabo *et al.*, 1995; Klabbers *et al.*, 1994; Vellinga *et al.*, 1995; Hisschemöller *et al.*, 1995).

Conclusively, the various notions embodied in the two-cultures theory of knowledge use serve to highlight some interesting and valuable insights on the further development and implementation of policy response options to possible changes in extreme weather. Basically, the recommendation is to establish an institution to link research on both climate change and disaster management to policy-making at different administrative levels. This institution might, by fostering a coherent knowledge system, contribute to knowledge use in policy-making. However, the two-cultures theory does not give an adequate recipe for the design and operation of such an institution. Hence the insights provided by this metaphor need to be deepened, adjusted and complemented by insights from other theoretical perspectives.

14.2.2 *Usable knowledge and the rational actor model of policy-making*

A second school of thought focuses on the specific characteristics of the policy process and its implications for using knowledge. It is assumed that analysis of organisational rules, bureaucratic procedure, conflicts of interests and specific organisational cultures is essential for understanding the dynamics of knowledge use (Rich, 1991). This perspective is not in all respects incompatible with the two-cultures theory, but it takes quite a different point of departure: the rational actor model. Policy-makers are expected to use information according to what they believe is in their own interest and ignore or abuse other information (Downs, 1967; Rich, 1981). Moreover, a bureaucratic institution is capable of handling a limited amount of information within the time available for a specific problem. This view corresponds to the theory of bounded rationality (Simon, 1976).

In the stage of utilisation, scientific information always competes with other sources of information, especially everyday observations, experiences, expectations and ideologies of policy-makers (Lindblom and Cohen, 1979). Policies are based on specific policy frames or frames of reference (Holzner and Marx, 1979; Dery, 1984), also referred to as policy-making frameworks (Lindblom and Cohen, 1979), belief systems (Sabatier and Jenkins-Smith, 1993; Hoppe and Peterse, 1993) or policy theories (Hoogerwerf, 1984; Hisschemöller, 1993). Policy frames comprise a broad range of assumptions with regard to the policy-maker's perception of causal relationships, relationships between goals and means, and relationships between norms and values (Hoogerwerf, 1984). Policy frames do not just relate to the narrow self-interest of policy-makers or policy-making institutions. Rather, they shape the framework in which interests and power relations are perceived. In competing with these policy frames, scientific information is often at a disadvantage, particularly if it contradicts policy-makers' beliefs and expectations.

Another factor of importance relates to the position of scientists themselves. Like policy-makers, scientists compete for research funds with institutions serving different policy interests. Depending on who pays for their work, scientists may disagree on the merit of scientific information. If this happens, the social and policy impact of science information is likely to decrease (Miller, 1983).

From these assertions, the following observations are made regarding the use of scientific knowledge (Rich, 1991: 326):

- The rational actor's information search will usually be quite limited.
- Institutions show a strong preference to confine the information search to their own organisation or, if necessary, to the policy network which is closely connected to their organisation.
- The use of information will be decided on the basis of whether or not the information confirms a policy position already held, or will be based on the intuition of policy-makers.

Two more observations, which fit into the rational actor model of knowledge use, may be added here:

- Policy-makers will not use scientific information unless this information yields recommendations that they consider politically feasible and easy to implement (van de Vall and Bolas, 1982; van de Vall, 1988).

369

- The impact of scientific knowledge on policy will be higher as scientists agree on the information disseminated to policy-makers (Miller, 1983).

What conclusions and recommendations can be derived from these observations? The first conclusion is that, given the way policy-makers operate, the chances for improving knowledge use are quite low. This is not to say that scientific information is not used at all. Policy-makers do utilise scientific information, but always as part of their general store of knowledge. The so-called *enlightenment function of science* (Weiss, 1980) is usually limited to confirming already prevailing values and beliefs. Scientific concepts and information about causal relationships may help structure policy beliefs and so may help to legitimate policies (Hutchinson, 1995). The second conclusion is that advising policy requires an attitude which fits in with the client's immediate needs; the scientist should, first of all, serve the interests of the policy-maker he or she works for. This means that the scientist should be most aware of the frame of reference of the policy-maker, especially with regard to the policy problem and the policy objectives that have been formulated. The researcher should not try to alter this frame of reference. Instead, the researcher should adjust scientific procedure to the needs of policy (van de Vall, 1988). Policy-makers have a primary interest in feasible and implementable action-oriented results, delivered on time, rather than in research that is good according to scientific standards. Therefore, as van de Vall (1988) points out, policy-oriented research has developed a paradigm of its own, distinct from rigid scientific procedure.

It is questionable whether these observations completely hold for the topic of this study. Researchers in the area of climate change and extreme weather events, too, should realise that the policy impact of their work increases as research reflects the values of policy-makers and fits in with policy options that are considered feasible. However, policies related to climate and extreme weather are in need of sound scientific information. A knowledge system too much engaged in everyday policy-making experience may become dangerous in the longer term, as the critical function of science becomes distorted. A politics-of-interest-based knowledge system, in which scientists are divided into policy networks closely linked to competing policy-making units, may safeguard this critical function of science. However, this supposes that the policy arena already has a structure which provides a more or less balanced representation of the interests involved, as is often the case with social policy. With regard to the response options for a possible change in extreme weather, such a policy arena does not (yet) exist, particularly at local and regional levels of administration where response options are to be implemented. As noted above, the knowledge system to serve the development and implementation of response options is also in a very early state of development. Therefore, the rational-actor-based solution, of a knowledge system which consists of competing applied research networks in a pluralist policy environment, does not hold for the problem at hand.

370

Rational actor theories do still contribute two valuable insights into the difficulties that are to be faced in the development and implementation of policy response options to address a possible change in extreme weather. First, a knowledge system will be confronted with the everyday practice of policy-making institutions at different levels. These institutions may be reluctant to consider scientific knowledge produced outside their immediate information-providing networks, as the knowledge producers will be perceived not to serve the immediate needs and interests of the respective policy-making agencies. Such a tendency might be greatest at the local and regional levels. Second, scientists from different disciplines, schools of thought, regions and institutions may show the same kind of behaviour as policy-makers do. They may be willing to co-operate with the well established, but they may exclude 'outsiders' from the knowledge system.

In conclusion, in establishing a knowledge system and a science–policy interface to support the further development and implementation of response options for a possible change in extreme weather, one must take into account the broad range of needs, interests, procedural rules and cultures of the different sectors and levels of administration in the regions affected, as well as the different information flows that have to be integrated into the knowledge system in order to make it effective. Neither the two-cultures approach nor the rational-actor-based theories of knowledge use provide adequate strategies for this, because of the specific character of the field under consideration.

14.2.3 Interaction-based models of knowledge use

A third school of thought within the knowledge-use tradition focuses on the different kinds of interactions between science and policy. This school of thought starts with the assumption that knowledge use is defined by a specific kind of interaction, depending upon the complexity of the problem that is addressed. Weiss (1977) distinguishes three models of policy–science interaction. Below, these models are discussed with reference to their implications for developing and implementing policy options to possible changes in extreme weather.

Model 1: Technocracy

In the first model, knowledge use is science-driven. Scientists tell policy-makers what to do. This model might be effective in the case of complex technical and engineering problems. An example of such a problem is standard-setting for construction in subsidence-prone areas (see Brignall *et al.*, Chapter 3, this volume). In general, however, science-driven policies are likely to meet with heavy public resistance, as political choice is replaced by a kind of technocracy. If diverging values are at stake, the science-driven model becomes incompatible with the type of democracy operating in most European countries.

Model 2: Ad-hocracy

In the second model Weiss observes, knowledge use is merely policy driven. Such a model, which shows similarities with the rational actor model presented above, would be most effective in a situation in which policy problems and objectives are clear to all involved. Policy-makers may still disagree; for example, about the cost-effective solution for installing air-conditioners in private houses to cope with heatwaves (see Gawith *et al.*, Chapter 11, this volume). In such situations, economists, for example, may be invited to provide a solution. It should be noticed, however, that much scientific knowledge would already have been provided long before this option was chosen. Policy-makers would have been aware about increased risk of heat stress, and where this would have been likely to have negative impacts. This is not yet the case for extreme weather events and climate change. Therefore, a policy-driven model is not allowed for. If this model is nonetheless employed, it will probably yield many *ad hoc* decisions and a fragmented research effort.

Model 3: Multi-stakeholder dialogue

In the third model, neither science nor policy is the driving force behind knowledge production and use. There is no straight path from research outcomes to the policy arena, or vice versa. Knowledge use cannot be a smooth process in which 'relevant' information is channelled from research institutions to decision-making arenas. This might be especially true for issues with a high degree of scientific complexity and uncertainty which are accompanied by conflicting values. With regard to these *unstructured problems* (Dunn, 1981, 1988; Mitroff and Sagasti, 1973; Mason and Mitroff, 1981), it is difficult or even impossible to identify 'scientific truth'. It is even questionable whether there is such a thing as truth, as scientists from different disciplines and policy-makers from different policy areas might ask different questions and yield to different findings and conclusions. This type of knowledge requires, as Funtowicz and Ravetz (1994) argue, 'post-normal' science. This type of science has done away with the boundaries of disciplines and the traditional dichotomy between 'facts' and 'values'.

The major assumptions that underlie this perspective are:

- Policy problems are socially and politically constructed, which means that different actors have different (subjective and biased) conceptions of the problem (for an elaboration of the concept of 'policy problem', see below).
- There is a plurality of *legitimate* perspectives. This legitimacy of perspectives relates not only to the multiple interests involved in policy-making, but also to the (epistemological) differences among scientists.

It is especially this third model that needs attention in institutionalising a knowledge system for extreme weather events. The (potential) policy and science communities both consist of many stakeholders, and may not be well defined yet. Therefore, knowledge use in this field does not simply imply processing science findings to policy-makers, as the two-cultures theory supposes. Neither can the policy–science interaction be limited to supporting already existing policy objectives, like the rational-actor-based theories suppose. If knowledge use goes beyond the level of implementing technical options, policy–science interaction should, first of all, provide clarification of the various assumptions, views, interests, obstacles and opportunities regarding the problem. Actors from different disciplines, working in separate networks and at different levels of public administration, should jointly work out the specific opportunities for responding to various kinds of extreme weather events. Establishing an institutionalised knowledge system may facilitate the agenda-setting of specific problems and solutions associated with land subsidence, agricultural drought, river floods, heatwaves, windstorms and tropical cyclones.

14.3 The agenda-building process: Barriers and opportunities

14.3.1 A model of the agenda-building process

The use of scientific knowledge may assume different forms in different stages of the agenda-building process. The theory of agenda-building assumes that a problem, in order to become a policy matter, has to pass various stages. In the model of Cobb and co-workers (Cobb and Elder, 1983; Cobb et al., 1976), a distinction is made between:

1   the *want* stage, in which a problem is sensed without being clearly defined;
2   the *demand* stage, in which a problem is defined and policy measures are being asked for;
3   the *issue* stage, in which various groups and agencies recognise the problem, whether or not it has acquired policy agenda status; and
4   the *political item* stage, in which the problem has reached the policy agenda, decisions are taken and – if the political item is not removed from the agenda – implemented.

It is worth noting that the stages distinguished here are not necessarily passed through in a fixed order. A policy problem can be recognised outside government as well as within it. A political item may not become a public issue, as government agencies have an interest in keeping it away from the public. A problem associated with extreme weather events may be found first by scientists and may become a political item later, without passing through the issue stage. Policy-makers may try to make it an issue afterwards, as they want to gain public support for a policy, for example if they need to raise money for certain investments. It often happens that a demand or issue never reaches the policy agenda at all.

The value of this agenda-building model is its focus on the barriers that are faced if a problem moves from one stage to another. Barriers may keep a problem from the political agenda but may also modify the (initial) meaning of the problem. In the next section, some specific barriers and opportunities are discussed which relate to the utilisation of knowledge in different stages of the agenda-building process.

14.3.2 The conversion from wants into demands for policy intervention

Each agenda-setting process starts with identifying a problem. The identification of a policy problem usually takes off with what Dunn (1981) refers to as problem-sensing. A person or small group, either inside or outside the actual decision-making agencies, feels uncomfortable with a certain phenomenon or situation that is perceived to exist. In the case of a possible change in the incidence of extreme weather, problem identification may very well start with a sense of insecurity. In this phase, the actor decides whether some basic needs will be harmed. If the answer is in the affirmative, the *want* may convert into a political *demand*. At the demand stage, an actual *policy problem* is identified. Hence for science to have an impact on policy, a policy problem should exist, or a science problem should become translated into a policy problem.

Extreme weather events as a policy problem

Policy problems do not just 'happen to be there'. They are always defined by people, who *frame* them from socio-economic, political and cultural perspectives. Hence, problems are not objective givens but social and political constructs (Dery, 1984). This is also the case for environmental problems (Hoppe and Peterse, 1993; Wynne, 1994; Liberatore, 1995; Hisschemöller and Hoppe, 1996). This is not to be interpreted as saying that problems are not real or exist only in human imagination. Problems refer to phenomena or situations which may become a problem if two conditions are met:

1   The observed phenomenon or situation is to be considered undesirable from some normative point of view (standard, value), which is not necessarily a majority view (Hoogerwerf, 1987).
2   There is a way of resolving the undesired situation, if only in the expectation of the actor who brings the problem forward. Problems thus suppose solutions (Dery, 1984).

In conclusion, a policy problem may be defined as a gap between a situation actually observed and one that is considered desirable. This gap can be bridged by collective (government) action. What do policy problem(s) associated with a possible change in extreme weather look like?

Natural disasters do not, by themselves, constitute policy problems. They may be considered as 'acts of God' which humans have to accept. Langen and Tol (Chapter 6, this volume) illustrate this with the behaviour of nineteenth-century farmer communities in the Netherlands. In the late twentieth century, natural disasters are considered a problem. Governments are held responsible if a weather-related disaster occurs; governments are expected to protect life and property in vulnerable areas. After the recent river floods in the European Union, people were highly frustrated, particularly in the Netherlands, as they were convinced that the impacts could have been anticipated. Socio-psychological studies provide ample evidence that people's reaction to risk differs according to whether the risk is considered unavoidable or it has been imposed by others – for example, a negligent government.

In the future, the scientific evidence with regard to anthropogenic climate change may further increase pressure on public policy. If changes in extreme weather are actually caused by human interference with the climate system, then weather-related disasters may no longer be felt to be natural phenomena, for which no human being is to blame. Disasters as such may become felt as social problems. Since the average citizen has quite high expectations of the protection that government can offer, ramifications for public policy in the EU may be far-reaching. There may be an increasing demand for reducing greenhouse gas emissions and demand for preparedness to prevent, mitigate or adapt to the impacts of extreme weather events.

At present, however, there seem to be two main obstacles for identifying policy problems in this respect. The first obstacle relates to the scientific uncertainties and diverging political views on the issue of *climate change*, which provides the context for a possible change in extreme weather. The second relates to scientific and policy uncertainties regarding the *change in extreme weather* as such. An adequate understanding of these obstacles is necessary for identifying opportunities to further the agenda-building process.

The climate change context

From a science perspective, a possible change in extreme weather is closely linked to the issue of climate change itself. However, not all policy-makers are convinced by the findings regarding climate change. And if they are, there may be other obstacles to prevent them from taking action. Improving scientific knowledge on the climate system will not necessarily yield a greater willingness to take action. The actual shortcomings of the science-driven view on policy-making are clearly illustrated by the findings from a study in six pilot countries on policy-makers' views (Bernabo et al., 1995) and from an ongoing multi-stakeholder study in the Netherlands (Klabbers et al., 1994; Vellinga et al., 1995; Hisschemöller et al., 1995; Akkerman et al., 1996).

Policy problems associated with climate change are partly based on scientific findings. They are also – and even more so – based on considerations with respect to the effectiveness of intervention. Of the policy-makers interviewed during the studies, many of those who trust the IPCC findings argued that action should be taken. Some of them, however, pointed out that even if the IPCC is completely right about climate change and its negative impacts, action on their part may not be needed. Some of them expect climate change to be mainly positive for their country. Others feel that their actions, however desirable, will not be effective. Because of economic and population growth in other parts of the world, their initiatives will be outweighed by inaction elsewhere and hence not make any difference. They are not prepared to accept costs on their own.

There are also policy-makers who do not believe that climate is actually changing, or that climate change constitutes a national priority. However, many of these people feel that action is nevertheless necessary, adhering to the precautionary principle. They feel that potentially disastrous effects of climate change make action unavoidable. Hence Bernabo et al. (1995) conclude that scientific information about climate change is only one, and not the most decisive, determinant for action. The other determinants consist of the range of values, ideas and interests held by stakeholders at the policy level. The policy and science dimensions that constitute the framing of the climate issue do not necessarily coincide.

Hisschemöller et al. (1995) point out that the policy-maker's inclination to act is not so much dependent on scientific evidence about climate change as on the policy-maker's confidence that action will produce visible results. Action is considered meaningful if problems that are considered urgent for the here and now will be solved. Both the international and the Dutch studies indicate that many policy-makers conceive of the climate issue as not very visible here and now. The findings confirm the observation made by the Second Assessment Report of IPCC Working Group III that actions to address climate change should simultaneously focus on goals other than climate. Such goals can relate to international competition, technological innovation, the quality of the physical environment and the quality of life in general (Hisschemöller et al., 1995: 63).

In conclusion, domestic actors in various countries frame the policy problem associated with climate change in very different ways, if a policy problem is recognised at all. Different values and interpretations of scientific facts constitute a major barrier for development and implementation of policy response options. Framing the climate issue largely depends on whether domestic actors see opportunities for effective action. A 'policy problem' is an opportunity to improve an undesirable situation through government action.

Extreme weather events and opportunities for policy intervention

Like policy problems associated with climate change, policy problems which relate to a possible change in extreme weather suppose the identification of specific responses to prevent, mitigate or adapt to disastrous impacts (Vellinga and Tol, 1993). At several places in this volume, recommendations are made about response options. A solution for subsidence is improved standard-setting for building construction and insurance (Brignall *et al.*, Chapter 3, this volume). Agricultural drought can be addressed by, *inter alia*, changes in crops and varieties or irrigation (Brignall *et al.*, Chapter 4, this volume). Solutions for river flooding include additional land use planning, physical protection, insurance and government relief (Handmer *et al.*, Chapter 5 this volume; Pittock *et al.*, Chapter 2, this volume). Heat stress can be addressed by air-conditioning (Gawith *et al.*, Chapter 11, this volume). For winter storms, solutions may be extended early warning and insurance (Dorland *et al.*, Chapter 10, this volume). Studies of tropical cyclones suggest improvement of early warning systems and government relief as well as aid from abroad (Olsthoorn *et al.*, Chapter 9, this volume) or adjustments in property markets (West and Dowlatabadi, Chapter 8, this volume).

It is interesting to note that the most specific actions are relatively simple: they require some technical accommodation and only a few decision-makers are involved. Standard-setting is a good example. In the case that many private decision-makers are involved, insurance policies may provide adequate incentives to implement precautionary measures. However, these simple options need not offer total protection or may leave some questions unsolved, especially those related to distribution of costs to protect people in highly vulnerable areas. More importantly, for certain extreme events, such options are not yet available, or need to be integrated into a comprehensive programme of action. For example, to take the case of agricultural drought (Brignall *et al.*, Chapter 4, this volume), fertilisers may be effective in addressing drought, but may conflict with other environmental requirements. Irrigation may be limited by competition for water. Langen and Tol (Chapter 6, this volume) discuss examples in which technical response options were counterproductive in the longer term. Therefore, Handmer *et al.* (Chapter 5, this volume) warn against the 'systemic bias in favour of engineering structures'.

14.3.3 Extreme weather events as an issue

If a demand does not meet with sufficient support from within the decision-making system (governments, environmental agencies, political parties or international bodies), it may be expanded to a wider public. If a demand receives a lot of attention, especially from the press, interest groups or the general public, it is transformed into an *issue*.

Issue linkages

The conversion of a demand into an issue may be accompanied by modification of the initial meaning of the problem. Policy stakeholders shape the problem in terms of their own interests and objectives. The issue becomes embedded in the prevailing dimensions of political conflict and hence linked to other issues.

However, it may happen that the actors pushing the issue forward do not rely upon existing interest groups or organisations. If the issue is considered very urgent, unique or novel, a new organisation may be founded. In such a case, a radical policy shift may be accompanied by a redefinition of vested interests in society, reflected in new issue linkages.

An instructive example of the creation of a 'new' issue linkage is the Dutch case of protecting the Wadden Sea (Waddenzee). After the flood of 1953, there was a national consensus that the Wadden Sea, like other big waters, was to be empoldered (Hisschemöller, 1993). The national physical plan (Nota Ruimtelijke Ordening), of 1966 estimated that the economic gains provided by the new land in the Wadden Sea would be tremendous. However, a counter-initiative by a grammar school pupil led to a national organisation for the protection of the Wadden Sea and resulted in a new policy vision. From 1973 onwards, the national consensus has been to maintain and strengthen the natural values of the Wadden Sea.

The Wadden Sea case not only provides an example of how a new linkage of issues can be brought about, but also focuses attention on the delicate and dynamic relationship between environment and security. On many occasions, environmental issues and people's safety go hand in hand (nuclear power, hazardous waste). However, in the Wadden Sea case, environment and security are opposites. At about the same time as the Wadden Sea case, a new, considerably more expensive dike was developed for the Oosterschelde, which left the tides intact. The ultimate government decision in 1977 marks a silent revolution which resulted in the recognition of environment as an issue with huge implications for water management (Hisschemöller, 1993).

It is interesting to note that the same dynamics of issue linkage have become apparent in recent river flood management policies. For the Dutch Rhine, the near-flood of 1995 resulted in local campaigns to improve the dikes against the will of environmentalists. The pendulum had swung back from environment to security.

It would be a misunderstanding to assume that issues are linked overnight. There are many organised groups in society looking for issue-linking opportunities. By issue linkage, they try to push forward their own goals and particular demands. As March and Olsen (1976) put it, many solutions are looking for problems. According to this view, problems and solutions looking for political attention constitute separate 'flows'. Once these flows meet, an issue gets access to the policy agenda (March and Olsen, 1976; Kingdon, 1984) and so-called win–win options may become visible. As Enloe (1978) shows, issue linkages may be embedded in national political cultures. On the basis of a historic overview of Dutch air pollution policy, Dinkelman (1995) suggests that if solutions do not address real problems, issue linkages may not be sustained.

What are the conclusions regarding the subject matter of this volume? First, it should be realised that policy problems related to possible changes in extreme weather will evolve over time. They are, or become, intertwined with other problems. Issue conflicts can be an obstacle in the agenda process, but issue linkage can serve as a vehicle to further clarify the problem, to identify opportunities for solutions, and to gain necessary social support. However, it should be realised that issue linkages may differ between countries, regions, sectors and levels of public administration. It should also be realised that the identification of a new policy problem may meet with much resistance, if the problem redefines vested interests. A knowledge system to cope with the manifold response options for extreme weather should therefore be extremely flexible and dynamic. It should facilitate learning about issue linkage across different regions and sectors.

Other issue characteristics

Apart from issue linkage, there are other strategies for mobilising public support. Cobb and Elder (1983) distinguish five factors which determine whether issues gain access to the policy agenda:

1   The degree of specificity of the issue: an issue defined in a very abstract way is less appealing to a larger public than an issue which is very concrete and specific. The link between global warming and extreme weather is not very specific (see Downing *et al.*, Chapter 1, this volume). However, extreme weather events are concrete. This observation can serve as an argument not to link the extreme weather events issue too closely to climate change models, but to start identifying specific and concrete issues in vulnerable areas.

2   Social significance: the larger the number of people affected by a problem, the more appealing the issue is likely to be. Depending on the hazard and the region, the number of people affected varies. Subsidence affects a relatively small number. Agricultural drought affects few people, but they can be highly organised people. River floods affect relatively many people, while heatwaves, winter storms and tropical cyclones affect virtually everyone.

3   Temporal relevance: some issues are relevant only within a short time frame, others are relevant for a longer period; the latter category is expected to gain more attention in the long run. By definition, extreme events last only for a short while. The window of opportunity for policy-makers to act is thus not too large.

4   Complexity: an issue can be defined as technically very complex or as very easy to understand; in the latter case, more people will get involved in the political conflict. The technical complexity of the link between the enhanced greenhouse effect and extreme weather is huge. The events and their impact are relatively straightforward. Possible responses vary from simple (air conditioning, early-warning systems) to complex (building standards, planting schemes).

5   Categorical precedence: issues may look routine or unique and relatively new; the more unique and new an issue looks, the more social impact it

may have. The weather hazards reported here are selected as to the present risk; they are thus relatively routine.

Strategies of issue expansion and issue perversion

Following Cobb and Elder (1983: 125), we can distinguish two main types of strategies with which either to expand or to marginalise issues related to a possible change in extreme weather: direct and indirect strategies. Most *direct* strategies aim at increasing the public awareness of the impacts of weather-related disasters. Most *indirect* strategies focus on the costs and benefits of policy (non)intervention.

WARNING FOR DISASTERS

Issues related to climate change suffer from the image of being very remote in time and place. This is the case in very different countries like the USA, Brazil, India, China, Poland and the Netherlands (Bernabo *et al.*, 1995). Research in the USA indicates that Americans have difficulties in seeing the difference between climate change and ozone depletion, although they are likely to believe that global warming has already occurred (Read *et al.*, 1994). Climate change does not have symbols like seals, which symbolised the Wadden Sea in the case discussed above (Hisschemöller *et al.*, 1995). However, the Greenpeace Climate Campaign draws on extreme weather events and nasty diseases, a strategy frequently suggested by policy and science experts (Bernabo *et al.*, 1995). It is assumed that people's willingness to act will increase if they become really worried about the impacts on their daily lives. There is, however, no convincing evidence for a positive relation. Too much concern may even lead to deference on the part of the public. An example is areas near nuclear power stations, where people express more than average faith in nuclear technology and sometimes even deny the risk (Hisschemöller and Midden, 1989). Although there is reason to believe that actual disasters may provoke government intervention to address the immediate threat, there is also evidence that the occurrence of disasters is not a sufficient condition to develop comprehensive anticipatory policies with a long-term perspective (Solecki and Michaels, 1994; cf. Langen and Tol, Chapter 6, this volume).

FOCUSING ON COSTS AND BENEFITS

Another strategy proposed and used is to stress the costs that follow from not addressing the issue at an early stage (e.g. Bruce *et al.*, 1996). Such a strategy may be effective, if it simultaneously increases awareness of the problem. It may also provoke a counter-strategy. In his instructive analysis of issue creation and disappearance, Downs (1972) pointed out that, after a period of widespread public concern, an environmental issue may lose priority as the costs of taking action are taken into consideration. In the case of weather disasters and other climate change impacts, the overall costs and benefits are still a matter of scholarly investigation. It is possible that pushing cost–benefit analysis too hard could be counter-productive at this stage, as observations in the Netherlands indicate (Akkerman *et al.*, 1996).

In summary, there are two main strategies for expanding issues related to a possible change in weather-related disasters to the public at large: stressing risks and stressing costs. Both have been tried out with regard to the issue of climate change, with mixed success. This does not, however, represent a serious obstacle to building a knowledge system in order to identify and address the impacts of specific extreme weather events in the European Union. Such a strategy can anticipate a possible future shift in public attention and get policy-makers prepared for implementing specific response options.

14.3.4 Extreme weather as a political item

Once it becomes a political item, a policy faces barriers in the implementation stage. Many of these barriers and opportunities to overcome them have already been mentioned. The crucial challenge for implementation is often to involve stakeholders in identifying the policy problem and opportunities for solution. This may not be easy, since these stakeholders have their own priorities and interests. Success depends on the open character of interaction between the stakeholders involved, from the science and policy communities alike.

14.4 The structure of policy problems and different types of science–policy interactions

Wants, demands, issues and political items reflect the social and political construction of problems. In this process, some problems are brought to the core of the public and political agenda, while others are marginalised. Different problem frames imply different solution strategies and hence mechanisms for including and excluding persons or aspects of the problem that, later on, may appear to be of the utmost importance (Hisschemöller and Hoppe, 1996). In order to enhance and improve policy–science interaction concerning response options for the types of extreme weather impacts dealt with in this study, one needs to understand the different problem frames and the mechanisms that underlie actors' views and interactions.

14.4.1 Four types of policy problems

Hisschemöller and Midden (1989), Hisschemöller (1993) and Hoppe and Peterse (1993) identify four approaches to environmental risks. The *technical* approach views problems associated with risk as being ones of technical expertise. Public attitudes are believed to be based on emotions and fears rather than on rational judgement. Following this approach, expert decisions may be frustrated by public involvement. The *market* approach considers the problem as a social dilemma; that is, a conflict between individual and collective interest. This conflict can be solved either by government coercion or by incentives in order to make people voluntarily follow the common interest. Victim compensation might be a way to address inequities. The *distributive justice* approach points at the lack of a coherent public policy, including coherent policy goals. Hence the most vulnerable interests are not adequately protected. The government must win public support by implementing integrated policies based on principles that are widely thought of as 'just'. The *public participation* approach assumes that citizens are capable of rational judgement on matters they feel personally involved with. If authorities tend not to take citizen participation seriously, valuable alternatives are excluded from the policy agenda.

None of these approaches can be observed in their 'pure form'. They are ideal types. Only in combination can they describe an actual policy problem as observed in practice. However, not all ideal types can be combined into actual policy theories:

1   The technical and the participation approaches give opposite answers to the question as to whether public participation is desirable.
2   The market and the distributive justice approaches are at odds concerning the kind of government intervention needed.

A typology of the problem shows four problem alternatives available to public policy (Figure 14.1). It is just a small step further to link the findings above to insights about problem structure and methodologies for problem structuring as developed in the policy sciences literature (Dunn, 1981, 1988; Mason and Mitroff, 1981; Schon and Rein, 1994). The rationale behind the distinction between problem types is that problems with a different structure require different solution strategies (Hisschemöller and Hoppe, 1996).

Problems that are mainly viewed as technical (the technical–market approach combination) are *structured problems*. Structured problems are characterised by standardised methods for solution, a great deal of expert involvement in the decision process, a hierarchical decision structure and one policy actor to make the final decision.

Problems that are characterised by a given set of values and goals but require different actors to negotiate an outcome (the market–participation approach combination) are termed *moderately structured problems (ends)*. This type of problem requires a policy strategy in which different actors are allowed to promote their interest by forming coalitions with like-minded actors. As policy ends are fixed, the negotiation space is limited.

Problems characterised by a persistent conflict between values and goals that at the same time enable a compromise about means (actions) are termed *moderately structured problems (means)*. These problems (the technical–distributive justice

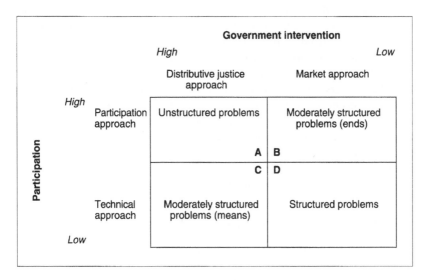

Figure 14.1 A typology of problems concerned with environmental risks

approach combination) require a depoliticisation of conflict. The compromise is usually formulated in rather broad terms in so far as the contested values are concerned, and may be characterised as 'symbolic policy'. Experts play a role in modifying the issue from a highly political to a technical one.

Problems characterised by disagreement about the values at stake and the kinds of knowledge that may contribute to a decision are termed *unstructured problems*. These problems (distributive justice–participation approach combination) require *problem structuring* followed by a deliberate *problem choice*. Learning is the key concept, because by interaction with others, who bring to bear conflicting views, interests and frames, participants gain new insights on the problem and options for solution. Stakeholders learn about the position taken by others and – ultimately – about their own particular interests. Scientists and other experts may play a remarkable role in this process, but they do not have rights distinct from those of lay-people. The unstructured problem in particular reflects the notion of 'post-normal' science (Funtowicz and Ravetz, 1994).

Multi-stakeholder studies carried out in the Netherlands (Klabbers *et al.*, 1994; Hisschemöller *et al.*, 1995) indicate that the problem types discussed above can be applied to different conceptions of the climate issue in Dutch society. The studies yield conclusions about what type of policy problem might be chosen to handle the climate issue in all its complexity and the kind of interaction which makes a dialogue productive.

Studies of climate change do not immediately offer a recipe for handling possible changes in extreme weather. All four types of problems have their respective advantages and disadvantages. The more structured a policy problem is when it enters the policy agenda, the more easily it is handled by public administration. At the same time, there is a risk that actors or parts of the problem become excluded from participation or political consideration. If this happens, policy controversies may become intractable (Hisschemöller and Hoppe, 1996). The more unstructured a policy problem is supposed to be, the more policy-makers will try to avoid biases reflected in the technical and market approach. Problem structuring may not produce total consensus, but the learning process will anyhow improve the choice for policy intervention. However, an unstructured policy process is time-consuming and there is no real guarantee that policy learning will evolve. The point is that parties cannot be forced to really participate in the process. What remains is the question whether there is more to say about the structure of problems related to specific types of extreme events.

14.4.2 The structure of problems for specific extreme weather events

Table 14.1 presents some impact features of land subsidence, agricultural drought, river floods, heatwaves, winter storms and tropical cyclones. These features indicate how the policy problems associated with these extreme events may be structured in Europe. In order to identify strategies for further developing policy response options, we may distinguish between the types of extreme events discussed in this study. Each type of extreme weather events has its own characteristics, which relate to scientific uncertainties, local and regional differences, technical options, the number of potential actors involved and the (potential) linkages with other problems. Some of them may create structured problems; some of them may confront policy-makers with moderately structured problems; others create unstructured problems or aggravate already existing ones. An institution for policy–science interaction which aims at encouraging the development of a knowledge system must take the different types of problems into account.

Land subsidence appears to be a reasonably structured problem. Scientific methodologies to identify the subsidence-prone areas are being developed (see Brignall *et al.*, Chapter 3, this volume). Although the impacts for individual households can be quite uncomfortable, standard-setting and insurance policies provide effective instruments to solve the problem. *Agricultural drought* is best treated as an unstructured problem (Table 14.1). Brignall *et al.* (Chapter 4, this volume) do not foresee major consequences from the enhanced greenhouse effect, but uncertainties are large. Moreover, the relation between drought management and environmental policies, competition for water, and European agricultural policy complicates the problem. Impacts of *heat waves* can be readily addressed by technical accommodation, thus the problem appears to be reasonably structured. *River floods* are a moderately structured problem (ends) or an unstructured problem, depending on local and regional circumstances. If a 'fair' trade-off between environment and security exists (as is the case along the Meuse; see Handmer *et al.*, Chapter 5, this volume), the issue might be solved by negotiation. If a trade-off is not possible, the only opportunity is to jointly and creatively search for win–win options. Such an option could include compensation of environmental values elsewhere. *Winter storms* do not seem to present a substantial problem for western European countries (see Dorland *et al.*, Chapter 10, this volume). As such, the policy problem is structured. *Tropical cyclones* may not create a new unstructured problem, but may aggravate existing ones, especially for the poorest. Development and aid is a highly unstructured problem at present, and climate change may exacerbate this.

14.5 Conclusions and recommendations for policy

This chapter offers a strategy for policy-makers at the European, national and regional levels involved in prevention of and adaptation to extreme weather events. The main recommendation is to strengthen collaboration between scientists and policy-makers from sectors and regions at risk.

In the short term, a change in extreme weather is not likely to lead to new policy demands or specific policies, because policy problems are not yet adequately defined. However, it is advisable to anticipate future demands and opportunities.

Policy–science co-operation can be improved by establishing and institutionalising a science–policy interface. This institution's primary responsibility would be to actively encourage, co-ordinate and evaluate interaction between different levels of administration and different disciplines that, so far, have worked in separation from one another. Such a knowledge system is required for adequate policy-making.

The Commission of the European Union is in a good position to initiate an institution for further development and integration of a comprehensive knowledge system. The primary contribution the Commission can offer is bringing together stakeholders. These are, from the science community, those working in the areas of climate impact, disaster management and engineering. From the policy community, it is advisable to involve national, regional and local officials, environmental and consumer organisations, and business representatives (including farmers' unions and insurance).

Interaction in an open-minded way is crucial for a long-term perspective on weather disaster policies; fine-tuning science and policy presupposes dialogue between all those involved, nationally and in a European context. The process of interaction will offer an opportunity periodically to evaluate science findings and policy opportunities, and hence reframe problems and solutions. It will contribute to formulating and clarifying problems for both policy and science. It will also offer an opportunity for identifying issue linkages that, in the long run, may create win–win options.

Table 14.1 Impact characteristics of six case studies of climatic hazards

Type of event	Spatial characteristics	Economic effects of changes in risks[a]	Current dominant type of risk management	Costs of risk management	Scope for disaster/risk management policies	Dominant policy-making institutions	Dominant perception of type of problem	Other observations
Subsidence	Local, related to soil types	Small to modest	Insurance (included in property insurance)	High for specific individuals	Risk-mapping, building reinforcement	Property insurance industry, building and construction	Structured	
Agricultural drought	Regional	Small to modest, adverse effects of droughts may be offset by beneficial effects of climate change	Government support	Probably low (for agricultural sector)	Intensified land, crop and economics management	Farmers' organisations, governments (regional to European)	Unstructured	Competition in demand for water with other sectors/linked to other problems in agriculture
Heat stress	Urban	Small, except for cooling costs	Health insurance, cooling	Medium	Air-conditioning, building design	Public health authorities, health insurance, construction	Structured/ moderately structured (ends)	Risk groups: elderly people and babies

Floods	Flood plains	Modest to large	Physical protection, (government) insurance and government relief	Medium (at societal level)	Non-engineering approaches next to physical protection	River management authorities, insurance, governments (local to national)	Moderately structured (ends)/ unstructured	Transboundary problem; structure may differ according to local situation
Winter storms	Regional	Modest to large	Insurance	Low	Building codes	Property insurance industry, building and construction	Structured	
Tropical cyclones	Regional to local	Large	Government relief and aid from abroad	High at societal level and for some properties	Physical protection (building codes)	Governments (local to international)	Unstructured	Linked with development

Notes
a   Economic effects from a national perspective; ranking is relative

Acknowledgements

The authors hereby thank Tom Downing, Angela Liberatore and Richard Tol for helpful suggestions. This chapter is a revised version of Hisschemöller and Olsthoorn (1996).

Note

1 The many conceptual difficulties related to questions such as how to measure these impacts, and whether knowledge use is necessarily intentional, are beyond the scope of this chapter.

References

Akkerman, A.E., Hisschemöller, M., Vellinga, P., Klabbers, J.H.G., Baede, A.P.M., Fransen, W., Komen, G.J., van Ulden, A.P., Berk, M.M., Leemans, R., and Swart, R.J., eds (1996) *Werkconferentie Klimaatverandering. De Tweede IPCC-rapportage: Indrukken en Reacties uit de Nederlandse Samenleving*, Amsterdam.

Bernabo, C., Postle Hammond, S., Carter, T., Revenga, C., Moomaw, B., Hisschemöller, M., Gupta, J., Vellinga, P., and Klabbers, J. (1995) *Enhancing the Effectiveness of Research to Assist International Climate Change Policy Development (Phase II Report)*, Washington, DC, and Amsterdam: Science & Policy Associates and Institute for Environmental Studies, Vrije Universiteit.

Bruce, J.P., Lee, H. and Haites, E.F., eds (1996) *Climate Change 1995: Economic and Social Dimensions of Climate Change*. Cambridge: Cambridge University Press.

Caplan, N. (1979) The two-communities theory and knowledge utilization. *American Behavioral Scientist* 22: 459–470.

Caplan, N., Morrison, A., and Stanbough, R.J. (1975) *The Use of Social Science Knowledge in Policy Decisions on the National Level*, Ann Arbor: Ann Arbor Institute for Social Research, University of Michigan.

Cobb, R.W., and Elder, C.D. (1983) *Participation in American Politics: The Dynamics of Agenda Building*, Baltimore: Johns Hopkins Press.

Cobb, R.W., Ross, J.K., and Ross, M.H. (1976) Agenda building as a comparative political process. *American Political Science Review* 70: 126–138.

Dery, D. (1984) *Problem Definition in Policy Analysis*, Lawrence: Kansas University Press.

Dinkelman, G. (1995) *Verzuring en Broeikaseffect: De Wisselwerking tussen Problemen en Oplossingen in het Nederlandse Luchtverontreinigingsbeleid (1970–1994)*, Utrecht: Van Arkel.

Downs, A. (1967) *Inside Bureaucracy*, Boston: Little, Brown.

Downs, A. (1972) Up and down with ecology: The issue attention cycle. *The Public Interest* 28: 38–50.

Dunn, W.N. (1980) The two-communities metaphor and models of knowledge use: An exploratory case-survey. *Knowledge: Creation, Diffusion, Utilization* 1: 515–537.

Dunn, W.N. (1981) *Public Policy Analysis: An Introduction*, Englewood Cliffs, NJ: Prentice-Hall.

Dunn, W.N. (1988) Methods of the second type: Coping with the wilderness of conventional policy analysis. *Policy Studies Review* 7: 720–737.

Dunn, W.N., and Holzner, B. (1988) Knowledge in society: Anatomy of an emergent field. *Knowledge in Society* 6: 6–26.

Enloe, C.H. (1978) *The Politics of Pollution in a Comparative Perspective: Ecology and Power in Four Nations*, New York: Longman.

Funtowicz, S.O., and Ravetz, J.R. (1994) Uncertainty, complexity and post-normal science. *Environmental Toxicology and Chemistry* 13: 1881–1885.

Hisschemöller, M. (1993) *De Democratie van Problemen*, Amsterdam: Free University Press.

Hisschemöller, M., and Hoppe, R. (1996) Coping with intractable controversies: Problem structuring and frame-reflective policy analysis. *Knowledge and Policy* 8: 40–60.

Hisschemöller, M., and Midden, C.H.J. (1989) 'Technological risk, policy theories and public perception in connection with the siting of hazardous facilities', in C. Vlek and G. Cvetkovich, eds, *Social Decision Methodology for Technological Projects*, Dordrecht: Reidel.

Hisschemöller, M. and Olsthoorn, A.A. (1996) 'Linking science to policy: Identifying barriers and opportunities for policy response', in T.E. Downing, A.A. Olsthoorn and R.S.J. Tol, eds, *Climate Change and Extreme Events*, Amsterdam: Institute for Environmental Studies, Vrije Universiteit, Oxford: Environmental Change Unit.

Hisschemöller, M., Klabbers, J., Berk, M.M., Swart, R.J., van Ulden, A., and Vellinga, P. (1995) *Addressing the Greenhouse Effect: Options for an Agenda for Policy Oriented Research*, Amsterdam: Free University Press.

Holzner, B., and Marx, J. (1979) *Knowledge Application: The Knowledge System in Society*, Boston: Allyn & Bacon.

Hoogerwerf, A. (1984) Beleid berust op veronderstellingen: De beleidstheorie. *Acta Politica* 19: 493–531.

Hoogerwerf, A. (1987) De levensloop van problemen: Definiering, precisering en oplossing. *Beleidswetenschap* 1: 159–181.

Hoppe, R., and Peterse, A. (1993) *Handling Frozen Fire: Political Culture and Risk Management*, Oxford: Westview Press.

Hutchinson, J.R. (1995) A multi-method analysis of knowledge use in social policy. *Science Communication* 17: 90–106.

Kingdon, J. (1984) *Agendas, Alternatives and Public Policies*, Boston: Little, Brown.

Klabbers, J., Vellinga, P., Swart, R.J., van Ulden, A., and Janssen, R. (1994) *Policy Options Addressing the Greenhouse Effect*, Bilthoven: Ministry of Public Health and Environment.

Lasswell, H.D. (1951) 'The policy orientation', in D. Lerner and H.D. Lasswell, eds, *The Policy Sciences*, Stanford, CA: Stanford University Press, 3–16.

Liberatore, A. (1994) 'Facing global warming: The interactions between science and policy making in the European Community', in M. Redclift and T. Benton, eds, *Social Theory and the Global Environment*, London: Routledge, 59–83.

Liberatore, A. (1995) 'The social construction of environmental problems', in P. Glasbergen and A. Blowers, eds, *Environmental Policy in an International Context: Perspectives on Environmental Problems*, London: Arnold.

Lindblom, C.E., and Cohen, D.K. (1979) *Usable Knowledge*, New Haven, CT: Yale University Press.

March, J.G., and Olsen, J.P. (1976) *Ambiguity and Choice in Organizations*, Bergen: Universitetsforlaget.

Mason, R.O., and Mitroff, I.I. (1981) *Challenging Strategic Planning Assumptions: Theory, Cases, and Techniques*, New York: Harper & Row.

Miller, J.D. (1983) *The American People and Science Policy: The Role of Public Attitudes in the Policy Process*, New York: Pergamon Press.

Mitroff, I.I. and Sagasti, F. (1973) Epistemology as general systems theory: An approach to the design of complex decision-making experiments. *Philosophy of the Social Sciences* 3: 117–134.

Olsthoorn, A.A., van der Werff, P.E., and de Boer, J. (1994) *The Natural Disaster Reduction Community and Climate Change Policy Making*, Amsterdam: Institute for Environmental Studies, Vrije Universiteit.

Read, R. S., Bostrom, A., Morgan, M.G., Fischhoff, B. and Smuts, T. (1994) What do people know about global climate change? Two survey studies of educated laypeople. *Risk Analysis* 14: 971–982.

Rich, R.F. (1981) *The Knowledge Cycle*, Beverly Hills, CA: Sage.

Rich, R.F. (1991) Knowledge creation, diffusion, and utilization: Perspectives of the founding editor of Knowledge. *Knowledge: Creation, Diffusion, Utilization* 12: 319–337.

Sabatier, P.A., and Jenkins-Smith, H.C. (1993) *Policy Change and Learning: An Advocacy Coalition Approach*, Boulder, CO: Westview Press.

Schon, D., and Rein, M. (1994) *Frame Reflection: Towards the Resolution of Intractable Policy Controversies*, Reading, MA: Addison Wesley.

Simon, H.A. (1976) *Administrative Behaviour: A Study of Decision-Making Processes in Administrative Organisations*, New York: Harper & Row.

Snow, C.P. (1960) *The Godkin Lectures*, Cambridge, MA: Harvard University Press.

Solecki, W.S., and Michaels, S. (1994) Looking through the post-disaster policy window. *Environmental Management* 18: 587–595.

van de Vall, M. (1988) 'De waarden-context van sociaal-beleidsonderzoek: Een theoretisch model', in M. van de Vall and F.L. Leeuw, eds, *Sociaal Beleidsonderzoek*, The Hague: VUGA, 35–56.

van de Vall, M., and Bolas, C. (1982) Using social policy research for reducing social problems. *Journal of Applied Behavioral Science* 18: 49–67.

Vellinga, P. and Tol, R.S.J. (1993) Climate change, extreme events and society's response *Journal of Reinsurance* 1, 2: 59–72.

Vellinga, P., Hisschemöller, M., Klabbers, J.H.G., Berk, M.M., Swart, R.J., and van Ulden, A.J. (1995) 'Climate change policy options and research implications', in S. Zwerver, R.S.A.R. van Rompaey, M.T.J. Kok and M.M. Berk, eds, *Climate Change Research: Evaluation and Policy Implications* Amsterdam: Elsevier.

Weiss, C.H. (1977) *Using Social Research in Public Policy Making*, Lexington, MA: Heath.

Weiss, C.H. (1980) *Social Science Research and Decision-Making*, New York: Columbia University Press.

Wynne, B. (1994) 'Scientific knowledge and the global environment', in T. Benton and M. Redclift, eds, *Social Theory and the Global Environment*, London: Routledge.

AUTHOR INDEX

AUTHOR INDEX

SUBJECT INDEX